EARTH AND UNIVERSE

Earth as seen from Moon. (NASA photo.)

EARTH AND UNIVERSE

BENJAMIN F. HOWELL, JR.

PENNSYLVANIA STATE UNIVERSITY

Charles E. Merrill Publishing Company
A Bell & Howell Company
Columbus, Ohio

International Standard Book Number: 0-675-09183-7

Library of Congress Catalog Card Number: 71-167500

1 2 3 4 5 6 7 8 9 10 — 77 76 75 74 73 72

Printed in the United States of America

79502

PREFACE

This book is primarily about Earth, the third planet from Sun, a small star in the Milky Way galaxy, which is a larger-than-average spiral galaxy located in the only universe we know anything about. A brief but comprehensive description of the structural framework of our universe is provided, to which the scholar or layman can add all knowledge of the physical environment he eventually acquires. Such a framework puts into proper perspective the scientific discoveries which are regularly reported to the general public by television, radio, newspapers and magazines and to specialists in science by more technical publications. The average citizen can understand, enjoy and use such information effectively only if he can relate it to previously acquired knowledge. Understanding of our place in nature is a basic human need. Concepts of size and rate of many unfamiliar phenomena are as necessary to such understanding as is familiarity with common events encountered in daily life.

The writing of this book was inspired by two exciting experiences of the author. The first of these was a thorough review and renovation of the Geological Sciences curriculum at Penn State. One of the things which the review showed was that many important discoveries have been made recently in branches of earth science largely neglected in typical geology courses. This led to the addition of a companion introductory course to physical and historical geology concerned with the interactions of the surface parts of Earth with the space around it and with its atmosphere, oceans and deep interior. This work is the text of that course.

The second experience was membership on the Steering Committee of the American Geologic Institute's Earth Science Curriculum Project. This brought home the responsibility of the scientist to serve not only his profession but also the general public. If our technological civilization is to provide abundance for everyone, or even if it is merely to survive, we must control the present rapid pollution of our environment and exploit more efficiently the non-renewable mineral resources on which it depends. This can occur only if our leaders in every field, not just our scientists, understand the problems involved. It will occur best if all citizens understand the problems. This means that every high-school graduate should know some earth science. The quality of their education is limited by the competence of their teachers. In writing this book, the author is making the best contribution of which he is capable to the education of these teachers.

This book is written with several themes running thru it. First of all, it has as its goal to enable the scholar or teacher to understand why we believe what we think we know about the universe. Therefore, it emphasizes the evidence for concepts as much as the concepts themselves. Some students would like to ignore the factual basis of sci-

v

ence. Facts tend to be dry. They are nonetheless necessary to real understanding. Concepts change as knowledge grows. The evidence remains, if it was any good originally, even after it has been reinterpreted.

A second theme is that change is universal. Even the most permanent-appearing features of our universe are changing slowly. The rates of change can usually be either measured or estimated, and both the future and the past predicted once the processes of change are understood. It is an amazing thing that so infinitesimal an organism as a man can comprehend the total organization of the universe, and from the brief view of one lifetime project its evolution from billions of years in the past to equally far into the future.

The third theme is the flow of energy. This is what causes and controls the changes. The process which dominates the evolution of the cosmos is the evening out of the energy. Our concepts of change are viewed in terms of rate of energy transfer. This provides a quantitative as well as a qualitative viewpoint. It is measurements of rate which enable man to make precise rather than vague predictions.

It would be impossible to cover adequately so large a subject as this work attempts if it were not assumed that the reader has some previous knowledge of science and its language. Since different students enter college with different backgrounds, a minimum previous experience has been assumed. The mathematics used in the book (aside from a few supplementary footnotes) can be followed by anyone who has mastered ninth grade algebra. Eighth grade general science should have made the chemical and physical terms used familiar. Most geologic terms are defined when they are first used. However, in order to avoid the necessity of discussing every simple concept encountered, it has been assumed that the student will have had a course in earth science or a general science course including earth science in high-school. Reference to geographic locations all over the world are abundant. The average reader should have no trouble finding these places on any good 1/40,000,000 scale world map such as that published by the National Geographic Society. He is urged to keep such a map, or even a globe, handy for quick reference.

An attempt has been made to keep technical jargon to a minimum. Familiar words have been used wherever possible even though corresponding words exist with more precise technical meanings. Thus, compressional seismic waves are referred to simply as sound waves. Only six rock compositions are recognized: shale, sandstone, limestone, granite, basalt and ultrabasic rocks. It is not possible to say all one might like with so unproliferated a vocabulary; but most important geological concepts can be introduced with only these simple terms.

To simplify algebraic notation, only English letters have been used throughout the text. This may irritate the advanced scholar who is familiar with other more conventional notations, but it will ease the task of many beginners by saving them from having to recognize unfamiliar symbols. Capital letters are used to represent distances, angles, times and other physical dimensions. Small-case letters are used to represent physical properties.

The earth is called Earth throughout the text to emphasize its nature as a planet. Similarly the sun is referred to as Sun. This notation has the advantage of compactness.

Since this is an elementary text, referencing to sources of the data presented has not been attempted. A list of books and articles on which the author has drawn is presented at the end of each chapter. These will, in general, serve as useful further reading in the subject. Some selectivity has been used in an attempt to include mostly works which can be easily and enjoyably read by the average student. The sources of previously published illustrations are given with the captions. Where a complete reference is not included here, it will be found in the bibliography at the end of the chapter.

More people have helped me in assembling this material than there is room to acknowledge here. I am grateful to all the friends and colleagues who have so generously aided me by supplying advice, illustrations and other information. One of the pleasures of writing a book like this one is the many kind letters I have received from both friends and strangers whom I have approached for permission to copy their photographs, diagrams or other works. Special thanks are due to Shelton S. Alexander, Roy J. Greenfield, Laurence H. Lattman, Peter M. Lavin, Jack Oliver, Hans Panofsky, Robert F. Schmalz, Karl V. Steinbrugge, Charles P. Thornton and Barry Voight for their helpful suggestions on parts of this text. Without the assistance and encouragement of all these people, this book would never have been completed.

B. F. Howell, Jr.

CONTENTS

1

INTRODUCTION

The scope of man's concern for his surroundings is continually growing. Ten thousand years ago a man was born, lived and died within a few tens of miles of the cave or hut he called home. His observations of the world were limited by how far he could walk from home to make them. His need for knowledge of his environment was restricted to the particular climatic zone in which he lived.

By two thousand years ago, the world had grown a great deal. Roman soldiers and merchants travelled from the Atlantic to Persia. The Chinese empire embraced most of eastern Asia.

In 1521 Magellan's crew sailed around the world, and the last stage of the exploration of Earth's surface was well under way. By 1911 Roald Amundsen of Norway had reached the south pole, and it could truly be said that a man could travel anywhere on Earth.

Today we are well launched on the exploration of our solar system. Governments are spending billions of dollars to find out what the universe beyond our Earth is like and how a man can live there. Some of you reading these words will eventually travel to Moon. What is it like, this universe in which we live?

In a civilization which is marked by rapidly expanding control of our natural environment and use of the resources it provides, every citizen needs an understanding of the universe in which he lives. In a democracy, everyone carries a share of the responsibility for the use or misuse of nuclear power, for the conservation or waste of our ore and oil supplies, for the pollution or protection of our streams. Failure to advise our legislators wisely of what we think should be done contributes to their errors in judgment. In the rapid pace of a nuclear world every citizen needs to understand the significance of new discoveries as they affect his life. To do so, he must understand the universe he inhabits, particularly the planet on which he lives.

In order to be able to comprehend the nature of the universe, it is necessary to know something of the nature of the material of which it is composed. Although this book is concerned primarily with large-scale structures, the biggest bodies are made up of small subdivisions, and the subdivisions are composed of even smaller parts down to the smallest particles known. To explain the phenomena to be discussed, it is necessary to assume a model for the nature of the matter of which the universe is composed. Although much remains to be learned about the nature of matter, the following simplified picture should prepare the reader for the discussions presented below.

1.1 ATOMIC MODEL OF THE STRUCTURE OF MATTER

Most matter in the universe occurs in the form of electrons and atomic nuclei. *Electrons* are small particles possessing a negative electrical charge. *Atomic nuclei* are relatively large bodies possessing a positive electrical charge. Atomic nuclei are composed of one or more particles. Some of the particles, called *protons*, have a positive electrical charge equal and opposite to that of the electron. Others, called *neutrons*, are of about the same mass as protons but possess no electrical charge. The mass of each proton is 1836 times that of the electron. Part of the mass of the nucleus is in a form which can be converted into energy if the nucleus is broken apart. Nuclei differ from one another in the number of protons and neutrons they contain. Because oppositely charged electrical particles attract one another, each nucleus tends to attract to itself electrons equal in number to the protons in the nucleus. Such a combination is called an *atom*.

The chemical behavior of an atom is controlled largely by the number of electrons attracted by the nucleus. There are 103 varieties of nuclei. These are the *elements*. Each has a name and a standard abbreviation (Table 1.1). The atom with one proton in the nucleus is called hydrogen (abbreviation, H); that with two protons, helium (He); that with three protons, lithium (Li); and so on to lawrencium (Lw) with 103 protons. Each species of atom except lawrencium has several sub-species depending on how many neutrons are contained in the nucleus. Such sub-species are called *isotopes*. Thus hydrogen has three isotopes determined by whether there are zero, one or two neutrons in the nucleus. Iron (Fe) has ten isotopes (26 protons

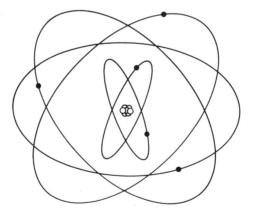

Fig. 1.1   *An atom consists of a compact nucleus with electrons in orbits around it.*

and 26 to 35 neutrons), uranium (U) has fourteen isotopes (92 protons and 135 to 148 neutrons). Different isotopes are represented by adding after the abreviation the sum of the number of protons and neutrons. Thus uranium's fourteen isotopes are designated by U227 to U240. Unlike the planets of our solar system, however, the electrons orbiting about the nucleus of an atom are not restricted to a single plane, but have a variety of relationships (Figure 1.1). To move an electron from one orbit to another or to separate the electrons from the nucleus takes a fixed amount of energy (Figure 1.2). This amount varies from one element to another and from one orbit to another for any element. Certain orbital paths are much preferred to others, and usually only certain orbital paths are stable for a given atom. Often a single orbit is much preferred over all others for each electron. The result is that every atom of a given species occurs in a finite number of conditions, one for each preferred arrangement of electron orbits.

Some nuclei have a greater attraction for electrons than others. Elements like oxygen have the ability to attract more electrons than the number of protons in the nucleus. Other elements have only a weak hold on the electrons in the orbits farthest from the nucleus. An atom with fewer electrons or more electrons than needed to balance the number of protons in the nucleus is called an *ion*. Ions are electrically charged.

Table 1.1 *The Elements. (Those marked with an asterisk are not found naturally on Earth. Those proceded by an n have one or more naturally occurring radioactive isotopes. Data largely from* Handbook of Chemistry and Physics *and* Geol. Soc. Am. Memoir 97.)

Name	Chemical symbol	Number of electrons	Neutrons in nucleus	Estimated abundance in universe (atoms per billion)
Hydrogen	H	1	0–2	9.2×10^8
Helium	He	2	1–5	7.5×10^7
Lithium	Li	3	2–6	1.1
Beryllium	Be	4	2–7	0.2
Boron	B	5	3–8	0.2
nCarbon	C	6	4–10	480,000
Nitrogen	N	7	5–10	87,000
Oxygen	O	8	6–12	840,000
Fluorine	F	9	7–12	29
Neon	Ne	10	8–14	84,000
Sodium (natrium)	Na	11	9–15	1,200
Magnesium	Mg	12	11–16	30,000
Aluminum	Al	13	10–17	2,600
Silicon	Si	14	12–18	29,000
Phosphorus	P	15	13–19	270
Sulfur	S	16	14–22	17,000
Chlorine	Cl	17	15–23	53

Table 1.1 (cont.)

Name	Chemical symbol	Number of electrons	Neutrons in nucleus	Estimated abundance in universe (atoms per billion)
Argon	Ar	18	17–24	6,900
nPotassium (Kalium)	K	19	18–26	85
Calcium	Ca	20	18–29	2,100
Scandium	Sc	21	19–29	0.84
Titanium	Ti	22	21–29	91
nVanadium	V	23	22–31	17
Chromium	Cr	24	22–32	350
Manganese	Mn	25	24–33	180
Iron (ferrum)	Fe	26	26–35	2,500
Cobalt	Co	27	27–37	20
Nickel	Ni	28	28–38	430
Copper	Cu	29	29–39	1.1
Zinc	Zn	30	30–42	5.8
Gallium	Ga	31	33–45	0.26
Germanium	Ge	32	33–45	3.6
Arsenic	As	33	35–52	0.13
Selenium	Se	34	36–53	0.54
Bromine	Br	35	39–55	0.11
Krypton	Kr	36	38–59	0.58
nRubidium	Rb	37	42–58	0.14
Strontium	Sr	38	42–57	0.61
Yttrium	Y	39	43–57	0.10
Zirconium	Zr	40	46–59	0.66
Niobium	Nb	41	48–60	.023
Molybdenum	Mo	42	48–60	.070
°Technetium	Tc	43	49–62	—
Ruthenium	Ru	44	49–64	.046
Rhenium	Rh	45	51–65	.0075
Palladium	Pd	46	52–69	.029
Silver (argentum)	Ag	47	55–70	.0075
Cadmium	Cd	48	55–73	.026
nIndium	In	49	57–74	.0032
Tin (stannum)	Sn	50	58–82	.038
Antimony (stibium)	Sb	51	61–84	.0043
Tellurium	Te	52	62–82	.087
Iodine	I	53	64–86	.013
Xenon	Xe	54	67–89	.091
Cesium	Cs	55	68–89	.0072
Barium	Ba	56	70–88	0.12
nLanthanum	La	57	69–87	.011
nCerium	Ce	58	73–90	.031
Praseodymium	Pr	59	75–89	.0046
nNeodymium	Nd	60	78–91	.020

Table 1.1 (cont.)

Name	Chemical symbol	Number of electrons	Neutrons in nucleus	Estimated abundance in universe (atoms per billion)
°Promethium	Pm	61	80–93	—
nSamarium	Sm	62	79–95	.0069
Europium	Eu	63	81–97	.0024
nGadolinium	Gd	64	81–98	.0095
Terbium	Tb	65	82–99	.0016
Dysprosium	Dy	66	83–101	.0095
Holmium	Ho	67	84–103	.0022
Erbium	Er	68	90–104	.0001
Thulium	Tm	69	92–107	.00092
Ytterbium	Yb	70	94–107	.0052
nLutetium	Lu	71	96–109	.00090
nHafnium	Hf	72	96–111	.0046
Tantalum	Ta	73	99–113	.00061
Tungsten (wolfram)	W	74	102–114	.0032
nRhenium	Re	75	102–116	.0016
Osmium	Os	76	105–116	.021
Iridium	Ir	77	105–121	.014
nPlatinum	Pt	78	106–122	.033
Gold (aurum)	Au	79	107–124	.0038
Mercury (hydrargyrum)	Hg	80	109–126	.0078
Thallium	Tl	81	110–129	.0032
nLead (plumbum)	Pb	82	112–132	.0064
nBismuth	Bi	83	113–132	.0040
nPolonium	Po	84	108–132	—
nAstatine	At	85	113–134	—
nRadon	Rn	86	118–140	—
nFrancium	Fr	87	125–137	—
nRadium	Ra	88	125–142	—
nActinium	Ac	89	132–142	—
nThorium	Th	90	133–145	.0020
nProtactinium	Pa	91	134–146	—
nUranium	U	92	135–148	.0012
°Neptunium	Np	93	138–148	—
°Plutonium	Pu	94	138–152	—
°Americium	Am	95	142–151	—
°Curium	Cm	96	144–154	—
°Berkelium	Bk	97	146–153	—
°Californium	Cf	98	146–156	—
°Einsteinium	Es	99	147–157	—
°Fermium	Fm	100	148–156	—
°Mendelevium	Md	101	154–155	—
°Nobelium	No	102	151–153	—
°Lawrencium	Lw	103	154	—

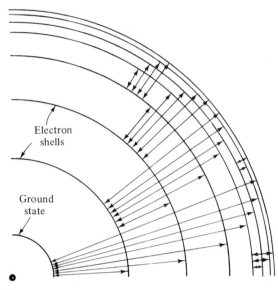

Fig. 1.2 *An electron can jump from any one stable orbit to any other, each jump being accompanied by the absorption or release of a characteristic amount of energy. The outermost orbits are so closely spaced that they become a continuum.*

There is a tendency for elements having a strong attraction for extra electrons to hold onto one of the electrons of an element with a weak attraction. Often the two atoms then share the electron, which may have a complicated orbit about the two nuclei. Two or more atoms thus held together are called a *molecule*.

The electrons moving about a nucleus protect the nucleus from the approach of any other atom. They form a shield which gives each atom an effective size and shape which is much larger than that of the nucleus itself and is related to the radii and positions of the orbits.

If two atoms are placed beside one another, they fit together best in certain orientations. Furthermore, matter tends to attract other matter. Gravity pulls it all together. Electrical, nuclear and other forces hold atoms together in molecular groupings. Thus, much of the matter of the universe is found assembled in dense concentrations. When molecules of one composition are gathered together, they tend to stack in regular arrangements, each molecule fitting against its neighbor in a certain orientation like eggs in a carton (Figure 1.3). Regular stackings of this sort are called *crystals*. Each variety of molecule in nature has this ability to stack compactly in a crystal lattice. A *mineral* is a stacking of a specific composition of molecules in a characteristic crystal form.

1.2 EFFECT OF PRESSURE.

If pressure is applied to a crystal, the atoms may rearrange themselves in a new more compact lattice. The smaller atoms will be found in the spaces

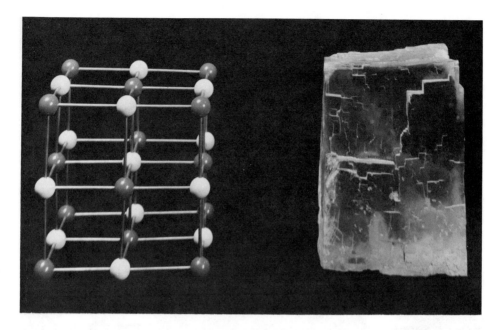

Fig. 1.3   *Regular stackings of atoms form crystals, in this case halite (NaCl, table salt).*

between the larger atoms. Thus, there may be many minerals with the same chemical composition, but with different crystal structures (Figure 1.4). The composition of a solid can be described by stating either the proportions of the different elements it contains or the proportions of the different minerals. The latter description is generally a more complete description.

If there is very little pressure on the material, the atoms will arrange themselves in an open lattice with lots of space around each atom. If the pressure is gradually increased, two things will happen. First, the atoms will snuggle up to one another just a little bit, keeping their same relative positions. The greater the pressure, the closer they will snuggle together. Pressure causes the density (amount of mass per unit of volume) to increase.

The second thing that can happen when the pressure is increased is for the atoms to rearrange themselves so that they fit together more tightly. Such a change is called a *phase change*. It might mean that one of the smaller varieties of atoms in the crystal lattice was squeezed into a hole that was really the wrong shape for its comfort but provided a better arrangement than being pinched by the big atoms when they moved closer together (Figure 1.5). Since atoms are of an irregular shape due to the various orbits of the electrons moving about each nucleus, there are many varieties of lattice in which the atoms can fit. As the pressure goes up, there is a series of phases through which a crystal of a given chemical composition will pass. Under increasing pressure, density increases slowly for any phase. At each phase change it jumps a finite step of increased density.

Fig. 1.4   *Two forms in which silica (SiO₂) is found. The molecular stacking on the left (tridymite) is more compact than that at the right (cristobalite).*

1.3 EFFECT OF TEMPERATURE.

Temperature has the opposite effect from pressure. The property of matter called heat corresponds to an agitation of an individual atom relative to surrounding atoms. If an atom possesses no heat energy, it is not moving with respect to nearby atoms. In this condition it will rest comfortably in a very compact crystal lattice, pressed tightly against adjoining atoms. As heat energy is supplied, the atom begins to vibrate slightly relative to its neighbors. It needs more space to do this. As temperature is increased, the atom is harder and harder to hold in even the most open crystal lattice.

If sufficient energy is supplied, the gravitative, and other forces holding atoms and molecules in the crystal lattice are no longer strong enough to keep them in place, and they break loose from their positions. When this happens, the substance is said to melt. The molecules are now occupying a volume which is large enough to allow each to move about freely with respect to its neighbors. If the material is not too hot, it may stick together to the degree that it occupies a fixed volume, settling under the influence of gravity into the lowest space available. The material is then said to be in the *liquid state.*

Fig. 1.5 *As pressure increases, the little atoms are squeezed more tightly into the holes between the big atoms.*

If even more energy is added, the molecules move about even faster, and the tendency to occupy a limited volume is lost. The material will expand to fill whatever space is available to it, being restricted only by any confining gravitational attraction. Fluids in this state are called *gases*. The temperature at which interatomic forces are no longer able to hold the material to a fixed volume is called the *boiling point*. Inside Earth, except very near the surface, the pressure is always so high that the distinction between gas and liquid is meaningless. There is never an empty space into which a gas can expand.

High temperature in a liquid or gas takes the form of a high velocity of each molecule with respect to its neighbors. The higher the temperature, the faster each molecule moves, and the more often and harder it collides with other molecules. If even more energy is added, the velocity of the atoms continues to increase. If the molecules are confined so that they cannot escape, as in the center of a star, this increased velocity causes them to bump violently into one another. Under such circumstances, the electrical forces holding molecules together are inadequate, and the atoms all separate.

As temperature rises further, collisions between atoms become so violent that electrons are knocked out of their orbits. The outermost, loosely held electrons are freed first. The gas now contains large numbers of both positively charged ions and loose electrons. As further electrons are knocked loose, the ions become more and more excited. The positively charged nuclei seek desperately to hold the free electrons. The high temperature of the gas is now partly in the form of high particle velocity of the atoms and electrons and partly in the form of the ionization energy used to separate electrons from the nuclei.

Finally, as the temperature is increased still further, the last of the electrons begin to be stripped from around the nuclei. The unprotected nuclei are now able to collide with one another. These collisions may cause nuclei to break up into their component parts or to combine, converting them into different species of atoms. A gas which has become so heated that it is completely ionized is called a *plasma*.

When the shielding screen of circling electrons is broken up, the atoms can be squeezed much more closely together than in the solid, liquid, or gaseous states so that plasmas can be very dense. The centers of stars are usually in the plasma state. In a few stars the atomic structure has collapsed so completely that the density reaches fantastically high values. Stars are known with average densities as great as 36,000,000 times that of ordinary water.

1.4 RADIATION OF ENERGY.

At very low temperatures, the heat is in the form of motion of the individual atoms and molecules. As temperature increases, part of the heat is in the form of excitation of the atoms thru loss of electrons or thru the shift of electrons from one orbit to another (Figure 1.2). For an electron to be in other than its most preferred orbit, it must have extra energy. It will also have a tendency to shift back to its preferred orbit. When this occurs, the extra energy is radiated as electromagnetic waves. The amount of energy in the radiation controls the type of electromagnetic waves which are radiated. Each gaseous element and molecule radiates primarily certain characteristic types of energy at any temperature. It is, therefore, possible to determine the composition and temperature of any gaseous body hot enough to radiate energy by studying the electromagnetic waves it gives off.

1.5 NATURE OF ELECTROMAGNETIC RADIATION.

Light is the most familiar form of electromagnetic energy. The eye recognizes different varieties of light as different colors. White light is a mixture of many colors. This can be shown by shining a beam of sunlight thru a prism, which separates the light into its component colors (Figure 1.6). An incandescent lamp produces a slightly different spectrum of colors, richer in yellow. Firelight is richer in red.

The eye is sensitive to only a small fraction of the types of electromagnetic waves radiated by excited atoms and molecules. Radio waves are a familiar type of electromagnetic waves which cannot be seen. The different varieties of electromagnetic waves can be distinguished from one another by a property called frequency. Imagine yourself as having a device which will measure the force exerted by passing electromagnetic waves on an electron. The force will be directed perpendicular to the direction from which the light comes and will reverse its direction at a frequency (number of times per second) which depends on the color of the light. Each color of light will be found to correspond to a different frequency of oscillation of the force. The eye can detect electromagnetic

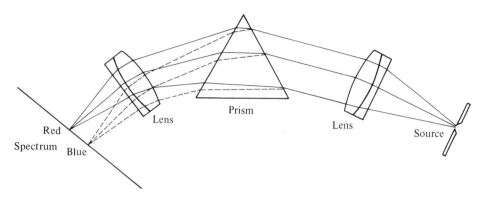

Fig. 1.6 *Light passed thru a prism can be split into different colors.*

energy with frequencies of oscillation from 4.3×10^{14} (red) to 7.5×10^{14} cycles per second (violet).° The time it takes one cycle of a wave to pass an observer is called its period. It is the inverse of the frequency, and hence is measured in units of time, such as seconds.

Other frequencies characterize other types of radiation (Figure 1.7). Heat radiation (infra-red) possesses frequencies just below those of light. Radio waves are electromagnetic energy of even lower frequency, 10^5 to 3×10^{11} cps. Frequencies above 10^9 cycles per second are known as microwaves. Radar, used in tracking rockets, is an example. Below this are ultra-high-frequency radio waves. Television and FM-radio are examples. The standard broadcast range is 5.5×10^5 to 1.6×10^6 cps. Lower frequencies, as well as certain higher radio frequencies, are reserved by law for special services such as aircraft, police and military communications.

Electromagnetic forces can cause electrical currents to flow in a metal wire. If this wire is part of a loud-speaker, some of the electrical energy is turned into sound energy. The human ear can hear sound vibrations in the frequency range from around 60 cps to 15,000 cps. These frequencies are, therefore, called audio frequencies. The range from 15,000 to 100,000 cps is called supersonic. Many animals (e.g. dogs) can hear sound at supersonic frequencies to which the human ear is insensitive. Frequencies below 60 cps are called sub-audio. There are electrical currents in the earth and in the atmosphere which can be detected by the electromagnetic forces they produce at all these frequencies.

Faster rates of oscillation than those of light correspond to other types of electromagnetic energy. The band from 7.5×10^{14} to 3×10^{16} cps is called ultraviolet light. (It is sometimes called black light, because it is invisible.) It is the ultraviolet component of sunlight which causes sunburn.

°When very large numbers must be written, as is often the case in scientific work, the exponential system will be used for convenience. In this system 10^6 means 1,000,000; 10^{27} means 1 with 27 zeros after it; 16,500,000 is written 1.65×10^7, equivalent to 1.65 times 10,000,000.

X rays are even more rapid vibrations. Low-frequency X rays are used by doctors to take shadow pictures of the bones within our bodies. Higher frequencies are used to make shadow pictures of the insides of metal objects. The electromagnetic energy radiated by atoms as they disintegrate is called gamma radiation. Its frequency is around 3×10^{19} to 3×10^{21} cps. Even higher frequencies are occasionally observed. These are called cosmic rays, and are still incompletely understood.

The different colors of electromagnetic energy can be described in another way. Imagine that, instead of standing in one place and counting the number of oscillations per second of the electrical force as light flows by, you were to move along the ray and observe the electrical force as a function of distance. Again, you would find that the direction of the force oscillated (Figure 1.8). The distance from one point of maximum force to the next is called the wavelength. If the electromagnetic energy is moving past an observer with a velocity of travel C_L, then wavelength, L, frequency of oscillation, F, and period, T, are related by the formula

$$C_L = LF = \frac{L}{T} \tag{1.1}$$

Experiment has shown that the velocity of light, C_L, in a vacuum is always the same. It appears to be one of the few fundamental constants of the universe. Numerically it has the value 3×10^8 meters per second°. Thus any color of light or of any other part of the electromagnetic spectrum can be equally well described by stating its frequency or its wavelength.

This property of frequency (color) of electromagnetic waves such as light provides one of the most powerful tools whereby we can study the universe. Almost all we know of the universe outside Earth comes from the study of the electromagnetic radiations reaching Earth from space. We will call upon frequency many times in explaining what goes on in stars.

1.6 LIGHT SPECTRA.

Light spectra are of three types called bright-line, dark-line and continuous spectra. A *bright-line spectrum* consists of the colors radiated by a hot gas. It consists of lines corresponding to all the energized reactions going on within the structure of the atoms and molecules present in the source (Figure 1.2). The hotter the source, the more numerous the reactions which will be stimulated.

°Throughout this text the metric system of units will be used since this is the language of science (as well as of most countries in the world). A table of English equivalents is included as Appendix I.

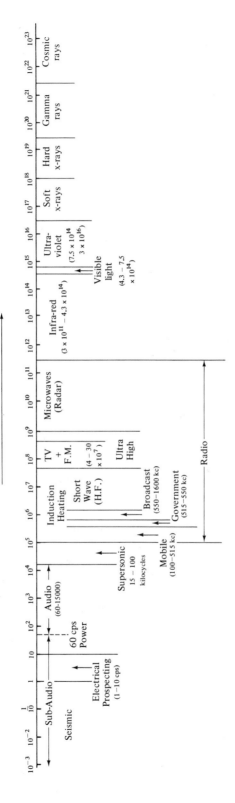

Fig. 1.7 *The electromagnetic spectrum.*

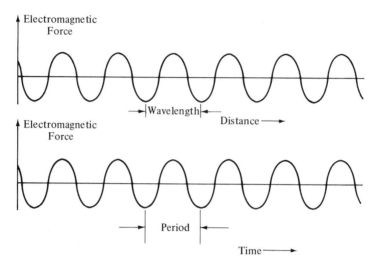

Fig. 1.8   *a. If electromagnetic force is observed at one time as a function of distance, it oscillates with a wavelength. b. If it is observed at one place as a function of time, it oscillates with a period. Velocity = $\dfrac{wavelength}{period}$.*

If a gas is very hot, the number of lines may be so great that the separate lines are indistinguishable from one another, and a *continuous spectrum* is obtained. Very high pressure has the effect of broadening the lines, assisting in the production of a continuous spectrum. The spectra of hot solids also are continuous.

Each line in the spectrum is produced by a particular reaction in the atomic structure of the radiator, such as an electron dropping from an outer to an inner orbit about the nucleus. When light of this frequency shines on a cool atom, the reaction tends to be stimulated in reverse. In jumping from an inner to an outer orbit, the electron absorbs energy from the light. Thus light received from a star contains dark lines formed as it passes thru the cooler, outer part of the star's atmosphere, thru any gas in interstellar space and, finally, thru Earth's atmosphere. Such a spectrum is called a *dark-line spectrum* (Figure 1.9). The missing lines are just as characteristic of the cool material thru which the light has passed as are the bright lines radiated from a hot source. Each line in a dark or bright spectrum corresponds to a particular atomic or molecular reaction. Since the number of radiated lines for a given atom increases with its temperature, the presence (or absence) of particular lines in a spectrum can tell not only whether a particular atom or molecule is present in a radiating (or absorbing) gas, but also the temperature of the radiating gas. The brightness (or darkness) of particular lines can be used to estimate how much such material is present. Studies of the spectra of stars are our main source of information about them.

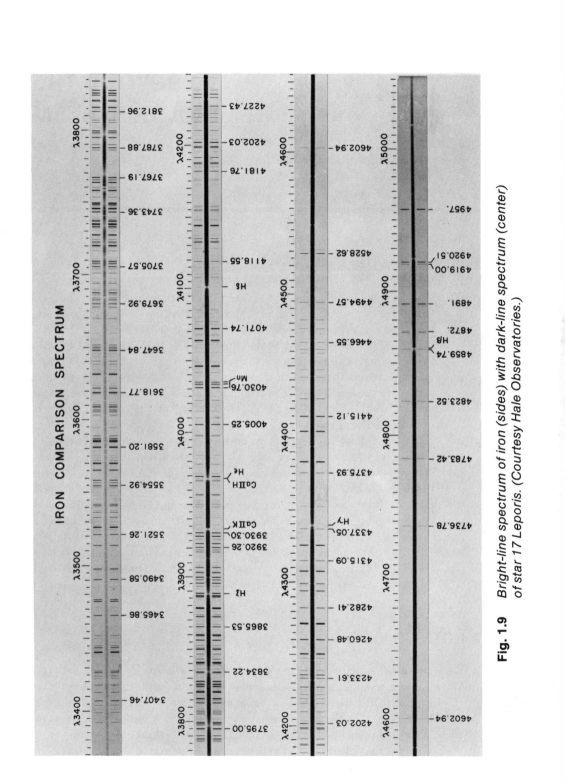

Fig. 1.9 *Bright-line spectrum of iron (sides) with dark-line spectrum (center) of star 17 Leporis. (Courtesy Hale Observatories.)*

2 BEYOND THE SOLAR SYSTEM

2.1 NOTHING AT ALL.

If on a cloudless night you look away from Earth upward at the sky, you will see the stars. At first glance it may seem that the sky is "full" of stars, but if you think carefully about what you are seeing it will become obvious that the stars are separately visible only because there are dark areas between them. If you consider what fraction of the sky is occupied by stars, you will realize that most of the sky is dark. Nearly everywhere there is nothing to be seen. This is the predominant characteristic of the universe—nothing at all. Almost all of the universe is empty space. Earth, Sun, each star is something special.

It can be demonstrated that space is empty in several ways. If it were not empty, you could not see thru it to Sun, Moon and the stars. Furthermore, as one goes upward from Earth, the density of air gets steadily less. At a few hundred kilometers elevation, there is less matter per unit volume than the best vacuum that can be created in the laboratory. However, space is not completely empty. Measurements made with rockets detect clouds of electrons and ions thrown off by Sun and surging past Earth. There are also fine dust particles. Near Sun these clouds can be seen during an eclipse by the sunlight they reflect (Figure 2.1). The concentration of this material decreases with distance from Sun.

2.2 NOTHING VISIBLE.

Outside the solar system, the density of matter in space is even smaller. If this were not the case, starlight would be dimmed and scattered more than it is. Actually, there is a measurable effect. Faint absorption and emission lines, especially certain frequencies in the ultrahigh radio frequencies, are most easily explained as due to interstellar gas. Estimates of the amount of such gas range from one to fifty percent of the mass of the universe. The material is predominantly hydrogen, but lines for heavier elements have also been detected.

A small part of interstellar material is in the form of dust. Aggregates of matter scatter light. If dense enough, they absorb or reflect light. Dust is much more irregularly distributed than interstellar gas. At places it is thick enough to hide parts of the sky. In Figure 2.2 the number of stars seen per unit area of sky in the lower half of the picture is much less than in the upper half. This difference seems to be caused by clouds of interstellar matter. The clouds below are thin enough that an occasional star shines thru. Wisps of the cloud extend into the less obscured area. Figure 2.3 shows another system of dark clouds partly obscuring part of an incandescent cloud. These obscured areas, however, are the exception rather than the general rule. Most dark areas of the sky appear to be dark because there are no light sources in that direction, not because there is any recognizable barrier to light reaching Earth. Elsewhere, dust is made visible as a thin nebulosity by the light it reflects from nearby stars (Figure 2.4).

Most often, dust is so diffuse that stars can be seen thru it. The density of the cloud in such a case can be estimated by the amount by which transmitted starlight is scattered. This scattering, the same process which makes

Fig. 2.1   *Sun's corona during an eclipse. (Courtesy Hale Observatories.)*

the sky blue by day on Earth, is frequency sensitive and affects the colors of the stars seen thru a thin cloud. The variability of the density of dust in the universe and the fact that it is a reflector and absorber rather than a radiator of light make its amount very hard to measure. That ten to twenty percent of the matter in the universe is in the form of dust is a reasonable estimate.

Fig. 2.2   *Horsehead nebula in the constellation Orion. (Courtesy Hale Observatories.)*

Fig. 2.3 *Bright and dark nebulosities near Gamma Cygni. (Courtesy Hale Observatories.)*

It is difficult to estimate the average density of total matter in the universe because of uncertainties in the amount of non-glowing material. Large bodies near Earth, such as Moon and the planets, are visible because of the sunlight they reflect. Small bodies even within the solar system do not reflect enough light for us to see them. Thus a star may typically have associated with it an unknown but possibly substantial amount of invisible matter. If Sun is typical, the associated mass is a small fraction of the total, but it is not known how typical Sun is.

2.3 DISTRIBUTION OF STARS.

If the distribution of stars in the sky is examined, it is found that there are many more stars to be seen in some directions than in others. Figure 2.5 shows a part of the sky known as the *Milky Way*. The Milky Way is a belt of stars tilted at 62° to Earth's equatorial plane. Detailed study of the bright areas reveals the presence of millions of stars so closely spaced that they cannot be separately seen in a small-scale photograph. This does not mean that the stars themselves are close together. Individual stars are actually far apart, some being much nearer Earth than others. The nearest stars (aside from Sun) are Alpha Centauri and Proxima Centauri, both about 4×10^{13} km away. It is easier to describe such large distances by the

Fig. 2.4   *Great Nebula in Orion. (Courtesy Hale Observatories.)*

time it takes light to travel to them. Light travels at 3×10^8 meters per second, which is 9.47×10^{12} kilometers per year, or roughly 10^{13} kilometers per year. Thus 4×10^{13} km is about four light years.

An analogy may help to visualize the density of stars in a distribution like the Milky Way. The English mathematician, James Jeans, has pointed out that if five apples were placed on Earth, one on the center of each of the five largest continents, their spacing in comparison to their size would be about that of stars in the Milky Way. If we assume that each apple has a diameter, D, of 7.5 cm (3 inches) and a density of 1.41 gm/cc (equal to that of Sun), and that half of each apple belongs to the volume inside Earth and half to the surrounding space, then the average density, d, is

$$d = \frac{\text{mass of 2.5 Sun-apples}}{\text{volume of Earth}}$$

$$= \frac{\frac{4}{3} \pi \left(\frac{D}{2}\right)^3 \times 2.5 \times 1.41}{\frac{4}{3} \pi (R_e)^3} \text{ gm/cc}$$

$$= \frac{7}{10^{25}} \text{ gm/cc} = 7 \times 10^{-25} \text{ gm/cc} \qquad \textbf{(2.1)}$$

where R_e is the radius of Earth. This is equivalent in density to about one spoonful of matter in a cube 1000 km (600 miles) on a side.

The Milky Way stars form a disk-shaped band over 100,000 light years in diameter. Sun is near the outer fringe of the disk. Within this disk stars are not evenly spread but lie in belts. There is a particularly dense grouping of stars toward the center of the disk (Figure 2.6). There are so many stars that the far side of the disk cannot be seen thru this bright concentration (Figure 2.5).

If one looks into space at right angles to the disk, where stars are less common, then other groups of stars can be faintly seen with the strongest telescopes. Such groupings of stars are called *galaxies*. The most common type of grouping consists of a dense central ball of stars surrounded by spiral arms (Figure 2.7). Between the arms are relatively empty belts, and along the arms there may be clouds of dark material. This is very like the Milky Way. Our solar system lies in one of the arms of a very large spiral galaxy.

Most of the stars of the universe are gathered into galaxies, of which there are a very large number. Galaxies tend to occur in groups (Figure 2.8). The galaxies are scattered far apart from one another in space, just as the stars are (though not proportionately as far apart considering their size). Thus the average density of matter in the universe is much less than the figure given above for density within the Milky Way. Interstellar dust and gas is thought to be thinner outside of galaxies than within them, perhaps about six percent of the universe's matter being outside of galaxies. This is a very rough estimate and may be greatly in error for the same reasons that the amount of interstellar material within galaxies is uncertain. Photographed by their radio-frequency radiations, galaxies frequently appear larger than when photographed using visible light. Considering the distribution of galaxies in space, it is estimated that the average density of matter in the visible universe is within several orders of magnitude of 10^{-30} gm/cc.

Among large galaxies, about three quarters are of the spiral type. The spiral galaxies can be further subdivided into those with a central bar and those without (Figure 2.7), by the presence or absence of a recognizable complete ring of matter outside the central nucleus, and by the degree of development of the spiral arms.

Most of the remaining galaxies are round or elliptical in outline (Figure 2.9). These galaxies also may be disc-shaped, the range in profiles being due to the degree to which we see them edge-on or face-on. In elliptical galaxies, the number of stars per unit area of sky increases more evenly toward the center than it does in spiral galaxies. There is no obvious appearance of rotation, although the disc shape suggests that ellipticals may also be rotating in their plane of flattening. They also lack the clouds of dark matter common in the arms of spiral galaxies.

The remaining galaxies are irregular in shape, lacking any clear organization of stars (Figure 2.10). At one time they were believed to constitute only a few percent of all galaxies; but recent studies have shown that there may be a great many small galaxies of this type, possibly more than there are of the larger, more highly organized galaxies.

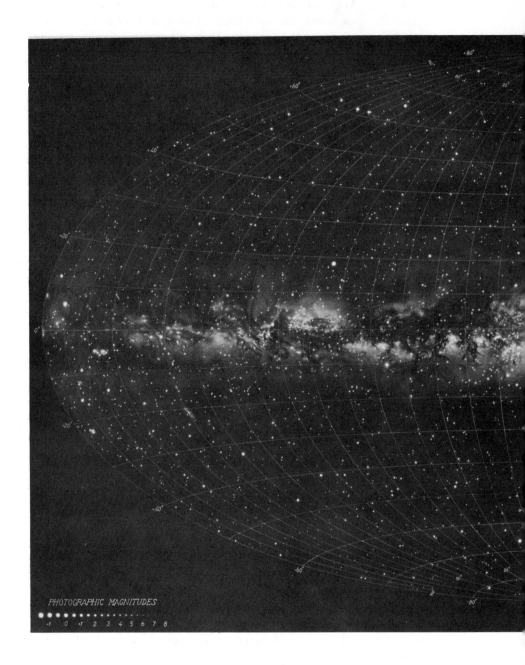

Fig. 2.5   *Panorama of Milky Way galaxy as it might appear from intergalactic space. (Prepared by Martin and Tatyana Kuskula. Courtesy Lund Observatory, Sweden.)*

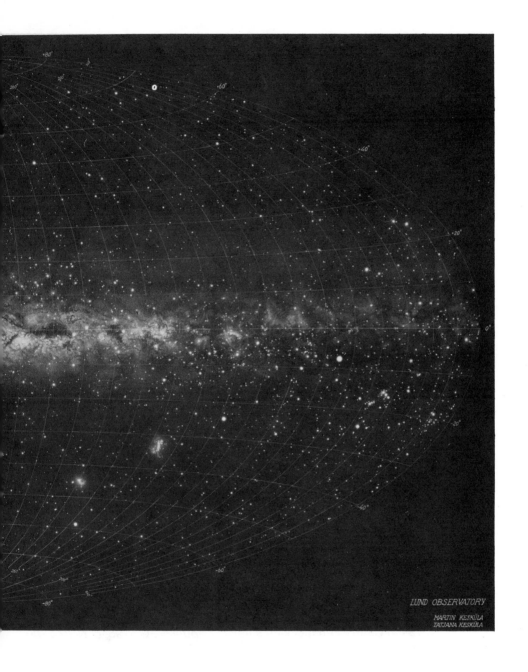

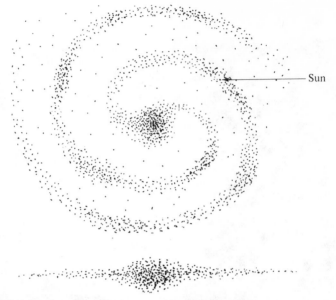

Fig. 2.6   *Distribution of stars in the Milky Way.*

Fig. 2.7   *Three spiral galaxies. a. NGC 4565 in Coma Berenices; b. NGC 5194 in Canes Venatici with satellite nebula NGC 5195; c. NGC 1300 in Eridamus. (Courtesy Hale Observatories.)*

2.4 TYPES OF STARS.

When individual stars are examined they are found to vary in size and in the types and amounts of light they radiate. Thus some stars appear to the

Fig. 2.8   *Cluster of galaxies in Hercules. (Courtesy Hale Observatories.)*

eye to be redder than others, e.g. Antares (in the constellation Scorpius) and Betelgeuse (in Orion). In general, the hotter the star, the richer its light is in higher frequencies, hence the bluer it appears to be (Figure 2.11). This is because it takes more energy to produce high-frequency than low-frequency radiation. Some stellar bodies are so cool that most of the energy they radiate is in the infrared and lower parts of the spectrum. Such stars are studied by examining photographs taken with film which is sensitive to lower frequencies than the eye can see, or by using "radio" telescopes which measure the intensity of energy at lower frequencies than can be resolved with a glass lens.

The color of a star is a measure of the temperature of the deepest part of the star which radiates light. The center of the star is hidden by the outer layers of matter. The outermost fringes of a star are thin enough to see thru them. It is common to call this transparent part of a star its atmosphere, even though there is no transition to solid or even liquid material beneath it. The upper part of this atmosphere may have levels where the gas is cooler than the lower portions, so that both radiation and absorption spectra can be observed from the same star, giving information on the temperature and composition at several levels.

The intensity of starlight also varies for a given color. This is due to several factors. First, the farther away a star is, the fainter it appears to be. The amount of light received from a radiating source falls off as the inverse

Fig. 2.9   *Elliptical galaxy. (Courtesy Hale Observatories.)*

square of the distance. A star half as far away is four times as bright. Second, the larger the diameter, the more light a star radiates in any direction. Lastly, the hotter it is, the more energy it radiates at all frequencies. Temperature can be determined from the relative amounts of light of different colors. If the distance to a star is known, then its size can be determined from the amount of the light it radiates. This is an important relation, because diameter cannot be measured by direct observation for bodies outside our solar system. All stars except Sun are so far away that their light comes from an unmeasurably small area of sky.

Stars are classified by types on the basis of two properties: color and luminosity. Luminosity is their brightness adjusted for the effect of distance. If the size (luminosity) of the 55 stars nearest Sun is plotted against surface temperature (predominant color) the vast majority of stars are found to fall close to a curved line running from upper left to lower right of Figure 2.12. Such stars are called *main-sequence stars*.

There is no reason to suppose that the main sequence ends with the small red stars. Smaller stars would be too small and faint to see. From this point of view, the planets of our solar system may be considered to be stars of the main sequence lying off the diagram of Figure 2.12 to the lower right. If it weren't for reflected and reradiated sunlight, Earth would radiate primarily the heat escaping from its interior. Its spectrum would peak somewhere far down in the infrared.

Fig. 2.10 *The Large Magellanic Cloud, an irregular galaxy. (Courtesy Lick Observatory.)*

Main sequence stars are classified in sequence as *B, A, F, G, K* or *M* stars on the basis of color. *B* corresponds to the bluest and *M* to the reddest stars. The sun is a medium-sized, slightly yellow star of the *G* type. The spectra of main-sequence stars differ from one another in color because of the relative intensity of different colors of light. Table 2.1 shows some of these temperature-produced differences. Stars of the main sequence or smaller are called dwarf stars in contrast to stars which are larger than the main-sequence stars. The latter are called giants. The designation of a star as a giant or a dwarf is thus not primarily on the basis of size, but refers to whether it is larger than a main-sequence star of its characteristic color or not.

The *B* stars are not the largest stars observed. There are a few slightly larger stars called *O* stars, or blue giants. These may be a continuation of the main sequence. The order of these stars can be remembered by the mnemonic "Oh, Be A Fine Girl, Kiss Me," in which the first letter of each word corresponds to a different type of star from hottest to coolest. Possibly even hotter than the *O* stars are the rare *W* (Wolf-Rayet) stars. In size these stars are below the main sequence but they are surrounded by a large atmosphere of ejected material.

In addition to the main-sequence stars, there are giant stars with the low temperature of the main-sequence stars but larger diameters. Very large

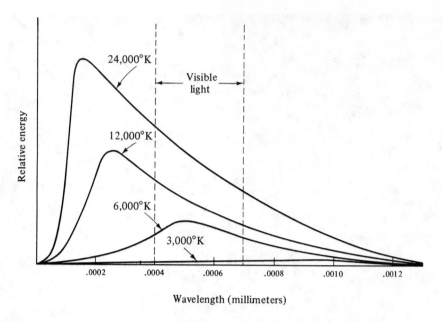

Fig. 2.11 *The spectrum of light radiated by a star depends on its tempera-*
ture. (After Ebbighausen, Astronomy, *Charles E. Merrill Books, Inc.)*

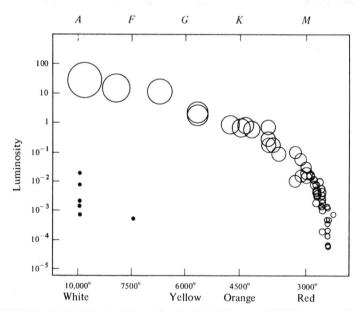

Fig. 2.12 *Distribution of spectral types of 55 stars nearest to Sun. The diame-*
ters of the circles are proportional to the diameters of the stars ex-
cept for the white dwarfs. (After van de Kamp, Astro. Soc. Pacific v.
58, Nr. 478 p. 9.)

Table 2.1 *Principal types of stars. B to M is main sequence*

Spectral Class	Surface temperature (°K)	Characteristic spectral lines
O (Blue giants)	over 25,000	Ionized helium, nitrogen, silicon
B	11,000 – 25,000	Helium, ionized silicon oxygen and magnesium
A	7,500 – 11,000	Hydrogen, ionized metals
F	6,000 – 7,500	Many metals
G (Sun)	5,000 – 6,000	More metals, especially calcium
K	3,500 – 5,000	Metals, molecules
M	less than 3,500	Molecules, especially TiO_2
R	3,500 – 5,000	Carbon, cyanogen (CN)
N	less than 3,500	Carbon and compounds
S	less than 3,500	Zirconium and lanthanum oxides

giant stars are called super-giants. There are also dwarf stars with the high temperatures of the *A* and *F* type stars but smaller in diameter than our Sun. Although the latter are small in size, they are not small in mass. Some of these "white" dwarf stars may be as small as Earth yet as heavy as Sun. The density of matter in such stars reaches values as great as 3.6×10^7 times that of water. This density is so large that the structure of the atoms inside such stars must have completely collapsed, with the electrons and nuclei forming a homogeneous plasma. For these stars to be so hot and yet dense, their processes of heat generation and radiation must be very different from those which operate in the main-sequence stars. Differences in generative processes can also be expected for the giant stars lying above the main sequence.

The size-temperature relationship of the main-sequence stars is the expectable result of a rate of energy loss from these stars which is roughly proportional to the cube of the mass. In other words, stars five times as large as Sun radiate 5^3 or 125 times as much energy per unit of time. The bigger the star in the main sequence, the faster it is consuming whatever fuel it has supplying the energy it radiates.

The stars lying off the main sequences can be expected to get their energy from different reactions than those common in most stars. This implies that the compositions of stars vary. The spectra of the main-sequence stars are all very similar except for the greater intensities of high-frequency lines corresponding to high temperatures in the larger stars. There are, however, differences in the relative intensities of different lines which lead to special categories of stellar types. *R* and *N* stars are dwarf stars differing from the main-sequence stars by being richer in the element carbon, in the carbon-nitrogen cyanogen (CN) and in the hydrogen-carbide (HC) molecules. *S* stars are unusually rich in the elements zirconium and lanthanum.

2.5 VARIABLE STARS.

Some stars vary in brightness. Stars whose luminosity changes measurably with time are called *variable stars*. In some cases this change may be due to the presence of two stars revolving about one another. Such stars are called binaries. If they are close together and far from Earth, they will not be separately visible even with a very large telescope. If one member of a binary passes in front of the other in the course of their revolution, then less light will be radiated toward Earth during this time than at other times. If one of the pair is too small to radiate visible light, then it will be invisible, and can be detected only by the small variations of the light of its companion. Jupiter may be considered such a binary with Sun. The fraction of stars which are known to have binary companions is so large that it is suspected that nearly all stars have small companions. Planetary systems such as ours may, therefore, be common in the universe.

Some stars appear to expand or to contract or to do both, alternately. One class of such stars is the cepheid variables named after δ-Cephei, the first such star discovered. There are an unusually large number of such stars in the nearby galaxy called the Small Magellanic Cloud. The cepheid variables vary in brightness regularly. The length of a cycle can be anything from 1.5 hours to over 45 days, but the period of each cepheid stays very nearly constant. The period increases with the luminosity of the star. A cepheid variable changes in color as well as diameter during its cycle. The color change implies a change in temperature, which leads to a possible explanation of the change in size. The process which provides the energy radiated by the star is presumed to be pressure sensitive. As the star contracts, the energy generation increases. This heats the center of the star, opposing further contraction. The increasing central temperature eventually stops the contraction. Expansion follows. This in turn reduces pressure so that energy generation slows, and the star can again contract, starting a new cycle.

The expansion and contraction are not observed directly, but are found from a study of the light spectrum. Not all colors of the spectrum are equally bright. Each line of color is due to a particular atomic reaction, and the reactions associated with the principal lines have been identified and reproduced in the laboratory. Many of these principal lines appear in the spectra of nearly all stars, with variations in relative strength depending on temperature and stellar composition. The frequency of each line in the laboratory is always the same. In most nearby stars the positions of the lines are exactly the same as in laboratory experiments, but in cepheid stars the whole spectrum of lines is displaced alternately up and down in frequency.

There are a number of things which can cause changes in the frequency of light. One of the simplest of these is a relative motion of the source of the light and the observer. Such a shift is called a Doppler effect. It is an expectable result of the oscillatory nature of light. The frequency of the light generated depends on the particular atomic reaction producing the radiation. It will be the same whether the atom is on Earth or in a distant star. Consider the case where 5.00×10^{14} oscillations are being emitted per second. The light leaves the star at a speed of 2.998×10^8 m/sec. The wavelength of the light is $2.998 \times 10^8/5.000 \times 10^{14} = 0.5996 \times 10^{-6}$ m, and one cycle of the light radiation is emitted every 0.2000×10^{-14} sec. If the star is moving away from earth at 3.000×10^5 m/sec (300 km/sec) then the next cycle of light radiation will have $3.000 \times 10^5 \times 0.2000 \times 10^{-14}$ meters further to travel to reach Earth than its predecessor, or its wavelength will be increased by 0.6×10^{-9} m to 0.6002×10^{-6} m. On Earth, all the spectral lines will have the same relative positions as for a star at a fixed distance, but the whole pattern of lines will be shifted to longer wavelengths (lower frequencies), which is toward the red end of the spectrum. Such a change is called a red shift. If the source is approaching Earth, the wavelengths will be correspondingly shortened and the frequencies increased.

In the case of the cepheid variables, during the time of maximum luminosity the spectral lines are raised in frequency indicating that the radiating surface of the star is moving toward Earth, presumably due to the rapid expansion of the star. During the minimum of luminosity, there is a maximum red shift of the spectrum, indicating that the surface of the star is receding and that the least energy is radiated during the contracting part of the cycle.

In addition to those of the cepheid stars, other periodic variations are observed, some as long as 1380 days (3 years, 9 months). The cycles of these long-period variables are not as constant as those of the cepheids, although the basic cause of their oscillation may be similar. Even longer periods have been observed, though the cycles of extremely long-period variables are only poorly known because of their rarity and the short time since their discovery.

Novas are variable stars which increase immensely in brightness, becoming as much as 50,000 times brighter than they were previously. After a brief interval of brightness they fade away again. Their sudden appearance and extremely high rate of radiation suggest an explosive process of energy release. This is supported by an upward shift of the spectral lines of their light, showing that they are expanding rapidly, commonly at rates exceeding 100 km/sec. Following the expansion, the rate of radiation falls off and changes in nature. The observed spectral lines then correspond to radiation from a diffuse gas cloud rather than a concentrated star. Apparently a nova throws off at least a part of its mass into surrounding space during its brief flare-up. Some novas have been observed to erupt more than once, the interval between eruptions being inversely related to the intensity of the event. The average period is estimated from this relation to be about 3500 years.

The most violently exploding stars are called *supernovas*. A supernova explosion may totally disrupt a star. Because of its extremely high rate of energy release, a supernova can be seen even in a distant galaxy. In our own galaxy, there was a supernova explosion in 1054. The star was one which had not previously been seen, and after the explosive phase it again disappeared until the development of powerful modern telescopes revealed its remains in the form of a distended, diffuse cloud of matter (Figure 2.13). Such rare clouds constitute a class called *planetary nebulas.* Many may be the remains of prehistoric supernovas. The Star of Bethlehem, seen at the time of the birth of Christ, may have been a supernova.

The variable stars generally are abnormal in that they lie out of the main sequence of stars. The novas in their more quiet stages are O and B spectral types which are smaller in size than the main-sequence stars but larger than white dwarfs. When they flare up, their spectral frequencies are displaced upwards, showing that their radii are increasing rapidly. Expansion velocity may reach several thousand kilometers per second, and luminosity may increase by as much as 50,000. Part of the outgoing matter may be permanently lost to the star, but probably all except a small fraction of one percent eventually falls back. During the cooling part of the cycle, the color shifts to the A or F type. Following its explosive phase, a nova gradually returns to approximately its pre-explosion condition.

A few variable stars show less systematic changes in the light they radiate. Some of these variations may be due to relative motions of Earth, the stars and intervening dust or gas clouds, to clouds within the atmospheres of the stars or even to internal processes producing large sunspots on the star's surface.

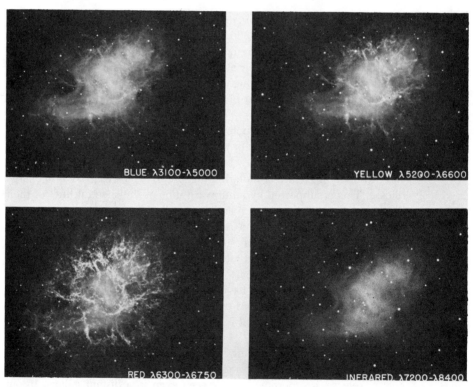

Fig. 2.13 *Crab nebula photographed using different colors of light. (Courtesy Hale Observatory.)*

2.6 LIFE HISTORY OF A STAR.

The fact that a star is rapidly radiating energy implies either that the star must eventually change due to exhaustion of its energy sources or that it must be refueled from some outside source. Something, the "origin" of the star, must have started the radiation process initially. The life of a star in the sense that it would be visible to an outside observer begins when it starts radiating with sufficient strength to be seen. This occurs when whatever heats the star has raised its surface temperature to the point where substantial electromagnetic energy is released. This implies the addition of a fuel or a change in state permitting the consumption of an already present fuel. Supernovas are the stars we see which are most obviously changing their pattern of radiation. After the explosive phase, the supernova appears to have been destroyed. This raises the question: is this the end state for all stars? The answer is probably "no"; but to see why this is the case, we must consider where stars get the energy they give off so generously.

If the energy radiated by a star had to come from a fixed reservoir of heat, a star such as Sun would use up its present reserve in a few hundreds of thousands of years. The geologic record shows that Earth has received radiation from Sun at a rate not very different from the present for several billions of years. Obviously, therefore, the heat comes from another source. The only reasonable known means of supplying the needed energy is the conversion of mass, M, to energy, E, according to Einstein's classic equation

$$E = Mc^2 \qquad\qquad (2.2)$$

where c is the velocity of light in meters/sec, M is in kilograms and E is in joules.

If Sun is losing mass in this way, then changes in its composition must be involved. Measurements of the strengths of the spectral lines suggest that at least its outer portions are composed largely of the element hydrogen, plus smaller amounts of most of the other elements (Table 2.2). This is true of all the main-sequence stars. Hydrogen is by far the commonest element in the universe, both in stars and, in so far as its composition can be found, in the material scattered thru interstellar space.

The simplest known means of converting mass into energy involve the conversion of light elements into heavier ones. Studies of the chemistry of atomic nuclei have shown that there are several ways in which four atoms of hydrogen can be converted into one atom of helium with the transformation of 0.7 percent of the mass of the original hydrogen atoms into energy. These processes cannot easily be reproduced in the laboratory because they involve collisions between nuclei, which are normally prevented from approaching one another by electrical repulsive forces. In the centers of stars, however, where all the atoms are highly excited and the pressures are so great that the atoms are tightly confined and in the plasma state, such encounters are expectable. If Sun started as almost all hydrogen 5–10 billion years ago, and has radiated heat since then at about its present

Table 2.2 *Composition of most abundant elements in Sun's atmosphere. (Largely after A.G.W. Cameron in* Geol. Soc. Am. Memoir 97 *pp 7–10.)*

Element	Estimated abundance in atoms per million atoms	
	Sun	Universe
Hydrogen	860,000	920,000
Helium	140,000	75,000
Carbon	450	480
Nitrogen	81	87
Oxygen	780	840
Sodium	1.7	1
Magnesium	21	30
Aluminum	1.3	3
Silicon	27	29
Sulfur	17	17
Calcium	1.2	2
Iron	3.2	2

rate, the energy required could have been provided by the conversion of hydrogen to helium. Theoretical studies predict that the rate of energy conversion would actually have increased slowly throughout this time (Figure 2.14). There should still be enough hydrogen left, however, to last several tens of billions of years.

This process suggests a reason why there are so few *O* and *B* type main-sequence stars. Normally, the larger the star, the hotter it is. This means that the very large stars, if their energy comes from the conversion of hydrogen to heavier elements, use up their hydrogen faster than smaller stars, and must have much shorter lives. Stars five times heavier than Sun would have used up their hydrogen in about five billion years. We should see such stars at the point in their lives where their hydrogen is about exhausted. The variable stars are of this sort. Novae are *O, B,* and *A* type stars. The cepheid variables are *F* and *G* type giants. Larger stars than this must have begun their lives less than five billion years ago.

The exact steps in the aging of a star still remain uncertain, but our limited knowledge of what happens in high-pressure plasmas suggests that something along the following lines is probable. Stars originate, somehow, by the gathering together of interstellar dust and gas, mostly hydrogen, thru their mutual gravitational attraction. This process may have begun from a random density distribution. In any continuous distribution of matter some irregularities must be expected. The density of matter would be greater at some places than the average for the whole of space. If at any point in such a mass, one region should ever become denser than a critical value, the distribution of matter would become unstable. Material would be pulled together faster than the general turbulence of the whole mass

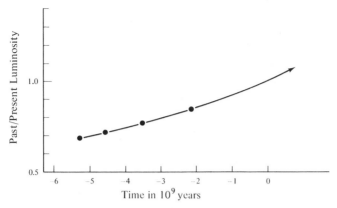

Fig. 2.14 *Predicted variation in solar luminosity with time. (Donn et al.,* Bull. Geol. Soc. Am. *v. 76, p. 289.)*

could scatter it apart again. Once such centers began to form, the process would accelerate. In the early stages, gas streaming into such a center would rapidly build it up. This would empty the surrounding space of matter, so that at present most of the matter of the universe is found concentrated into stars.

As the gas streams in, the inflow of dense material may prevent the escape of radiant energy. In the early stages of formation, a star may be hard to observe. As material falls to the growing center thru the gravitational field, it gains energy. Thus, contraction heats the star. The rise in temperature contributes to the increase in pressure in the plasma. Contraction goes on at each level until the presssure is able to support the weight of the overlying layers. In a large star, this weight is very great, so the temperature will be correspondingly high, especially at the center.

Eventually, the material becomes so dense and hot that nuclear reactions begin. Hydrogen-to-helium reactions are the easiest ones to produce, because they are possible at the lowest temperatures. As nuclear reactions begin, the star becomes even hotter, and begins to radiate enough energy to be visible. After a time, the star will reach an equilibrium where its size and temperature provide a near balance between energy produced by the mass loss involved in the nuclear reactions and the electromagnetic energy radiated away. This is the state in which we see the main-sequence stars today.

Ultimately, the hydrogen will approach exhaustion. The percentage of helium in the star will have risen. The hydrogen content will fall faster at the center of the star than near the surface, because the reaction takes place faster where the temperature is greatest. As the hydrogen is converted into helium, the average compactness of the atoms composing the star increases, because one helium atom is smaller than four hydrogen atoms. The inner layers of the star thus become more closely packed with a consequent increase in temperature. The increase in temperature speeds up the rate of conversion of the remaining hydrogen into helium causing the active layers to become hotter and to expand. More heat is radiated from the star because it is bigger. The surface temperature, however, is less because the star is larger, so that the spectrum of the radiation is shifted slowly to the right in the mass-color diagram (Figure 2.15). The star is turning into a red giant and now occupies a position to the right of the main sequence on a luminosity-color plot.

As the hydrogen becomes exhausted, some stars become unstable, especially the very largest ones. The cepheid variables are such stars. It is probable that many stars lose material during the final stages of hydrogen consumption. The increased size means that the gravitational field at the surface is less than it was earlier. Material is thrown far out from the star's surface in stellar storms. Some is given enough velocity to escape entirely.

Meanwhile, the continued increase in temperature in the star's center allows reactions to occur which form heavier elements than helium. The spectra of many of the giant stars are rich in lines corresponding to metals such as strontium, barium and zirconium. That such materials are actually being produced in the stars is proven by the presence of the radioactive element technetium. Technetium decomposes relatively quickly, and would long since have disappeared in these stars if more were not continuously being made.

Ultimately even the giant stars must run out of fuel. Rearrangements of neutrons and protons in the nucleus offer only limited possibilities. Iron

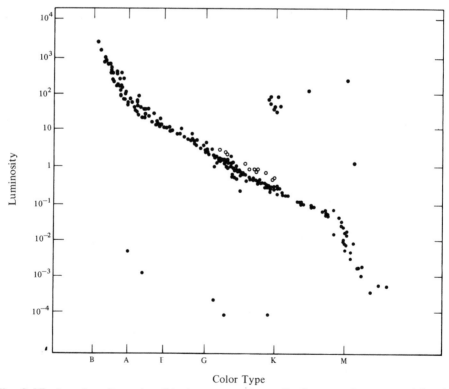

Fig. 2.15 *Luminosity-color (Hertzsprung-Russell) diagram for normal (main-sequence) stars. (After Johnson and Morgan,* Astrophys. Jour. *v. 117, p. 338.)*

has the nucleus containing the least concentration of energy. Smaller nuclei can be combined to form iron with the release of energy; but because of the electrical repulsion of protons for one another, larger nuclei than iron can be formed from smaller only by adding energy. As a result, iron is one of the more common elements in the universe.

Finally the core of a giant star will become so dense that the whole star will collapse under its own intense gravitational pull. As the star shrinks, the surface temperature rises again. But now the star may have become so small in size that it is a white dwarf. Figure 2.16 is a luminosity-temperature diagram of the stars in a galaxy which seems to have many stars in the process of taking the final step from giant to white dwarf. The latter stars are very hot, but are so small that their luminosity is low.

It is difficult to know much about stars which are smaller than the main-sequence stars, because they cannot be seen unless they are near us. There may be varieties of stars in this range which have not yet been detected. One of the newest discoveries is the *pulsar*. A pulsar radiates a large amount of radio-frequency energy at a single frequency, yet is so faint that it cannot be seen in photographs. One theory is that the radiation comes from very rapid expansion and contraction of a superdense star which is as small as or smaller than Earth yet has a mass roughly equal to that of Sun. In stars of such great density, the structure of the component atoms may be so disrupted that even the nuclei are crushed.

What happens beyond this point is still a mystery. Some stars may disrupt themselves thru explosions, scattering themselves in space to provide components for later accumulation in other stars. Others may continue to collapse. How far the process can go is unknown. There are, however, places in the universe where the nature of the electromagnetic radiation is best explained by a steady influx of matter under vastly greater gravitational fields than any known star possesses. If this means that there can be sinks into which the exhausted matter finally flows, then the channels into which these sinks vent are still unexplored regions of the universe.

2.7 DISTANCE MEASUREMENTS.

In order to establish the classification of stars given above, it was necessary to measure distances to many stars. It is easy to measure distance between two points near enough to one another to stretch a tape measure between them. Larger distances between easily reached points can be determined by measuring the number of lengths of the whole tape measure between them. This soon becomes tedious for large distances. A simple method for measuring moderately large distances was introduced by Willebrord Snell in Holland in 1617, and is called triangulation (Figure 2.17). If the length of one side of a triangle and the angles at either end of the side are measured, then the lengths of the other two sides can be calculated using trigonometric relationships.°

° The formula used is the law of sines

$$\frac{\sin b}{CA} = \frac{\sin a}{BC} = \frac{\sin c}{BA} \qquad\qquad \textbf{(2.3)}$$

where the sine of the angle b (abbreviated sin b) is the ratio of the perpendicular distance between the line BC and the point A to the length AB.

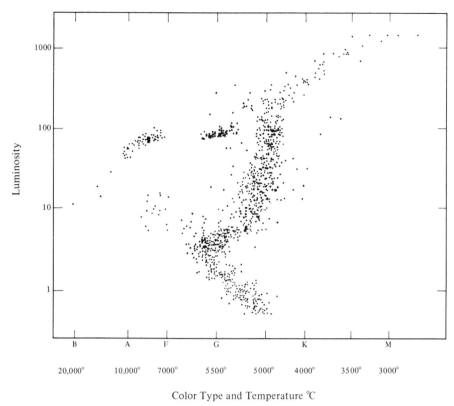

Color Type and Temperature °C

Fig. 2.16 *Luminosity-color diagram for stars of the globular cluster M3. (After Johnson and Sandage,* Astrophys. Jour. *v. 124, p. 356. Copyright 1956 by the University of Chicago.)*

The distance from Earth to Sun can be found in this way (Figure 2.18). The directions to one edge of Sun are measured simultaneously from two points a known distance apart on Earth.

To measure the distance to any other star, sightings are made one half year apart from opposite sides of Earth's orbit around Sun. This method is good out to points where the distance subtended by Earth's orbit (*P* in Figure 2.19) becomes too small to measure accurately. It is useful to about 1.4×10^{15} km (150 light years). At this stellar distance the angle subtended is only 1/20 of a second of arc.

The distances to more remote stars have to be found by other means. One method is to take advantage of the motion of Sun thru space. If it is assumed that the average velocity of all stars in the vicinity of Sun is zero, and their relative motions with respect to Sun observed, then it is found that Sun has a velocity of about 20 km/sec (6×10^8 km per year) toward a point in the constellation of Hercules. Over a period of years this provides a base line which is twice as long per year as the diameter of Earth's orbit. With this base line, the distances to other groups of stars in the Milky Way can be found.

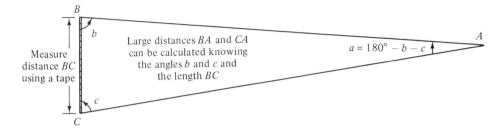

Fig. 2.17 *Principle of triangulation.*

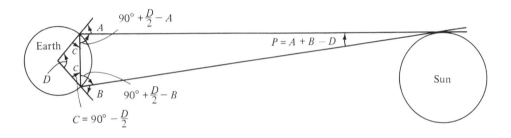

Fig. 2.18 *Distance to Sun can be found using a known length on Earth as a base line.*

This motion of Sun relative to nearby stars should not be confused with the rotational motion of the Milky Way galaxy as a whole, which is larger. If the position of Sun with respect to distant galaxies is observed, it is seen to move at 280 km/sec (9×10^9 km/year) about the center of the galaxy. Stars nearer the center of the galaxy revolve faster. More distant stars revolve more slowly. Assuming that groups of stars are neither falling in toward Sun nor escaping into space (which is probably not quite true), it can be calculated from this pattern that the center of the Milky Way is 3×10^4 light years away. Sun would take 2×10^8 years to make one revolution about this center.

From this velocity, V, and distance, R, it is possible to calculate the mass, M, the galaxy would have to have, if it were all concentrated at its center, to hold Sun in orbit. The gravitational force of the galaxy on Sun exactly overcomes the centrifugal tendency of Sun to fly out into intergalactic space. Equating Newton's universal equation of gravitational attraction to the centripetal force needed to hold Sun in a circular orbit:

$$G\frac{MM_s}{R^2} = \frac{V^2M_s}{R} \tag{2.4}$$

where M_s is the mass of Sun and G is a constant known as the *Newtonian constant* or as the *universal constant of gravitation*. Solving this for the mass of the galaxy

$$M = \frac{V^2R}{G} \tag{2.5}$$

Using mks units (velocity, V, in meters/sec, radius, R, in meters and M in kilograms)

$$M = \frac{(2.8 \times 10^5)^2 \times 3 \times 10^4 \times 9.47 \times 10^{15}}{6.67 \times 10^{-11}} \tag{2.6}$$

$$= 3.3 \times 10^{41} \text{ kilograms}$$

which is the mass of nearly 200 billion (2×10^{11}) Suns.

The Milky Way appears to be a large galaxy, but not exceptionally so. The average spiral galaxy is thought to be about half this size, or to contain about 100 billion stars.

To extend distance scales further, it is necessary to use even more approximate methods. The first of these depends on the observation that the luminosity of the cepheid variables is a function of the period of variation of their light intensity. This was first discovered by plotting apparent luminosity against period in a distant star cluster where all the stars are at approximately the same distance from Earth (Figure 2.20). Assuming this to be a universal relationship, it is only necessary to measure the period, luminosity and distance of a nearby cepheid to obtain a formula which can be used to measure the distance to any cepheid. This is possible because the apparent brightness of a star falls off as the inverse square of its distance, whereas luminosity is defined as its brightness at a standard distance of 33 light years. Therefore

$$\frac{\text{Apparent brightness}}{\text{luminosity}} = \left(\frac{D}{33}\right)^2 \tag{2.7}$$

$$D = 33 \sqrt{\frac{\text{Apparent brightness}}{\text{luminosity}}} \tag{2.8}$$

where D is distance in light years, luminosity is determined from Figure 2.20 and the period of oscillation, and apparent brightness is measured with a telescope equipped with a meter for measuring the amount of light received from the star.

The Cepheid-period method is useful for measuring the distances to galaxies near enough that individual stars can be seen thru the telescope. The nearest galaxy to the Milky Way is the Large Magellanic Cloud, at a distance of about 170,000 light years. In galaxies further away than about 5×10^6 light years, however, only a few very bright stars or no stars at all can be separately distinguished.

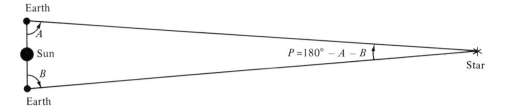

Fig. 2.19 *Distances to stars other than Sun are found using a diameter of Earth's orbit as a base line.*

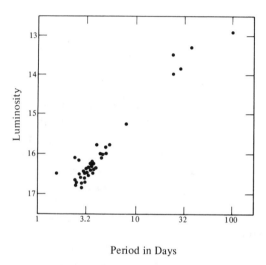

Period in Days

Fig. 2.20 *Luminosity-period dependence for cepheid variable stars in the Large Magellanic Cloud. (After Hodge,* Galaxies and Cosmology, *p. 66. Courtesy McGraw-Hill Book Co.)*

The distances to some remote galaxies can be estimated roughly by assuming that their brightest stars, the O-type stars, are of about the same luminosity as O-type stars in a nearer galaxy at a known distance. Their distances are inversely proportional to their apparent brightness. This extends the distance scale to about 2.5×10^7 light years.

The method for measuring larger distances depends on comparison of the spectra of galaxies. If the spectrum of light from a distant galaxy is compared to the light from a nearby star, it is observed to contain the same arrangement of spectral lines, but the whole sequence exhibits a displacement toward the low-frequency end of the spectrum (Figure 2.21). This *red shift* is most easily explained as a result of the galaxies moving away from the observer. The amount of the red shift is a measure of the velocity of recession. For velocities which are much less than the speed of light, wavelength is increased by the distance the galaxy travels in the time it takes to radiate one wave

$$L_0 = (1 + \frac{V_S}{V_L}) L_S \qquad (2.9)$$

where L_0 is the observed wavelength and L_S is the wavelength generated at the source; V_S is the velocity of recession of the galaxy and V_L is the velocity of light. This can be solved for the velocity of the galaxy where ΔL is the red shift.

$$V_S = (\frac{L_0}{L_S} - 1)V_L = (\frac{L_0 - L_S}{L_S})V_L = (\frac{\Delta L}{L_S})V_L \qquad (2.10)$$

If the velocity is not small compared to the velocity of light, there is a correction introduced by Einstein's relativity. Equation 2.9 becomes

$$L_0 = \frac{1 + \dfrac{V_S}{V_L}}{\sqrt{1 - (\dfrac{V_S}{V_L})^2}} L_S \qquad (2.11)$$

The denominator becomes increasingly important as V_S approaches V_L.

For galaxies whose distances are measurable, the amount of the spectral shift is proportional to the distance of the galaxy. By extrapolation of this relationship the distance to any galaxy can be found from the displacement of its spectral lines (Figure 2.22). The most distant galaxy observed has a red shift of about 46 percent of its normal wavelength. Much of the light from this galaxy is so lowered in frequency that it is outside the visible spectrum in the infra-red. This galaxy appears to be at about 4×10^9 light years from Earth.

What may be even more distant objects have been observed using radio telescopes. Certain quasi-stellar sources (quasars) appear to have even larger red shifts corresponding to recessional velocities of eight tenths the velocity of light and distances of 5×10^9 light years. Their light is shifted all the way out of the visible spectrum into the infra-red.

This brings out a peculiar situation. Since light travels at a finite velocity, one can never see an event as it happens, but only as it has already happened at some previous time. In the laboratory, this time delay is so small that it can be ignored. In looking at very distant galaxies, however, one can see them only as they existed billions of years ago. Thus, as one looks into space, one simultaneously looks back into time.

There are other ways in which light can be changed in color. In the case of very dense stars, there appears to be a loss of energy, and hence a decrease in frequency, of all light leaving the star. This causes a slight shift of all the spectral lines toward the red end of the spectrum. The density of visible matter in galaxies is far too low to cause the observed red shift.

In the presence of a strong magnetic field, the frequency radiated can be both increased and decreased, producing a splitting of each spectral line into two lines, one a little higher and one a little lower than normal (Figure 2.23). This is called the Zeeman effect after its discoverer. It has been used to recognize the presence of strong magnetic fields around some stars. Double lines rather than shifted lines are produced.

RELATION BETWEEN RED-SHIFT AND DISTANCE

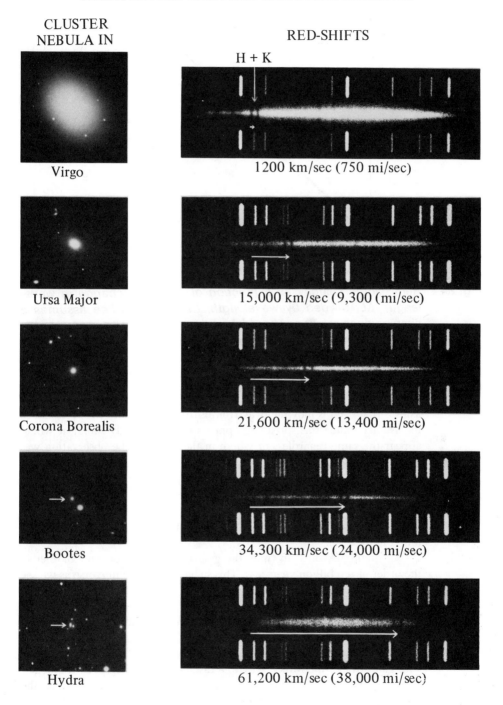

CLUSTER NEBULA IN	RED-SHIFTS

H + K

Virgo — 1200 km/sec (750 mi/sec)

Ursa Major — 15,000 km/sec (9,300 (mi/sec)

Corona Borealis — 21,600 km/sec (13,400 mi/sec)

Bootes — 34,300 km/sec (24,000 mi/sec)

Hydra — 61,200 km/sec (38,000 mi/sec)

Fig. 2.21 *Six galaxies and their spectra with the* H *and* K *absorption lines of calcium showing red shifts. (Courtesy Hale Observatories.)*

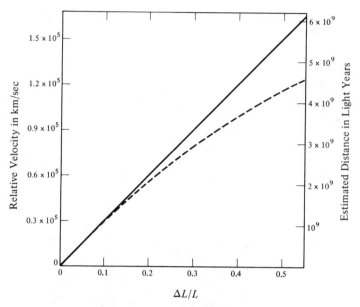

Fig. 2.22 *Recessional velocity as a function of red shift (change in wavelength, ΔL, divided by wavelength, L). The solid line assumes a simple linear relation. The dashed line includes allowance for the effect of Einstein's general theory of relativity.*

2.8 ORIGIN OF THE UNIVERSE.

By far the simplest explanation of the red shift of the light from distant galaxies is that each is moving away from Earth at a velocity proportional to its distance. Since this spreading appears to be primarily radial, it would appear the same regardless of where in the universe the observation was made. All galaxies appear to be receding from all other galaxies. It is as though a man stood on one coil of a long, spiral spring whose ends were too far away to see, and he watched the other turns of the spring receding from him. No matter on which turn of the spring he stood, the turns on either side would move away at a rate proportional to their distances from him (Figure 2.24).

Projected backwards in time, the galaxies converge on one another in the order of ten billion years ago. This suggests that the universe as we see it is undergoing an orderly progression from a former state of great compaction.

It was pointed out earlier that galaxies are much more closely spaced, relative to their size, than stars. The distance between galaxies is on the average only about 100 times their diameters. Suppose the universe is filled with galaxies like the Milky Way to a distance of 10^{10} light years, one galaxy for every part of this volume equal to a sphere of radius 100 times the radius from Sun to the center of the Milky Way, which is 3×10^4 light years. Then, 10 billion years ago the density of matter in this assemblage

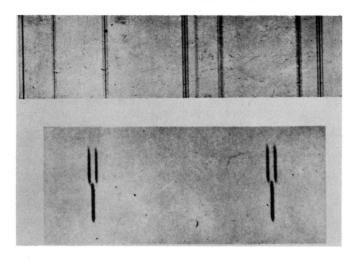

Fig. 2.23 *Spectrum showing Zeeman effect. (Yerkes Observatory Photograph, Univ. of Chicago.)*

would be $\left(\dfrac{10^{10}}{3 \times 10^{6}}\right)^{3}$, or 4×10^{10} times as great as in the average galaxy at present.

Under this degree of concentration, the average density of matter in space would still be very low, only 3×10^{-14} gm/cc. Even if the present universe extended to 10^{12} light years, the original condensed density would be only 3×10^{-8} gm/cc. Such a concentration of matter, however, would be expected to collapse under its own gravitational attraction rather than to spread outwards. Something must have occurred to supply the kinetic energy which allows the galaxies to move apart.

One theory to explain this postulates that all the matter was originally concentrated into a volume small compared to the size of even a galaxy. If the matter is allowed to become sufficiently concentrated, the amount of energy gained by all this matter condensing in its mutual gravitational field eventually becomes as great as the energy represented by its mass. (Remember, energy = mass $\times V^{2}_{light}$.) This condition is so far beyond our present experience that we cannot even predict the laws of nature which might apply. It does not seem unreasonable, however, that such a mass would be unstable and would explode. What we see today is the debris of this explosion still spreading thru the universe.

This explosion is for us the origin of the universe. This does not mean that this is the beginning of the material of the universe. It is merely the earliest stage for which any evidence remains. Earlier stages are forever beyond our ability to observe them.

Another possibility is that the material of the universe alternately expands and contracts. We are at present in an expanding phase. If this is so, then there should be a thinning of material at large distances from the

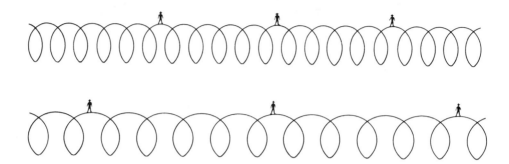

Fig. 2.24

point in space where the matter was once concentrated. But no such thinning of matter has been detected out to the distances we can see.

Other explanations require that the red shift not be the result of expansion. It is possible that some physical process, about which we are ignorant, would cause a shift in spectral color. There is no reason to suspect this to be the case; but this is not evidence. It is also conceivable that the laws of nature as we now know them are only approximations, good for small dimensions and times, such as the interiors of galaxies and recorded history, but poor for large dimensions, such as the whole universe and billions of years. It is assumed that the velocity of light is a constant because all measurements of it have so far given the same value. We cannot prove it was not different 10^9 years ago. Indeed, it is possible to postulate a system of natural laws which will permit the red shift without the galaxies being required to retreat from one another with time. According to one such theory all dimensions change with time, and physical constants, which appear fixed from recent laboratory measurements, are time-dependent. Unfortunately, it has so far been impossible to find a means to check whether such laws are better than the simpler ones we normally assume. Expansion of the universe is the easiest way to explain the red shift of the light from galaxies.

If the universe were formerly more concentrated than it is at present, this would place a maximum age on its parts. It has already been pointed out that individual stars go thru an evolutionary process. The theory of the evolution of stars presented above is consistent with all observed stars having been formed at the same time as the universe, about 10^{10} years ago, or more recently. Only stars about five times heavier than Sun or larger are commonly found to the right of the main sequence on a luminosity frequency chart.

On the other hand, there is ample evidence that stars are still forming. O-type stars are common in the arms of spiral galaxies. They are rarer in the central parts of spiral galaxies and in elliptical galaxies. The latter, on the other hand, are richer in unstable stars and in red giants. This suggests that the arms of spiral galaxies are younger than the centers. Apparently such galaxies are still in the process of formation, with interstellar material still being gathered together out of space.

How it will all end is not yet known. There is no evidence for a sequence of stages of development of galaxies as there is for stars. There is some suggestion that elliptical galaxies may be older than spiral galaxies in terms of the stage of development of the average star. This could be because they develop faster in the absence of the rapid rotation of the spirals.

The relation of the irregular galaxies to the others is not clear. Are they young galaxies still in the process of formation, or are they old galaxies which have somehow broken up, or are they formed entirely independently?

One argument favors an explosive origin for the universe a few billion years ago. This is the existence of the radioactive elements like uranium and thorium. These are unstable, a constant fraction decomposing every year. To exist at all on Earth they must have formed somewhere at some time in the past. Known nuclear reactions likely to take place in the main-sequence stars will not produce them. It is more likely that they are formed only at the tremendous temperatures and pressures encountered in the hearts of old stars after they have used up their hydrogen, or in the original dense parent which blew up to produce the universe. But which? Do the fragments of supernovae find their way into other stars, or is decomposition and death by disruption or collapse a one-way street down which all the stars are marching?

The vast majority of matter in the universe is hydrogen, which is slowly being converted into heavier elements. Unless there is somewhere a process whereby this hydrogen can be renewed to keep on forming new stars and new galaxies, the universe is evolving in one direction. The mass of the universe is constantly being spread thinner and thinner by expansion. The chemical composition is shifting to a mixture rich in stable iron nuclei. The electromagnetic energy is being radiated away at the speed of light.

An alternative possibility is that new hydrogen atoms are continuously being created in space so that the average density of volumes much larger than galaxies is constant. But this is contrary to our most cherished physical rule—the law of conservation of energy. It is ridiculous to think that this most cherished principle could be wrong. Or is this so impossible?

BIBLIOGRAPHY

Abell, G., 1966, *Exploration of the Universe*. Holt, Rinehart and Winston. 646 pp.

Azimov, I., 1966, *The Universe from Flat Earth to Quasar*. Avon Books. 315 pp.

Brandt, J. C., 1966, *The Sun and Stars*. McGraw-Hill Book Co., Inc. 161 pp.

Hodge, P. W., 1966, *Galaxies and Cosmology*. McGraw-Hill Book Co., Inc. 179 pp.

Hoyle, F., 1962, *Astronomy.* Doubleday and Company, Inc. 320 pp.

Krogdahl, W. S., 1962, *The Astronomical Universe.* The MacMillan Company. 599 pp.

Kruse, W.; W. Dieckvoss, 1957, *The Stars.* University of Michigan Press. 202 pp.

Mehlin, T. G., 1968, *Astronomy and the Origin of the Earth.* Wm. C. Brown Co., Publ. 131 pp.

3

THE SOLAR SYSTEM

3.1 SUN

Sun is the star about which we know the most. It is a G-type star. Its distance from Earth can be determined in several ways. One is the triangulation method described in Section 2.7. This is not highly accurate because of the difficulty of measuring precisely the angles involved. A better method is to measure the velocity of Earth in its orbit by observing the spectral shift of the light of nearby stars. Any part of the shift due to the motion of the star relative to Earth can be removed by measuring the shift when Earth is moving at right angles to the direction to the star. The circumference, C, of Earth's orbit is

$$C = VT \qquad (3.1)$$

where V is the velocity around the orbit and T is the time Earth takes for one revolution around Sun. Actually V varies throughout the year. A more accurate formula is

$$C = \sum_T V \triangle t \qquad (3.2)$$

where $\sum_T$ means "take the sum over the total time interval T" of all the component parts, $\triangle t$, times the appropriate values of V. The average radius of the orbit is

$$R = C/2\pi \qquad (3.3)$$

The average distance from Earth to Sun is 1.49599×10^8 km (1.49599×10^{11} meters). This distance is called an astronomical unit.

From this distance and the angle Sun subtends in the sky, its diameter is known to be 1,391,000 km. Knowing the period, T, of Earth's motion around Sun (once per year, or 3.156×10^7 seconds), its mass, M_s, can be calculated from the force required to hold Earth in its orbit around Sun. The gravitational force of Sun on Earth just overcomes the centrifugal acceleration of Earth to the approximation that Earth's orbit is a circle (compare equation 2.3):

$$G\frac{M_e M_s}{R_{es}^2} = \frac{4\pi^2}{T^2}R_{ec} M_e \qquad (3.4)$$

where G is the universal constant of gravitation, R_{es} is the distance from Earth to Sun, R_{ec} is the distance from Earth to the center of mass of Earth and Sun, and M_e is the mass of Earth. Solving for M_s and neglecting the very small difference between R_{es} and R_{ec}

$$M_s = \frac{4\pi^2 R_{es}^3}{T^2 G}$$

$$= \frac{4\pi^2 \times (1.496 \times 10^{11})^3}{(3.156 \times 10^7)^2 \times 6.67 \times 10^{-11}} = 1.99 \times 10^{30} \text{ kg} \qquad (3.5)$$

From its mass and diameter, D, its average density can be found

$$d_s = \frac{M_s}{\frac{4}{3}\pi\left(\frac{D}{2}\right)^3} = 1.41 \text{ gm/cc} \qquad (3.6)$$

The force of gravity (attraction per unit mass) at Sun's surface can be calculated using the universal law of gravitation

$$g_s = \frac{GM_s}{\left(\frac{D}{2}\right)^2} = 273 \text{ newtons/kg} \qquad (3.7)$$

This is about 28 times as great as gravity on Earth. It is because of this high gravity that Sun has so sharp an outline. In spite of its being entirely gaseous, most of its matter is tightly held together.

The amount of electromagnetic energy radiated per unit of time from Sun has been very nearly constant within the time available to measure it. The amount flowing past a point at the average distance of Earth is 20.0 kg-calories/m²/min. This is called the *solar constant*. This is a large amount of energy. One kilogram-calorie is the amount needed to raise the temperature of a liter (1.06 U. S. quarts) of water 1°C (1.8°F). This solar energy is enough to raise a layer of water 2mm (.08 inches) thick from 20°C (68°F) to boiling and to evaporate it every hour. If Earth did not radiate away the heat received from Sun just as fast (on the average) as it arrived, Earth would soon be too hot to live on.

The total amount of heat radiated from Sun can be found by multiplying the solar constant by the area of a sphere whose radius is equal to the distance to Sun. Sun radiates 9.4×10^{22} kg-calories per second. This is equivalent to a 3.9×10^{23} kilowatt generator.

Figure 3.1 shows the relative amounts of energy at different frequencies in sunlight. Most of the energy above and near the top of the visible spectrum is filtered out by Earth's atmosphere. This is fortunate, as this radiation is damaging to living material. Astronauts in space have to be protected from it or they would quickly get a lethal sunburn. By comparing the frequency spectrum of this sunlight with theoretical predictions of the dependence on temperature of the radiated spectrum, the temperature at Sun's surface is calculated to be $5750°K$ $(0°K = -273.16°C)$. Calculations based entirely on theoretical speculations of what Sun's interior may be like indicate that the temperature rises to $20,000,000°K$ at the center.

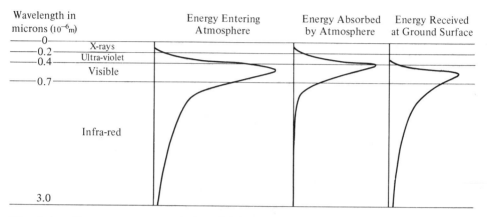

Fig. 3.1   *Energy spectrum in sunlight and at Earth's surface. (After Strahler, The Earth Sciences, p. 194. Courtesy Harper and Row Publ.)*

From the observed lines in Sun's spectrum, its composition is estimated to be that given in Table 2.2 (p.35).

The surface temperature of Sun is not the same everywhere. Enlarged photographs show that the surface is broken into cells giving it a granulated appearance (Figure 3.2). At times, relatively large dark spots appear on its surface (Figure 3.3). These sunspots are areas where the temperature is lower than in adjoining areas. They appear to be areas cooled by upwelling of gas from the interior. As it rises, the upwelled material cools enough by expansion to have a lower temperature ($3700–4600°K$) than normal areas.

Sunspots move (viewed from Earth) from east to west across the face of Sun. The rate of rotation of Sun can be calculated from their motions. At its equator, Sun rotates once every 25 days. The polar regions rotate more slowly, once every 34 days. The motions of the sun-spots also show that Sun's equator is inclined 7° to Earth's orbit. Most sunspots have a life of only two weeks, but a few persist as long as 4.5 months.

Seen in profile as they pass from front to back, sunspots are often accompanied by eruptions of gas from Sun's surface (Figure 3.4). Although most of this material falls back to Sun in a few minutes to hours, some of the ejected material is thrown out at least as far as Earth. Around sunspots, the spectral lines are often split (Zeeman effect), showing that there are strong associated magnetic fields. The spots tend to come in pairs with one member of the pair being of opposite magnetic polarity from the other. The number of sunspots present at any one time fluctuates, reaching maximum every 11.1 years. They are more common near the equator than near the poles. At latitudes below 16° they generally drift slowly equatorward during their lifetime. At larger latitudes they drift poleward. The sunspot cycle is tied to the growth rate of living organisms on the earth in a manner still not understood. Tree-ring growth cycles with the same period as the sunspot cycle have been traced back to 1000 B.C.

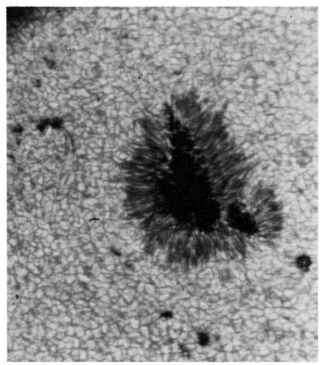

Fig. 3.2 *Portion of Sun's disc photographed from a balloon at an altitude of about 24 km (80,000 ft.). (Courtesy Project Stratoscope of Princeton University sponsored by National Aeronautics and Space Administration, Office of Naval Research and National Science Foundation.)*

The polarity of the magnetic fields of sunspots is tied also to the eleven-year cycle. Between one sunspot minimum and the next in the northern hemisphere of Sun the magnetic polarity of the westernmost of each pair of sunspots is almost always the same. In the southern hemisphere the polarity is the reverse. During the next cycle this arrangement is reversed, the polarity of the westernmost of any pair of sunspots in the southern hemisphere being that of the westernmost spot in the northern hemisphere in the previous cycle. Magnetically, therefore, the cycle is 22 years long.

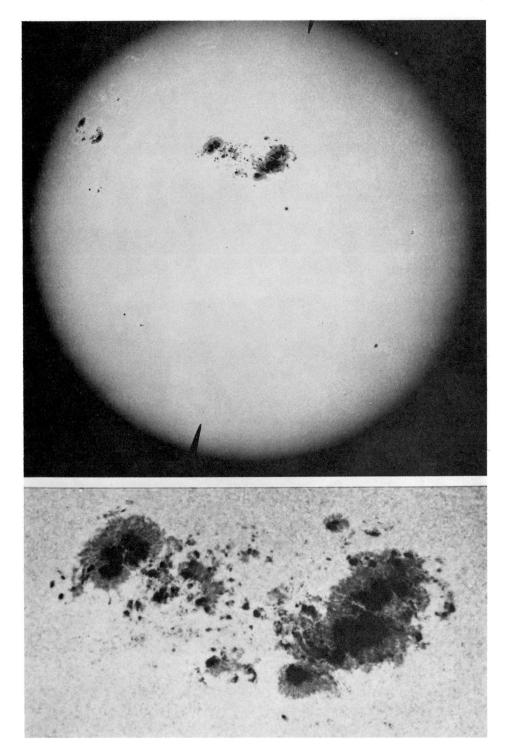

Fig. 3.3 *Large sunspot group on 7 April 1947. (Courtesy Hale Observatories.)*

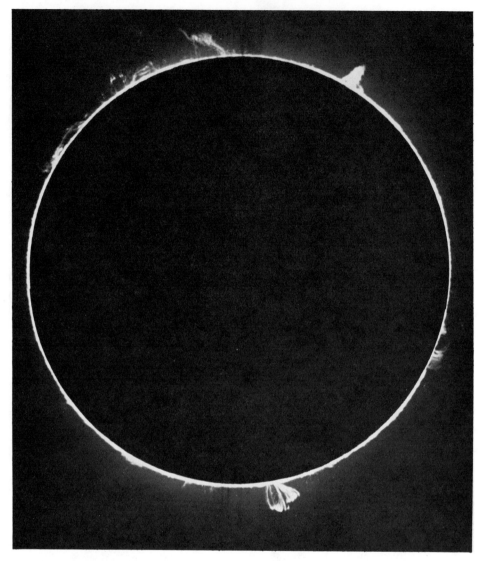

Fig. 3.4 *Solar prominences on 9 December 1929. (Courtesy Hale Observatories.)*

There appear also to be even longer-term variations in the numbers of spots and intensity of sunspot activity. What this means with respect to the radiation received by Earth is unknown. It has been suggested that there may be slow variations in solar radiation which could produce climatic and weather cycles on Earth, possibly even explain the occurrence of large-scale continental glaciation at times in the past. Until the details of the sunspot cycles and their effects on solar radiation are better known, such ideas are mere speculations. The lack of adequate data on how uniform Sun's radiations are points up the danger of extrapolating observations taken over a few decades to millenia or millions of years. What seems to be a constant today may only be so slowly varying that its changes have not yet been detected. For instance, recent fluctuations in the solar constant are suspected, but not proven, to be as great as one percent. Such variations are very hard to measure because so much of the energy is removed from sunlight as it filters thru Earth's atmosphere. The amount removed varies with cloudiness and with other, more subtle variations in the transparency of the atmosphere. To measure the solar constant, the radiation received at Earth's surface is observed and a correction made for losses in passing thru the atmosphere. The exact size of the correction is uncertain. The largest variations are known to occur in the radiation at frequencies which contribute only a small percent of the total energy. The solar energy flux is one of the first quantities being monitored on Moon, because here it can be observed without the interference of an atmosphere.

Sunspots occur within areas of exceptionally high temperature in Sun's surface, and are often associated with eruptions. There are several types of these eruptions. One type is called a prominence. Prominences are large bodies of luminuous, hot gas rising from or falling back to Sun. The material typically is seen to flow along curved paths (Figure 3.4). This can be explained as the motion of ions controlled by magnetic fields. Another type is called a flare. Flares also generally are associated with sunspots, though they originate in the bright areas accompanying the spots, not in the spots themselves. They radiate energy largely at frequencies above and below visible light. They can best be observed with radio telescopes. They also produce X rays and cosmic rays. A stream of ions and electrons is also ejected at high velocities.

These various types of activity throw a great deal of matter into the space around Sun, some of it as far out as Earth. The inflow to Earth of charged particles may interact with Earth's magnetic field. The resulting disturbance is called a magnetic storm. Magnetic storms tend to occur about one day after one of a group of sunspots has passed across the side of Sun directly facing Earth, showing that the sunspot is somehow associated with the source of the ejected material, and that something travels to Earth at velocities far less than the velocity of light.

One can think of the clouds of ejected matter as part of Sun's atmosphere and can consider Earth to be immersed in this atmosphere. The flow of matter past Earth is called the solar wind. The variations in conditions in interplanetary space which result are called space weather.

3.2 PLANETARY ORBITS.

There are a great variety of solid bodies held to Sun by its gravitational pull. Many of these have motions which are so strongly tangential that the bodies are able to maintain separate identities for long periods of time.

The nine planets are the largest bodies orbiting Sun (Table 3.1). Each moves along a nearly elliptical path in a counterclockwise direction as observed from a point in space north of Earth. The paths of the planets about Sun can be described by three laws first pointed out by Johannes Kepler. Kepler obtained these laws by analyzing data gathered painstakingly over many years by Tycho Brahe. Kepler's laws are:

I. Each planet moves around Sun in an elliptical orbit with Sun at one focus (Figure 3.5). Such a path can be described by the formula

$$\frac{x^2}{A^2} + \frac{y^2}{B^2} = 1 \tag{3.8}$$

where x and y are the coordinates of a plot of the path of the planet and A and B are the largest and smallest radii of the ellipse. The ratio $\frac{A-B}{A}$ is called the ellipticity (flattening) of the orbit. The ratio of the distance between the center of the ellipse and the focus to the longest radius, $\frac{Ae}{A} = e$ in Figure 3.5, is called the eccentricity. It is related to the ellipticity by

$$\text{ellipticity} = \frac{A-B}{A} = 1 - \frac{B}{A} = 1 - \sqrt{1 - e^2} \tag{3.9}$$

$$\frac{B}{A} = \sqrt{1 - e^2} \tag{3.10}$$

$$e = \sqrt{1 - \frac{B^2}{A^2}} \tag{3.11}$$

The more nearly equal A and B are, the smaller are the ellipticity and eccentricity. For all the planets except Mercury and Pluto, the ellipticities of the orbits are small, i.e., the orbits are nearly circular.

II. Each planet moves about Sun at a speed such that if a line is drawn from the planet to Sun, this line sweeps out equal areas in equal times (Figure 3.5). When the planet is nearest Sun it moves along its orbit faster than when it is farther away. Looked at another way, as the planet moves out from Sun along its orbit, it slows down because Sun's gravitational pull is holding it back. As its speed along the orbit slows, it is losing kinetic energy of exactly the amount gained by climbing in the gravitational field. Eventually its radial velocity drops to zero. But Sun continues to pull on it, so it starts to fall back. Falling back, its speed increases until eventually it is going so fast that it is carried past Sun and outward again on the opposite side of its orbit.

III. The square of the period of time it takes any planet to revolve about Sun is proportional to the cube of the largest diameter of its orbit. For two planets with periods T_1 and T_2 and largest diameters D_1 and D_2

$$\left(\frac{T_1}{T_2}\right)^2 = \left(\frac{D_1}{D_2}\right)^3 \tag{3.12}$$

Table 3.1 *Data on the planets.*

Body	Distance from Sun in Astr. units°	Average radius in km	Average density in gm/cc	Number of natural satellites	Sidereal period of rotation	Inclination of orbit to Earth's orbit	Tilt of axis to planet's orbit	Sidereal period of revolution about Sun in days	Eccentricity of orbit	Escape velocity (km/sec)	Mass (Earth = 1)	Princ. gases in atoms.
Sun	—	696,000	1.42	—	—	—	7°16'	—	—	619	332,500	H,He
Mercury	0.387	2,420	5.4	0	58.65 days	7°0'	7°0	88	.2056	4.2	.0543	—
Venus	0.723	6,100	5.16	0	241.3 days	3°24'	6°	224.7	.0068	10.4	0.814	CO_2,O_2
Earth	1.00	6,370	5.51	1	23.934 hr.	0°	23°27'	365.26	.0167	11.19	1.000	N_2,O_2
Mars	1.52	3,430	3.95	2	24.623 hr.	1°51'	25°12'	687	.0934	5.03	0.108	CO_2,H_2O
Asteroids	2.8	—										
Jupiter	5.20	71,800	1.34	12	9.93 hr.	1°18'	3°7'	4,333	.0484	59.7	318	NH_3,CH_4
Saturn	9.53	60,300	0.69	9	10.4 hr.	2°29'	26°45'	10,759	.0557	35.4	95.0	NH_3,CH_4
Uranus	19.2	26,700	1.36	5	10.8 hr.	0°46'	98°	30,686	.0472	21.6	14.9	CH_4,H_2
Neptune	30.2	24,900	1.30	2	15.8 hr.	1°46'	29°	60,188	.0086	22.8	17.6	CH_4,H_2
Pluto	39.5	2,900	—	0	6.39 days	17°10'	—	90,885	.2485	—	0.8	—
Moon	.00255†	1,740	3.36	—	27.32 days	5°9'	6°41'	27.32†	.0549	2.38	.0123	—

° One astronomical unit = 1.49599×10^8 km.
† Referred to Earth. Mass of Earth = 5.977×10^{27} gms.

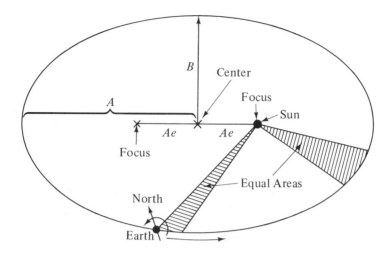

Fig. 3.5   *Earth moves around Sun in an elliptical orbit.*

Kepler's third law provides the most accurate means of measuring the distance from Earth to Sun. The distance to Venus can be measured by noting the time it takes a radar pulse to travel to Venus, be reflected and return to Earth. The distance is simply half this interval times the velocity of light. If this distance, D_v, is measured at a time when Earth, Venus and Sun are all in line (Figure 3.6), then

$$R_1 = R_2 - D_v \qquad\qquad (3.13)$$

Remembering that

$$\left(\frac{D_1}{D_2}\right)^3 = \left(\frac{2R_1}{2R_2}\right)^3 = \left(\frac{R_1}{R_2}\right)^3 \qquad\qquad (3.14)$$

where the R's are the radii of the orbits, and substituting into equation 3.12 (neglecting the ellipticity), an equation is obtained in which R_2 is the only unknown.

$$\left(\frac{T_1}{T_2}\right)^2 = \frac{(R_2 - D_v)^3}{R_2^3} = \left(1 - \frac{D_v}{R_2}\right)^3 \qquad\qquad (3.15)$$

$$\frac{D_v}{R_2} = 1 - \sqrt[3]{\left(\frac{T_1}{T_2}\right)^2} \qquad\qquad (3.16)$$

$$R_2 = \frac{D_v}{1 - \sqrt[3]{\left(\frac{T_1}{T_2}\right)^2}} \qquad\qquad (3.17)$$

Venus' distance can be used even if the three bodies are not in line at the time D_v is measured, but then the calculations become more complicated because they also involve measurements of angles. In practice, a correction must be made for the ellipticity of the planets' orbits.

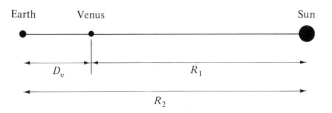

Fig. 3.6

The distance to Sun cannot be measured directly by radar because Sun does not have a solid surface to serve as a distinct reflector and because Sun is a source of a large amount of radiation at radar frequencies.

Isaac Newton used Kepler's laws to derive the *universal law of gravitation.* He argued that if the three rules applied to all the planets, they should apply to all bodies. He was able to show that motions of the type observed were the expectable result if every body exerted a force, F, on every other body of amount:

$$F = G \frac{M_1 M_2}{R_{12}^2} \tag{3.18}$$

where M_1 and M_2 are the masses, R_{12} is the distance between them, and G is a constant of proportionality whose value depends on the units used. G is called the *Newtonian constant* or the *universal constant of gravitation.* To determine the gravitational force near a large body, it is necessary to sum the pull of all its parts. For bodies as widely separated as Sun and the planets, the formula can be conveniently applied treating each body as though all its mass were concentrated at its center.

Newton's formula implies that each body such as Sun or a planet exerts a force on and moves relative to every other body. This is exactly what is observed. Moon and Earth obey this law in the same way that the planets and Sun do. If the paths of Earth and Moon are observed relative to Sun, it is found that each moves about a common point whose distance from Sun follows a nearly perfect elliptical course. Both Earth and Moon move on paths which fluctuate in distance and in velocity relative to Sun (Figure 3.7). The small departures of the center of gravity of the Earth-Moon system from a perfect ellipse can be shown to be very nearly what would be expected as a result of the gravitational pull of the other planets.

The radius of the elliptical path of a planet about Sun depends only on the tangential velocity of the planet and not on the planet's mass. Thus a small rocket would have a velocity with respect to Sun exactly the same as that of Earth to stay in an orbit of the same size and shape. If it left Earth with a low tangential velocity, it would fall inward toward Sun until it gained enough energy to allow it to move outward again. Its final orbit would have a smaller average radius than Earth's. If it started out fast, it would move outward until it lost speed, and its final orbit would have a

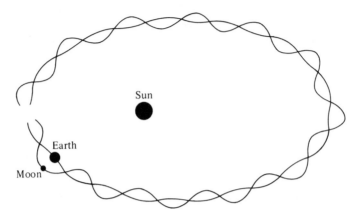

Fig. 3.7   *The paths of Earth and Moon about Sun showing mutual per-*
turbations (exaggerated).

larger average radius. Because the gravitational pull is inversely proportional to the square of the distance from Sun, a planet near Sun must move
tangentially faster than a planet at a larger distance to keep from falling
inward. This is why the period of Mercury is shortest and that of Pluto
longest among the planets (Table 3.1).

All the planets move about Sun in the same sense that Sun rotates on its
axis (counterclockwise when viewed from the north). Furthermore, all the
planets except Mercury, the innermost one, and Pluto, the outermost one,
move nearly in a single plane thru Sun. The tilt of Mercury's orbit is only
7° from this plane, and that of Pluto is 17°19′. When one examines the rotation of the planets on their own axes, the variation is much greater.
Many of the planets have tilts of their axes of rotation of over 20° with respect to the plane of the orbit. The axis of Uranus lies only 8° from the
plane of its orbit, and the direction of its rotation is the reverse of that of
its revolution about Sun (Figure 3.8). It is as though the axis of Uranus
were tilted 98° with respect to the plane of its orbit. Venus also rotates in
the opposite sense from its revolution, although very slowly.

The masses of those planets having satellites can be found from the
forces necessary to hold the satellites in their orbits at their observed velocities (Equation 3.4). The mass of the satellite need not be known for this
calculation. For those planets lacking satellites, their masses are calculated
from their gravitational effects on the orbits of other planets. Venus per-
turbs the path of Moon around Earth and of Earth around Sun a measurable amount.

The masses of the various moons of the planets (Table 3.2) are much
harder to find than the masses of the planets themselves. Of help in determining satellite masses are the motions of some of the smaller planetary
bodies. In addition to the nine large planets there are many smaller *planetoids* moving about Sun. (They are often, but less appropriately, called as-
teroids.) Most have orbits whose average radius is greater than that of Mars
but less than that of Jupiter. The largest have names, and their paths thru

Table 3.2 *Data on the satellites*

Planet	Name	Period (in Earth days)	$M_s SM_p$	Largest radius of orbit (thousands of km)	Direction of revolution compared to rotation of planet	Density (gm/cc)	Radius (km)
Earth	Moon	27.322	.0123	406	same	3.36	1,740
Mars	Phobos	0.3189	—	9.45	same	—	16
	Diemos	1.262	—	23.7	same	—	8
Jupiter	V	0.4982	—	18.2	same	—	120
	Io	1.769	3.81×10^{-5}	424	same	4.0	1,600
	Europa	3.551	2.48×10^{-5}	675	same	3.8	1,500
	Ganymede	7.155	8.17×10^{-5}	1,080	same	2.4	2,500
	Callisto	16.69	5.09×10^{-5}	1,890	same	2.1	2,200
	VI	250.6	—	11,500	same	—	50
	VII	260.1	—	11,800	same	—	10
	X	260	—	11,780	same	—	10
	XII	617	—	20,800	retrograde	—	8
	XI	692	—	22,500	retrograde	—	7
	VIII	735	—	23,400	retrograde	—	8
	IX	758	—	23,800	retrograde	—	6
Saturn	Mimas	0.924	6.7×10^{-8}	188	same	0.5	260
	Enceladus	1.370	1.3×10^{-7}	241	same	0.7	310
	Tethys	1.888	1.1×10^{-6}	298	same	1.2	610
	Dione	2.737	1.8×10^{-6}	382	same	2.8	520
	Rhea	4.518	—	533	same	2.0	780
	Titan	15.95	2.4×10^{-4}	1,230	same	2.4	2,600
	Hyperion	21.28	—	1,500	same	—	210
	Iapetus	79.33	—	3,600	same	1.3	610
	Phoebe	550.4	—	13,100	retrograde	—	160
Uranus	Miranda	1.413	—	147	same	—	160
	Ariel	2.520	—	216	same	—	1,300
	Umbriel	4.144	—	300	same	—	650
	Titania	8.706	—	493	same	—	1,300
	Oberon	13.46	—	659	same	—	1,300
Neptune	Triton	5.877	1.3×10^{-3}	395	retrograde	—	2,100
	Nereid	359.4	—	6,210	same	—	160

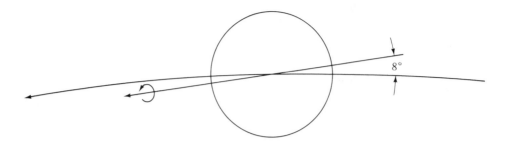

Fig. 3.8   *Uranus' axis of rotation lies only 8° from the plane of its orbit.*

the solar system are well established. Many have much larger ellipticities than Earth, and at their minimum radii are nearer Sun than Earth. These bodies as they pass near a moon are deflected from an elliptical path by the moon's gravitational pull. The moon tries to pull the planetoid into an elliptical path about itself instead of about Sun. By observing the amount of deflection the moon causes, the moon's mass compared to Sun's mass can be found; and since Sun's mass is calculable from Earth's orbit, the mass of the moon deflecting the planetoid can be found.

The diameters of the planets are found from the angles in the sky they subtend and their distances. Knowing their diameters and masses their average densities can be calculated. The densities vary from 0.69 gm/cc for Saturn to 5.51 gm/cc for Earth. The density of Pluto is not known. The figure obtained from the best available estimates (Table 3.1) of its diameter and mass is so high that one or both of these figures must be in gross error. (Its calculated density of 47 gm/cc is more than four times that of lead, twice that of the densest element, osmium.) The mass is hard to measure since it must be found from the planet's very small effect on the orbit of Neptune. The diameter is hard to measure because Pluto is so faint. It may be that Pluto is larger than it appears to be. Perhaps the edge of Pluto where it is tilted away from Sun reflects so little light that it is invisible.

3.3 PHYSICAL APPEARANCE AND COMPOSITION OF THE PLANETS.

The light reflected from any planet is controlled primarily by three factors. First, the composition controls the relative amounts of energy reflected and absorbed at any frequency. A body with a relatively poor ability to absorb blue light will look blue because it reflects this color. The absorbed part of the energy received from Sun is eventually reradiated, but almost entirely at infra-red frequencies which are not visible. Only hot bodies radiate most strongly at visible or higher frequencies. Burning materials often radiate enough red to be said to be "red hot," and many materials, e.g. a lamp filament, may become "white hot" before or after melting.

The second factor affecting reflected light is the shape of the surface. A rough surface reflects in all directions, a smooth surface reflects with less scatter. If the particle size on the surface is small enough, it may affect various colors differently. A surface may reflect differently with respect to the direction of vibration of the electromagnetic force, resulting in differences in the polarization of light.

Lastly, and of great importance, the atmosphere of the planet will absorb some frequencies more efficiently than others. The atmosphere may be a reflector also. This is likely if there are clouds in the atmosphere. These clouds may be condensed droplets (liquid) or crystals of one or more of the gaseous components of the atmosphere or they may be dust clouds picked up by winds from the planet's surface or thrown into the air by volcanic or some other (unknown) process.

3.3.1 MERCURY

Radar reflections from Mercury's surface show it rotates once in about 59 days. The rate is determined from the broadening of the energy spectrum due to the fact that one edge of Mercury is approaching Earth while the other edge is moving away. The scatter of radar reflections shows that Mercury's surface is rougher than Mars's.

From careful measurements of the positions of surface features, the period is more accurately set at 58.65 days. Determinations of the period from surface markings is not as easy as one might expect, because the planet is so close to Sun that it can be seen easily only twice per revolution, when it is neither in front of nor behind Sun, and it is most clearly seen only once per year when the distance to Earth is a minimum. Because Mercury's period of rotation is two-thirds its period of revolution, it usually presents the same face to Earth each year at the time of best observation. As a result, previous to the radar measurements, astronomers believed Mercury always kept the same face toward Sun. Either a one-to-one or a three-to-two locking of the periods of rotation and revolution can occur as a result of the gravitational pull of Sun on any density irregularities in Mercury. Mercury is too far away for us to observe the surface in detail. Mottled markings seem to be permanent in position, and so are presumably the result either of compositional or surface-roughness factors.

From the degree of polarization of reflected light from Mercury, it is thought that, if it has an atmosphere at all, the thickness is not over .003 that of Earth (measured in mass per m^2 of surface area). Because of the weak gravity at the surface of Mercury, it is likely that any atmosphere consists primarily of elements which are heavier than those common in Earth's atmosphere. Light gases will have a tendency to escape from an atmosphere. This is because the temperature of the atmosphere is a result of the average heat of the component atoms and molecules. In a gas at a low or moderate temperature, heat is largely in the form of particle velocity. In a large body of gas the individual atoms are continually colliding and exchanging momentum. Some particles will be moving faster and some slower than others. There will be a range of distribution of particle velocities centered on the velocity corresponding to the temperature of the gas

(Figure 3.9). The directions of motion will be random, that is motions in every direction will be equally likely. Some particles will be moving straight up. In the lower part of the atmosphere such a particle will sooner or later collide with another particle and be deflected down again. In the upper part of the atmosphere, however, the particles are far apart, and the upward-moving particle may not have such an encounter. If it is moving slowly, it will soon fall back to the planet due to the pull of gravity. If it is moving fast enough, however, the planet will be unable to stop its upward motion. The velocity a particle must have to be able to keep going upward indefinitely is called the *escape velocity.*

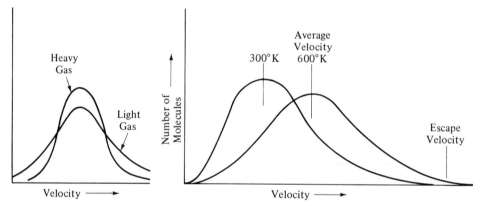

Fig. 3.9   *Relative particle velocities in a gas. (After Gamow, 1958,* Matter, Earth and Sky. *Courtesy Prentice-Hall, Inc.)*

The broadness of the particle-velocity profile (Figure 3.9) is greater for light gases than heavy gases. (This is because the amount of energy it takes to give a unit increase in velocity to a particle is inversely proportional to the square root of its mass. Therefore, the lighter and hotter the gas, the greater the proportion of that gas at very high velocities.) Mercury, being nearest Sun, has the highest surface temperature. Being the smallest planet it also has the lowest gravity. Any hydrogen or helium it may have had in its atmosphere a few billion years ago would by now have escaped, along with a large fraction if not all of its oxygen and nitrogen. It is likely, there-fore, that whatever atmosphere Mercury has is largely or entirely com-posed of heavy elements such as argon.

The composition of Mercury cannot be precisely determined. Mercury is the densest (5.4 gm/cc) of the planets except for Earth (5.51 gm/cc) and possibly Pluto. Therefore, Mercury is too dense to be composed of the type of rocks exposed on Earth, all of which have densities of 3.5 gm/cc or less. The only known bodies in the solar system with densities as great as Mer-cury are the metallic meteorites, which sometimes fall thru Earth's atmo-sphere. These are composed largely of iron. The surface of Earth is com-posed largely of compounds of oxygen, silicon, aluminum and iron, plus lesser amounts of almost all the other elements. Of these atoms, oxygen oc-cupies the largest space in crystalline compounds. One theory for the great density of Mercury is that it is simply low in oxygen, which would easily escape from its atmosphere, leaving the whole planet enriched in other elements commonly found in the solid part of Earth. Whether or not the proportions of these heavier elements are similar to the proportions on Earth is unknown. Certainly Mercury must be on the whole richer in heavy elements.

3.3.2 VENUS

Venus is covered at all times by clouds which obscure its surface. In detail, these continually change appearance, but they have so few recognizable features that no pattern of atmospheric circulation or planetary rotation has yet been identified by observing them. Radar reflections from Venus's surface show that it rotates about once every 250 days in the opposite sense from its revolution about Sun.

The Russian Venus probe found that its atmosphere was composed of about 90% carbon dioxide with 0.4% oxygen. Spectroscopic studies have also indicated the possible presence of nitrogen tetraoxide (N_2O_4), water, molecular nitrogen, carbon monoxide (CO), hydrofluoric acid (HF) and hydrochloric acid (HCl). The clouds themselves could be N_2O_4, ice crystals or even dust. The temperature is too high for them to be carbon dioxide.

Venus is a little smaller and less dense than Earth. This suggests that the overall composition of the solid part of the planet may be similar, although surface conditions will be made very different by the different atmosphere and high temperature. The surface temperature of Venus can be estimated from the strength of radio-frequency radiation, and must be around 400°C. With so high a temperature, Venus may be very windy. Wind erosion rather than water erosion can be expected to dominate its land forms. This could result in thick dust layers over low-lying areas and very rough or even jagged rock outcrops wherever the surface has not been worn smooth. Radar reflections from Venus exhibit reversed polarization, presumably due to double reflections from the planet's surface. From this it is known that there are areas of substantial relief.

The Mariner-II space vehicle, which passed within 41,000 km of Venus, detected no magnetic field. Later probes indicate that any field which may be present must be no greater than .0003 of Earth's field. This could mean that Venus lacks a fluid core such as is believed to be the source of Earth's magnetism.

The possibility of a fluid "ocean" on Venus' surface is harder to evaluate, particularly since the composition of the atmosphere is incompletely known. At 400°C there could not be bodies of water. On the other hand, the low rate of rotation of Venus is most easily explained by supposing that a more normal rate of rotation once existed but has been transferred to orbital motion by tidal damping (see Section 3.5). If the interior of Venus is not liquid (now or formerly), then the possibility of some sort of liquid surface layer is enhanced. The dielectric constant of the surface of Venus can be found from analysis of radar reflections, and is about that of common rocks and dry soils on Earth. It is much too low for water. It is possible, however, that lakes or seas of some liquid hydrocarbon might be present.

On Earth, such a body would probably combine with oxygen to form water and carbon dioxide, but Venus' atmosphere is believed to contain very little free oxygen.

3.3.3 Earth

Earth will be omitted in this chapter, because it is treated in detail later.

3.3.4 Mars

Mars is the planet whose surface environment differs least from that of Earth. Mars has a thin atmosphere, so that its surface is visible. Faint markings enable its period of rotation of 24.623 hours to be measured accurately (Figure 3.10). Parts of its surface are good reflectors of red and orange light. This could be due to a surface composition rich in iron oxides and silicates. Elsewhere the surface color is bluish and greenish brown, grey and black. These areas are less reflective than the red and orange colored areas. Pictures of the surface taken from space vehicles show it to be scarred by craters assumed to be produced by meteorite impact (Figure 3.11).

During the Martian cold season, white caps appear on the polar ends of Mars (Figure 3.10). These vanish every spring, their disappearance being accompanied by the development of a dark fringe suggesting some chemical variation in Mars's surface as the cap melts. Temperatures are lower and cover a wider daily range between extremes than on Earth. At the equator, temperatures may rise to about 20°C (68°F). At night they may drop to –90°C (–130°F). Polar temperatures as low as –130°C (–202°F) are expected. Mars's orbit is more eccentric than that of Earth, and the tilt of its axis to the plane of the orbit is slightly larger, producing greater seasonal variations in climate than on Earth.

With such low average temperatures, the surface material on Mars would be perpetually cold at a few centimeters depth. There could be no soil of the type found on Earth except for a thin veneer. Earth's soil properties depend greatly on its water content. At low temperatures, Earth's soil freezes and is as impermeable as rock. Mars could have thick dust layers, and in the absence of moisture these could provide dust clouds which, blown by wind, would be a major factor in modification of the surface features of Mars. The prevalence of meteorite craters, however, suggests that erosion rates are slow on Mars compared to Earth.

By recording the strength of radio waves received from Mariner IV as it passed from sight behind Mars, the density of the planet's atmosphere was calculated. At the surface this atmosphere has about one hundredth the pressure of Earth's atmosphere. The low pressure is due both to the lesser amount of mass per unit area and to Mars's lesser gravitational pull. Spectroscopic studies indicate the atmosphere is composed largely of carbon dioxide, with a small amount of water vapor. No other gases have been detected, although there could easily be considerable nitrogen and carbon

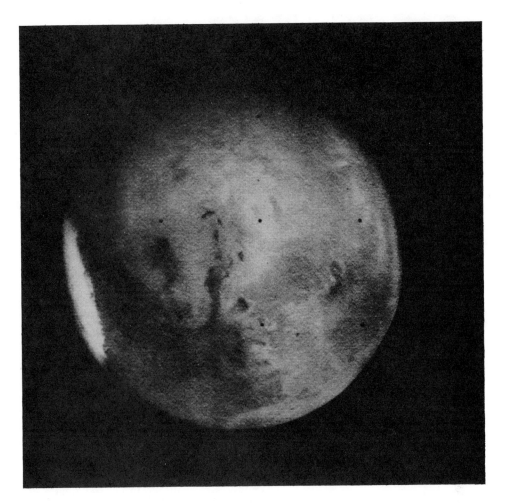

Fig. 3.10 *Mariner 7 photograph of Mars. (Courtesy U.S. Natl. Aeronautics and Space Administration.)*

monoxide without either being found, as well as small percentages of the heavier inert gases. There is some spectral evidence (weak absorption lines) for the presence of nitrogen tetraoxide, N_2O_4, and methane, CH_4. There cannot be more than 0.1 percent of free oxygen or it would be detectable. The white polar caps could be either ice or dry ice (frozen carbon dioxide), N_2O_4 or mixtures of these materials. Haziness at the Martian dawn is believed to be due to condensation in the atmosphere.

Telescopic observations of Mars's surface under the best atmospheric conditions on Earth have detected faint linear markings. These were named "canals" on the hypothesis that they might represent cultivated valleys. It is now believed that any life which may exist on Mars must be much more primitive than that on Earth. The extreme cold implies a much slower rate of evolution. The lack of liquid water requires that any life which exists center on different chemical processes than those of greatest importance on Earth. It is quite possible that Mars is lifeless.

Fig. 3.11 *Surface of Mars photographed by Mariner 6 spacecraft. (N.A.S.A. photograph.)*

Mariner IV approached within 13,000 km of Mars. As in the case of Venus, no magnetic field was observed. Mars has only a little more than one tenth the volume of Earth, and its average density is only 3.95 gm/cc. This low density suggests that the average composition of Mars is much poorer in heavy elements such as iron than either Earth or Venus. The lack of a magnetic field suggests, again as in the case of Venus, that there is no liquid core. With a density of only 3.95, there may be no core at all.

Mars is not perfectly round, having an equatorial diameter 1/192 larger than the polar diameter. This is a greater equatorial bulge than that of Earth, and indicates that there is a lesser increase in density with depth than in Earth. This is consistent with there being no dense core in the center of Mars such as that found in Earth.

3.3.5 Jupiter

The four planets from Jupiter to Neptune contrast with the inner ones in size and in density. Jupiter is the largest. Its density is only slightly less than that of Sun, suggesting that it is composed largely of similar materials. Its upper-atmosphere temperature, estimated from spectrographic studies and the amount of radiant energy reaching it from Sun, is around −140°C. A body of the observed volume and temperature of Jupiter but with the same composition as Sun would be expected to be less massive than Jupiter. Therefore, Jupiter must be richer in some heavy elements than Sun. The amount of extra material in Jupiter beyond an equal volume of solar composition is roughly equal to the mass of Earth. This much rocky material scattered throughout the whole of Jupiter's volume would constitute little more than scarce impurities.

The surface of Jupiter is completely obscured by clouds in its atmosphere. Hydrogen, ammonia (NH_3) and methane (CH_4) have been identified spectrographically and helium is presumed also to be an important constituent. (The spectrograph cannot detect helium at low temperatures.) At the estimated temperature, the clouds are likely to be composed largely of crystals of ammonia. In detail, the cloud patterns of Jupiter's surface are constantly changing, but the overall pattern is remarkably constant. There are dark belts parallel to the equator which are thought to be the result of some sort of permanent current pattern in the atmosphere (Figure 3.12). These cloud belts are deep red to brown in color. The color is thought to be due to traces of some minor constituent of the atmosphere, possibly sodium.

Bright and dark spots which form and fade away on the surface of Jupiter are believed to represent atmospheric current cells of some sort. Most notable of these is the *Great Red Spot* which has existed with varying prominence at least since 1831 (Figure 3.12). It rotates around Jupiter once every 9.929 hours. The rate of rotation of Jupiter's atmosphere varies with latitude, being greatest at the equator. It also fluctuates with time. On the average the equatorial region of Jupiter gains one rotation on the polar region every 103 Jovian days. This type of circulation, where the equatorial rotation is greater than the polar, also occurs in Sun.

The rate of rotation of the solid portion of Jupiter cannot be observed directly. However, Jupiter has a magnetic field linked to the upper fringe of the atmosphere, which is ionized. This ionized region radiates radio-frequency energy. From variations in the intensity of this radiation it is found that the direction of the magnetic field is not parallel to the axis of rotation of Jupiter's atmosphere, but is tilted between 9 and 10°. The magnetic axis rotates in 9.925 hours (13.5 seconds per day faster than the Great Red Spot). Assuming that the magnetic field is created beneath the atmosphere and that it is not moving with respect to the axis of rotation of the solid portion of Jupiter, then 9.925 hours is the period of rotation of this core. This is inconsistent with the hypothesis that the Great Red Spot is a result of an updraft or downdraft in Jupiter's atmosphere due to a topographic irregularity on Jupiter's surface. An alternative is that the magnetic field is moving about 0.14° of arc with respect to Jupiter's solid portion every rotation. Such a magnetic drift would be very rapid for so large a field.

Jupiter rotates fastest of all the planets (once every 9 hr. 55 min.). The rapid rotation of Jupiter produces a large equatorial bulge. The equatorial diameter is about 1/15 larger than the polar diameter.

The presence of a magnetic field also suggests that at least part of Jupiter's core is fluid, a convective core being the most likely source of the magnetic field. The diameter of the solid part of the planet is not known.

Fig. 3.12 *Jupiter. (Courtesy Hale Observatories.)*

At the temperatures and pressures calculated for Jupiter's interior, hydrogen would freeze at a depth of a few hundred kilometers from the top of the observed clouds. Assuming, as seems likely, that the solid part of Jupiter is largely hydrogen, the temperature at the center would rise to several thousand degrees. It is further postulated that the center of Jupiter is liquid. Liquid hydrogen at these temperatures and pressures is thought to behave very much like a metal and to have high electrical conductivity. It is in this liquid inner core of Jupiter that the magnetic field is believed to be generated.

3.3.6 Saturn

Saturn has the distinction of being the only planet whose density is less than that of water. It also has the greatest equatorial bulge, 1 part in 9.5. This is due in part to its low density, which leads to a gravitational attraction only a little greater than Earth's, and in part to its high rate of rotation: 10.23 hours at the equator, 10.63 hours near the poles. Like Jupiter,

Saturn rotates faster at the equator than at the poles. It also has similar bands of clouds in its atmosphere and occasional dark spots, from which its rate of rotation is determined. Both bands and spots are less prominent on Saturn than on Jupiter. This may be due to its lower temperature (-150°C).

Both ammonia and methane are present in Saturn's atmosphere, but the amount of ammonia is less, as would be expected at a lower temperature. Almost all of the ammonia should be frozen out of the atmosphere on Saturn. Its size and density are consistent with its being composed largely of solid hydrogen. It can be presumed to have lesser amounts of the other elements, though the proportions are impossible to determine.

3.3.7 Uranus

Uranus is a smaller, colder edition of Saturn. It shows faint markings in its atmosphere and spectrographic lines for methane and hydrogen. In reflected light it has a greenish tint, believed due to the tendency of methane to absorb the yellow and red from sunlight. At times a whitish equatorial zone and other faint bands can be seen. Its atmospheric markings are too indistinct to be used to determine its rate of rotation. This can be found, however, from small variations in the amount of light reflected by the planet, and from the spectral shifts of light reflected from the approaching and receding sides.

Its rotation is Uranus' most unusual feature. Its axis lies only 8° from its plane of revolution, and its sense of rotation is the reverse of its direction of motion around Sun (Figure 3.8). The two peculiarities are likely related. Its reverse direction of rotation would be a consequence of its axis being tilted 98°, just 8° more than at right angles to the plane of revolution. That this is the correct way to view the sense of the rotation is supported by the fact that Uranus' five satellites all revolve in the same sense that Uranus rotates, as is the usual case in other planetary systems.

3.3.8 Neptune

Neptune is slightly smaller than Uranus. It also appears greenish due to methane. Its period of rotation is determined by the same methods used for Uranus.

3.3.9 Pluto

Pluto is so small and distant that little can be said of its nature. Even its density is unknown. It is the only planet whose orbit does not lie close to the plane of the other planetary orbits, and its eccentricity is the largest of that of any planet. This eccentricity is so great that Pluto can pass between Sun and Neptune.

This raises the question of what might happen if Pluto and Neptune passed very close to one another. Pluto is smaller than several of the planetary satellites. Calculations of its motion show that it could pass between

Neptune and Triton, that planet's inner satellite. Triton is a peculiar satellite in that its direction of revolution is retrograde, the reverse of Neptune's direction of rotation. Should Pluto pass between Neptune and Triton, both could be deflected from their present orbits. Pluto could become a satellite of Neptune and the direction of revolution of Triton would be reversed to the prograde (normal) sense. The reverse could have happened in the past. A close encounter between Pluto and Triton when both were satellites of Neptune could have ejected Pluto and reversed the direction of Triton.

3.4 ORIGIN OF THE PLANETS.

The relation of Pluto to Neptune raises the question whether the orbits of other satellites are similarly unstable. It is possible that the solar system once consisted of more bodies than at present, and that some of these have been lost to space or have combined with those that remain. The present great stability of the planetary orbits may only mean that these are the bodies whose motions are so symmetrical that they are stable. Bodies with less symmetrical motions may have had only a limited lifetime.

The reverse is also possible. If a closer encounter occurred between Sun and some other small body, the latter could have been captured.

A close encounter between Sun and any other large body would very probably have so disrupted the whole solar system that the present arrangement of the parts would bear little resemblance to the previous arrangement. At one time it was believed that the origin of the solar system involved such an encounter. There is no known evidence that such an event has taken place; nor, considering the distances between stars in our galaxy, is it likely that such an encounter ever occurred, unless it took place during the primeval concentration of all galaxies in a tiny part of space.

It seems more likely that the origin of the planets is a normal corollary to the origin of Sun, and that planets are composed of the same parent matter as Sun. The great similarity of planetary orbits and their close approximation to a single plane of motion strongly suggests a common origin for all of the larger planets and their satellites. The principal differences in composition are easily explained in terms of the sizes of the various bodies. The surfaces of the major planets from Jupiter thru Neptune are cold because of their distance from Sun. As a result, they lack the tendency of the inner planets to lose their light gases, and are hence large and of low density. Beyond Neptune, Sun's gravitational force is so weak that the gas cloud from which all the material is postulated to have condensed would at best have been more tenuous than further in, and the material available for forming planets correspondingly less plentiful. There is a general but not uniform decrease in planetary mass moving outward from Jupiter. The possibly special origin of Pluto has already been discussed.

There is no way of being sure that there is not another planet beyond Pluto; but, equally, there is no reason to suppose that there is one. Whatever the process of formation, it must explain the concentration of the planets into a narrow planar zone, and this restraint begins to break down with Pluto. A more distant planet with a much more eccentric orbit would be in danger of escaping completely due to some internally or externally caused perturbation of its orbit.

A great deal has been written concerning the probability and improbability of bodies such as the planets having condensed from interstellar matter. The fact that the possibility of condensation is controversial means that the actual origin of the solar system is likely to remain unsolved for some time. The abundant evidence that stars do form, evolve and die is clear proof that they, at least, can condense. If a wide range of sizes of stars can condense, then it seems reasonable that bodies as small as the planets can also form from dust or gas clouds. The only question is whether the right conditions for this to happen ever did exist in the vicinity of Sun.

Once the process is started, it is easy to picture the gradual drawing together of matter into a cloud whose density is steadily increasing under its mutual gravitational attraction. A rotating cloud would of necessity condense fastest parallel to its axis of rotation. The final stages of contraction of the cloud might be violent, with all of the cloud passing thru a stage of instability in which only the roughest sorts of regularity would be expected. This is exactly the degree of regularity which the solar system exhibits: general but not complete uniformity in the directions of rotation and revolution; a predominant direction to the axis of rotation and revolution with much scatter about this direction; and nearly circular but not perfectly circular orbits. With so little information on such critical points as the compositions of the planets, it is hard to say more than that condensation of the planets from a cloud at the same time that Sun formed seems to be a reasonable theory of the origin of the whole solar system.

3.5 MOON.

In addition to the nine planets there are an uncounted number of smaller bodies in the solar system. The best known of these are the 31 satellites of the planets (Table 3.2). Earth's Moon is the most accessible for study.

Moon is unique among planetary satellites in that its mass is more than one per cent of the mass of the planet it circles. Moon is also the best known body outside Earth and Sun. Data on Moon's mass and orbit are given in Tables 3.1 and 3.2. The given dimensions and period of Moon's orbit are only averages, because Moon's motion is complicated by a number of factors. First, the attraction of Sun on Moon is greater than the attraction of Earth on Moon. This makes the ellipse of Moon's revolution rotate around Earth once every 8.85 years (Figure 3.13). Because Earth's orbit is elliptical, and Moon is carried by Earth around Sun, the amount of Sun's gravitational pull on Moon varies with the time of year. The period of Moon's revolution is altered by this, varying by as much as two hours. There are also variations in the eccentricity of its orbit and slight fluctuations in its inclination. Finally the whole plane in which Moon moves rotates in space once every 18.6 years.

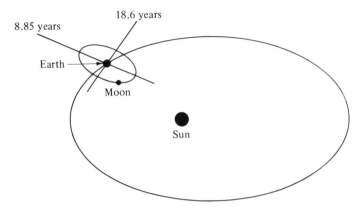

Fig. 3.13 *The line of intersection of Moon's and Earth's orbital planes rotates once every 18.6 years. The long axis of Moon's orbit rotates once every 8.85 years.*

As Moon moves about Earth, occasionally its shadow passes across the face of Earth. This is called an eclipse of the Sun. If the solar disc is completely invisible the eclipse is said to be total. The area of total shadow is small compared to the area of Earth. Only if Moon is near perigee, the point in its orbit where it is closest to Earth, can the shadow be total (Figure 3.14). Most eclipses are only partial. There are at least two solar eclipses annually, and in some years as many as five.

When Moon moves into the shadow of Earth, a lunar eclipse occurs. Either a solar or a lunar eclipse occurs every time Moon intersects a cone tangent to both Earth and Sun (Figure 3.15). Because the cone is wider between Earth and Sun than behind Earth, solar eclipses are commoner than lunar eclipses. There can be at most three lunar eclipses per year. There are never more than seven eclipses altogether in one year, lunar plus solar, nor fewer than two solar eclipses. The average number of all eclipses is four. Lunar eclipses can be seen from anywhere on the dark half of Earth's surface, whereas solar eclipses are visible only from a much smaller area. As a result, lunar eclipses are more commonly seen by the average person than solar eclipses, even though they occur less often.

Moon rotates on its axis in exactly the average length of time it takes to revolve once around Earth. As a result it always keeps the same side toward Earth. This is not likely to be mere coincidence, and is believed to be the result of a transfer of rotational angular momentum to the orbital angular momentum of Moon. The system is particularly liable to such exchanges because of the large relative size of Moon and the presence of a large amount of liquid on Earth's surface in the form of the oceans. The gravitational forces exerted by each on the other are not everywhere the same in either body, being greatest on the near sides and least on the far sides. Since neither body is completely rigid, these forces deform both bodies.

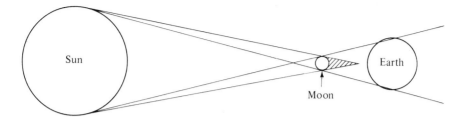

Fig. 3.14 *If Moon is too far away in a solar eclipse, the point of its cone of total shadow will not reach Earth's surface.*

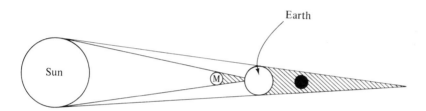

Fig. 3.15 *An eclipse occurs whenever Moon passes thru all or part of a cone tangent to both Earth and Sun.*

In the case of Earth, this results in a bulging of the surface toward Moon, where the pull is strongest, and away from Moon, where the pull is least. The bulge takes a finite time to form; and, hence, tends to peak somewhat after the Moon is directly overhead. Visualized in another way, the crests of the tidal bulges are not found facing and opposite Moon, but are dragged past this location by Earth's rotation (Figure 3.16).

The height of this bulge in the case of the solid part of Earth depends on the rigidities of the rocks of which Earth is composed. Solid-earth tides as great as 45 cm (18 inches) have been calculated from measurements of the variations in gravitational force which result from the change in distance of the surface from the center of Earth (see Equation 3.18).

In the oceans and atmosphere, the effect is greater. The oceanic tides are complicated by the shape of the ocean basins. Because Earth rotates, the tidal bulges as seen from a point on the surface appear to progress in the opposite direction from the direction Earth turns. But these waves are deflected by the borders of the oceans. Resonances due to the shape of the

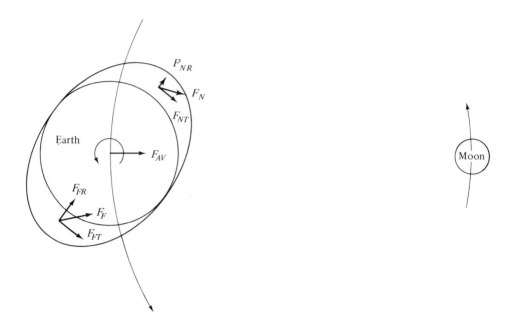

Fig. 3.16 *The oceanic tidal bulges are dragged eastward by Earth's rotation.*

ocean basins change the shapes of the tidal bulges, causing peaks and nodes in a wide range of patterns. The situation is further complicated by the pull of Sun, which produces a solar component of tide about half as strong as the lunar tide.

The eastward drag of the tidal waves produces a torque opposing the rotation of Earth due to the friction of the waves on the ocean bottom as they progress around Earth. Another way to see this is to note that the facing and opposition bulges of the tides are at different distances from Moon (Figure 3.16). The gravitational force per unit mass F_F, on the far side is less than the average gravitational pull for the whole earth, F_{AV}. The force per unit mass on the near side, F_N, is larger. The tidal force at any point can be thought of as being composed of two parts, one radial to Earth's surface and one tangential. The heights of the tides result from the fact that the radial component F_{NR} is larger and F_{FR} is smaller than average. There are also tangential forces, F_{NT} and F_{FT}, with F_{NT} being larger than F_{FT}. The difference between these last two forces acts as a torque opposing the rotation of Earth on its axis, slowing it down. Comparison of the times of solar eclipses in recorded history with those predicted from current motions of Earth and Moon show that the time it takes for Earth to rotate once on its axis is being increased by about 10^{-5} sec every year. This increase in the period of rotation has been going on for a long time. Counts of the number of daily growth rings per year of corals show that 400 million years ago Earth rotated on its axis about 400 times per year compared to the current 365¼.

There are corresponding forces acting on Moon. The result is a very small acceleration of the revolution of Moon and Earth about one another. This causes them to move farther apart at about 2 meters per century. As separation increases, the tides will be weakened. Ultimately, after some tens of billions of years, Moon will have received so much of Earth's rotational momentum that Earth will rotate once on its own axis in the time it takes for Moon to revolve about it. As this stage is approached, the lead of the tidal bulge over Moon's position will decrease to zero, and the transfer of angular momentum will cease.

Before this state is reached, the lunar tide will have become smaller than the solar tide. A similar process of transfer of momentum between Earth and Sun will further decrease Earth's rate of rotation. The tidal bulge will then lag behind the position of Moon. Moon will be drawn in and Earth's rotation will be accelerated. The nearer Moon comes the more the tide will lag Moon's position, and the faster Moon will be drawn in, until finally it will come so close to Earth that tidal forces will pull Moon apart. Pieces of Moon will rain onto Earth in a gigantic meteorite shower, destroying everything on the surface. This will be the end of the geologic record as we know it today. It may never happen, because long before then Sun will have used up its hydrogen, and other equally extreme changes will have taken place in the solar system.

This process of tidal exchange of momentum must be occurring for every planet which has a satellite. Phobos, the inner satellite of Mars, is believed to be late in its life. Phobos revolves about Mars in less than a Martian day. It is losing momentum, and moves slowly in toward Mars. In about forty million years, Phobos will get so close that it will break up, and its pieces will fall on Mars as huge meteorites.

The other satellites are all too small and distant from their planets to produce large tides. Their rates of transfer of momentum are too small to measure. There may, however, have been low-level satellites in the past which were drawn into one or more of the planets.

The slow rate of rotation of Venus could have resulted from a tidal exchange of energy. Venus may have had a satellite, now lost, whose tidal effect slowed Venus, making its rotation take longer than one revolution about Sun.

Moon has the further peculiarity that its radius is longer by about half a kilometer on the side facing Earth than elsewhere. This amount of bulge is the size that would be produced in a fluid Moon at one-third its present distance from Earth, which is where it would have been a few billion years ago, about the time the solar system is estimated to have been formed. It is a larger bulge than Earth's pull would produce at Moon's present distance. In addition to suggesting that Moon was once nearer Earth, this implies that Moon's internal strength was then much less than at present. It may even be that Moon was largely molten when it was at this distance. A liquid Moon would relatively quickly transfer its rotational momentum to its revolution about Earth thru the mechanism described above because of the damping caused by tides. If it then solidified by cooling, with the bulge frozen in, this means that the material of which Moon is made is strong enough to have held its shape undeformed throughout its life. This would be an amazing feat considering how impermanent most features of the solar system are. On Earth, geologic history has recorded again and again the rise and destruction of mountain ranges and vast plains. If the lunar bulge is an old permanent feature, then the interior of Moon must be very different from the interior of Earth.

Moon appears to have almost no atmosphere. There is not enough gas to refract measurably the light of stars as Moon passes between them and Earth, so any atmosphere must be less than 10^{-4} as dense as Earth's atmosphere. Studies of radio-frequency radiation suggest, however, that there may be a very tenuous layer of gas about 10^{-9} as dense as on Earth. It is hard to tell how much of this is a real atmosphere permanently held by Moon, and how much of it is a temporary concentration of solar wind, continuously pulled in by Moon's gravitation but quickly lost again.

The strength of Moon's gravitational pull is too small to hold for long any of the common gases except possibly krypton and xenon. Under the direct ultraviolet and X-ray radiation of Sun, even these gases may be slowly lost. There may, however, be some volatile matter trapped in dust formations beneath Moon's surface. Even water might be present in the form of ice at a few centimeters depth. If volcanic processes have, during Moon's life, released gases from the interior, some of this material could be trapped for long periods in undisturbed dust layers. Such trapped material would have a tendency to escape each time a meteorite impact moved the dust about.

The surface of Moon is marked by craters, mountains, rilles, rays, and maria (frontispiece, Figures 3.17 and 3.18). The craters range in size from a few microns across, visible only under a microscope, to the crater Clavius, which is 235 km in diameter. Each crater is a depression surrounded by a rim rising above the surrounding territory. In the larger craters there is often a peak in the center (Figure 3.19). The ratios of the diameters of the craters to their depths and to the rim heights are similar to what is found on Earth for explosion craters, and are different from what is common for volcanic structures. The lunar craters are believed to be all or nearly all the result of the impact of meteorites striking Moon's surface. Craters are scattered all over Moon, both on the side toward Earth and on the back side, which has been photographed from orbiting spacecraft.

The larger craters are frequently surrounded by patterns of rays—lightly colored bands showing up against the more generally dark background of Moon's surface (Figure 3.17). The rays almost never are associated with large enough topographic variations to cast an observable shadow, so it must be assumed that their appearance is caused by differences in the exposed rock types. One possible explanation is that they represent ribbons of pulverized rock thrown out of the impact craters. Fresh, powdery dust would be more reflective than a surface composed of large masses of solidified rock. Very old dust is believed to become darkened by changes produced by Sun's radiations and by bombardment with solar-wind particles.

Fig. 3.17 *Moon. (Courtesy Hale Observatories.)*

Some crater walls intersect other craters, and are thus obviously younger. The craters with the best-developed ray systems are always relatively young ones.

The composition of Moon can be estimated from studies of the samples brought back by the Apollo astronauts and from its overall density. The samples have a composition resembling that of terrestrial basalt, but richer in titanium (TiO_2) and such trace elements as scandium, zirconium, hafnium and yttrium (Table 3.3). Surface rocks on Moon are composed of fragments of a variety of compositions, including pieces of meteoritic iron. Some of these fragments may have come from considerable distances away. They suggest that Moon is composed of a variety of rocks of different compositions (provided that all are not fragments of meteoritic material). Water is notably scarce. The average density of Moon is 3.36 gm/cc. This is low enough that Moon may be composed throughout of material of the same chemical composition as the surface samples.

Fig. 3.18 *"Alpine Valley" area of Moon between Mare Frigoris and Mare Im-brium. Taken by Lunar Orbiter 5. The photograph shows an area about 460 km (286 miles) across. (N.A.S.A. Photo.)*

Much of the surface of Moon has a very low thermal conductivity, de-termined by the rate it reradiates infrared energy after sunset, suggesting that it is composed of powdered rock. At one time it was feared that much of the surface of Moon might be covered with a deep layer of rock so finely powdered that a man would sink into it as into quicksand. This was not the case at the sites of the first landings of spacecraft on Moon (Figure 3.20). The foot on which Surveyor I rested sank only a few millimeters into Moon's surface, about the depth it might have sunk into beach sand on Earth. Neil A. Armstrong and Edwin E. Aldrin were able to walk about on Moon's surface. One possible reason is that solid grains on the Moon's sur-face tend to cement themselves together more easily due to inter-atomic forces than on Earth, where surfaces tend to adsorb layers of gas or liquid. These adsorbed surface layers hold particles apart. Since Moon has very

Table 3.3. *Composition (by weight) of Moon's surface rocks compared to two typical Earth basalts. (After A.E.J. Engel and C. J. Engel, 1970, Science v. 167, p. 527.)*

Composition	Lunar gabbro fragment	Lunar diabase fragment	Lunar dust	Earth oceanic basalt	Earth alkali-olivine basalt
SiO_2	42.01	39.79	41.50	50.01	48.01
$FeO + Fe_2O_3$	17.98	19.35	15.68	9.39	11.43
Al_2O_3	11.67	10.84	14.31	16.18	15.97
CaO	12.18	10.08	11.84	11.33	9.04
TiO_2	8.81	11.44	7.50	1.37	2.92
MgO	6.25	7.65	7.95	7.71	5.26
Na_2O	.48	.54	.48	2.79	3.73
K_2O	.11	.32	.16	.22	1.89
H_2O	.00	.01	.01	.87	1.33
P_2O_5	.08	.17	.10	.13	.42

little or no atmosphere, dust grains would contact one another intimately, and there would be a tendency for recrystallization to occur at contacting points, linking the particles together. This process is called vacuum welding.

In addition to the crater walls, which sometimes rise to heights as great as 6000 meters, there are ten conspicuous mountain ranges on Moon (Figure 3.21). There are also what appear to be fault scarps, sometimes several hundreds of meters in height. The mountains are irregular in outline and do not possess the linear pattern common in terrestrial mountains. Valleys often cross the mountains.

The mountains are usually found along the edges of large, relatively smooth plains called maria (Figure 3.18). The surfaces of the maria are less often broken by craters than the rest of Moon's surface. When sunlight strikes them at very low angles, it reveals many low, bubble-shaped swellings, shallow hollows, and long sinuous ridges either in rings parallel to the borders or crossing the maria from side to side. The rounded outlines of the maria suggest that they, like the smaller craters, are produced by meteorite impact. In the case of meteorites fifty kilometers or more in diameter, the total energy involved is so great that large amounts of molten rock may be formed on impact. The smooth surfaces of the maria thus might be the result of resolidification of this melted pool or of lava which flowed into the large crater from the interior.

The latter possibility would imply that the interior of Moon is or once was close to or above the melting point at relatively shallow depths. No confirmed observations of volcanic eruptions have occurred, although there have been reports of such events seen telescopically by individual observers.

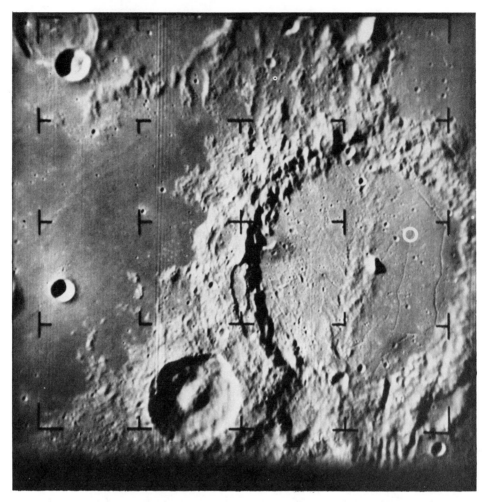

Fig. 3.19 *Ranger 9 photograph of Moon's surface, taken from 430 km (265 mi) elevation. The small circle marks the spot where Ranger IX struck Moon's surface. (N.A.S.A. Photo. Courtesy Calif. Inst. Technology Jet Propulsion Lab.)*

The *rilles* are grooves or furrows in the surface of Moon (Figure 3.18). They are commonest on the maria. They are long, narrow channels with steep walls from a few to hundreds of kilometers in length and up to several hundred meters deep.

If meteorite impact is the explanation of the great majority of surface features of Moon, as seems likely, then a very large amount of fragmental material must have been formed. This would now be scattered all over Moon's surface. Lacking both running water and an atmosphere, this material will be relatively undisturbed (compared to Earth) on all except the steepest slopes. Also, the process of repeated impacts must have produced

a.

b.

Fig. 3.20 *a. Part of landing gear of Surveyor I on Moon's surface. (Courtesy Calif. Inst. of Technology Jet Propulsion Laboratory.) b. E. A. Aldrin setting up first seismograph on Moon's surface in 1969. (N.A.S.A. Photo.)*

a mix of all varieties of lunar surface materials. The different colors and textures of the lunar surface, aside from steep ridges, must therefore result primarily from the fineness and degree of weathering of the rock fragments.

Weathering will take place in several ways. Rocks on the surface of Moon are exposed to intense radiation ranging from ultraviolet to cosmic rays. These, particularly the highest frequencies, can cause changes in crystal structure modifying the surface appearance of the rocks. Moon's surface is also directly exposed to the solar wind, which consists in large part of protons (hydrogen nuclei). It is not known to what degree the variations in the surface appearance of Moon are due to length of exposure and to what degree they represent differences in chemical or mineralogical composition or texture of the surface rocks. In addition, the rapid and large temperature variations between day and night on Moon will cause thermal stresses which may tend to break rocks into smaller and smaller fragments. Whatever the processes are, weathering is probably slower on Moon than on Earth, where it is aided by chemical changes as the rocks interact with air and moisture.

Five periods of formation of Moon's surface features have been recognized from their cross-cutting relationships. All may be billions of years old. One theory is that all the planets formed from the gathering together of material from a dust cloud, and that in the final stages of this accumulation moderate-sized chunks of coagulated matter rained in on the larger bodies. Because Moon lacks any but the most minimal erosional processes, it still presents the scarred surface left by this clearance of loose debris from the solar system. On the other hand, meteorites are still observed falling to Earth, and it is equally possible that the impacts which molded Moon's surface have occurred throughout the duration of its existence.

3.6 METEORITE CRATERS ON EARTH.

It is certain that material is still being gathered in by the larger bodies of the solar system. Three large meteorite falls and many small ones have been observed on Earth in historic times. The largest of these occurred in the Tunguska River area of northern Siberia seventeen minutes past midnight on June 30, 1908. This meteorite appears to have exploded in the air without creating a recognizable crater, but trees were knocked down over an area 35 km in diameter. In the central part of this, the heat generated was so intense that the trees were seared out to distances as great as 10 km. The uprooted trees lay with their trunks pointing away from the impact site (roots toward it). The impact was recorded on the seismograph at Irkutsk and by barometers over much of the world. The fall was seen as far away as 750 km from the impact site as a "blindingly" bright meteor travelling from southwest to northeast. A few metallic fragments have been collected from the impact area; but it is believed from the amount of material found in relation to the energy calculated from the seismic and air waves produced that most of the meteorite must have been in the form of rock-like minerals and frozen gases. From this composition it is suspected that the falling body was a comet.

So much dust was added to Earth's atmosphere by this one event that a measurable darkening of the sky occurred on the following day, and the night sky was brightened by scattered light for several days. Much of this dust may have come from the tail of the comet rather than from its main mass.

On February 12, 1947 another body landed in Siberia, this time in the Sikhote-Alin Mountains. Metallic meteorite fragments weighing up to several tons each have been recovered from an area of 1.6 sq. km. Shortly before striking the ground, the meteorite is believed to have broken into pieces, producing 122 individual craters from 0.4 to 30 meters in diameter. As a result of the fragmentation, estimated to have occurred within 6 km of the surface, there was not a single large impact explosion to disturb seismographs or barographs. The small area over which the individual pieces struck indicates there was only one body on entering the atmosphere. That there were many pieces at the time of impact is proven by the fused appearance of the surfaces of many of the fragments. Most of the large pieces were broken up by the final impact. The amount of meteorite material picked up in the area amounted to 23 tons. The total mass is estimated to have been about 70 tons. From the path of fall, it is calculated that the meteorite was travelling at about 14.5 km/sec when it entered the atmosphere.

About 250 years ago two large meteorites were seen to fall near Murgab in the Tadzhik S.S.R. The craters were still identifiable in 1951. One was 80 meters in diameter and 15 meters deep. The other, 250 meters away, was only 15 meters in diameter.

Small meteorites are observed to fall frequently, but only rarely are pieces of the meteorite recovered. Numerous prehistoric meteorite craters have, however, been identified. In the United States, there is one in Arizona (Figure 3.22), a group of five near Odessa, Texas, and one near Haviland, Kansas. In northern Quebec there is a meteorite crater nearly 3.5 km (2 miles) in diameter. The giant Ries Kessel crater in Bavaria, Germany, is 27 km (16 miles) across. Other craters have been identified in Australia, Argentina, Arabia, Esthonia, West Africa and Algeria. Many other crater-like structures have been observed. Some are in remote localities like the Sahara Desert and have not yet been studied. Others are so old that their features have been weathered and eroded beyond certain identification.

Truly ancient craters have been identified on purely geologic evidence in a few cases. A crater formed in late Devonian or early Mississippian time (about 260 million years ago) has been identified near Gainsboro, Tennessee, U.S.A. This crater has the typical features—rim, basin and fractured rocks—of a modern crater but it has been buried beneath later sedimentary rocks. Identification of such a structure as meteoritic is made possible by the presence of distinctive small structures such as shatter cones and by high-pressure mineral phases.

Shatter cones are conical fractures produced in rocks by the sudden application of very high pressures (Figure 3.23). Slowly applied force leads to yielding along lines of weakness related to grain patterns in the rock. Sudden pressures crush the rock in patterns independent of any previously present structures.

Certain minerals are also found only at meteorite impact sites. Stishovite and coesite are high-pressure forms of silica (SiO_2). They are formed only at pressures vastly greater than those which normally exist within twenty

Fig. 3.21 *Lunar orbiter photograph of a part of Moon's surface. (N.A.S.A. Photo.)*

kilometers of Earth's surface. When a large meteorite strikes, the pressure at the impact point is so great that rocks are greatly compressed. Some of the material transforms to dense phases such as coesite characteristic of the high pressures involved.

3.7 SATELLITES OF OTHER PLANETS.

Mars has two satellites. Phobos, the inner one, is only about 16 km in diameter and is so close to the planet that it circles Mars more than twice daily. No other satellite in the solar system revolves around its planet in less time than it takes the planet to rotate once. This produces the unusual (to Earthmen) situation that the Phobian "month" is shorter than the Martian "day." Diemos, the outer satellite, is only about half the diameter of Phobos.

Phobos is so close to Mars that its motion is strongly affected by the ellipsoidal shape of the planet. For purposes of calculating the effect of the gravitational force on the satellite, all of Mars's mass cannot be treated as though it were concentrated at its center. The orbit is perturbed a measurable amount by Mars's equatorial bulge. The problem of Phobos's motion

Fig. 3.22 *Meteorite Crater, Arizona, U.S.A. (Courtesy Meteor Crater Museum.)*

is also complicated by the fact that it seems to be in an unstable orbit, slowly spiralling in toward Mars.

This may explain why other planets do not have satellites near their surfaces. Low-flying satellites are very sensitive to perturbations due to the bodies to which they are held gravitationally. Other planets may have gathered in satellites that once circled in low, unstable orbits.

The remaining planetary satellites are so far from Earth that much less is known about them. Jupiter has twelve satellites, of which three are larger than Moon. Io, Europa, Ganymede and Callisto have surface markings which make it possible to tell that they rotate at the same rate that they revolve around Jupiter. Io and Europa are both denser than Moon (Table 3.2) and may have compositions similar to surface rocks on Earth. Ganymede and Callisto are lighter, but still more than twice as dense as Jupiter. They are likely, therefore, to be a combination of condensed gases and rocky or metallic materials. Indeed, this is possible for all four of the satellites. Ganymede is large enough that it could hold an atmosphere, although none has been detected on it. The outer satellites have such eccentric orbits that they move in and out past one another at times, although because of the inclinations of their orbits they never approach closer than 3×10^6 km from one another. The outer four satellites are retrograde. They revolve around Jupiter in the opposite sense from its rotation, like Triton around Neptune.

In addition to nine recognizable satellites, Saturn is surrounded by three bright rings of matter (Figure 3.24). These are believed to be swarms of small solid bodies. The thickness of the rings is so small that they are invisible when viewed edge-on from Earth. This indicates that these belts must

Fig. 3.23 *Shatter cones, a fracture structure produced by violent impact. From the Sierra Madera, Texas meteorite crater. (Courtesy F. Dachille.)*

not be over 30 km thick. The inner ring is much less dense than the outer two. All lie in the plane of Saturn's equator. The total mass of each ring can be calculated from its perturbing effect on Saturn's moons. The combined mass is less than the mass of Earth's Moon. The average particle size can be estimated from the intensity of the light the ring reflects and the mass. The particles are thought to lie between the size of dust grains and ordinary sand or fine gravel, but to be larger than flour. The first and third rings are diffuse enough that stars can be seen thru them. The rate of rotation of the rings decreases from the inside to the outside, as is necessary for individual particles to maintain their orbits. Saturn's rings are so close to Saturn that a single Moon revolving at this distance would be broken up by the gravitational pull of the planet. The minimum distance for a satellite is called Roche's limit. For a satellite of the same density as the planet it orbits, Roche's limit is 2.44 times the planet's radius.

The large satellites of Saturn all lie outside of the rings. The outermost one, Phoebe, revolves retrogradely. Titan is the largest moon in the solar system, and is the only satellite of Saturn larger than Earth's Moon. It is known to have an atmosphere. Spectroscopic studies have detected methane. It is reasonable that Titan should have an atmosphere and Moon none both because of Titan's greater size and its lower surface temperature. Titan does not reflect light as well as Saturn's other satellites, which are suspected to be coated with frozen water or ammonia.

Iapetus is peculiar in that it is five times brighter when it lies east of Saturn than when it lies to the west. This shows first that it always keeps the same face toward Saturn and second that the surface exposed on one half is composed of different materials than that on the other half, or that the shapes of the surfaces are different.

A tenth satellite was tentatively identified and named Themis, but has since disappeared. It is not known whether it was originally misidentified, is moving in an orbit different from that predicted, or has escaped from Saturn.

Uranus has five satellites moving in nearly circular orbits almost in the plane of Uranus' equator. Besides Triton, discussed earlier, Neptune has a second, normal satellite, Nereid.

3.8 COMETS.

In addition to the planets which travel in nearly circular orbits about Sun there are a large, unknown number of bodies travelling in highly eccentric orbits—the comets. These bodies are so small that they cannot be seen most of the time. Their paths must be determined from observations made during the brief periods when they are near enough Sun to be visible. A few have appeared repeatedly at regular intervals, and their paths are reasonably well established. Others have been seen only once, and in some cases it is uncertain whether they are permanent members of the solar system or are transients.

Compared to the planets and their satellites, comets are small bodies of low density. A typical comet consists of three parts: a fuzzy coma, or head; a nucleus embedded in the coma; and a tail extending from the head radially outwards away from Sun (Figure 3.25). There is great variety to the appearance of comets, and one or two of the three characteristic features may be missing.

Over 1640 comets have been observed. On the average six or seven comets are observed annually, mostly new ones never seen before. Only about once every ten years is a comet visible without the aid of a telescope. The orbits of about 200 comets are known to be elliptical, so that the comets' return dates can be predicted. Others have either unknown orbits or they are known to be parabolic or hyperbolic (Figure 3.26). A comet with a parabolic or hyperbolic orbit will not remain in our solar system.

All the measured orbits are so shaped that it is probable that perturbations by the larger bodies of the solar system are changing cometary orbits in such a manner that there is a steady loss of these bodies into space. No parabolic or hyperbolic comet has ever been observed which has

Fig. 3.24 *Saturn. (Courtesy Hale Observatories.)*

an orbit so eccentric that it could not have once been elliptical with only a small perturbation in its path. It is also possible that orbits which closely approximate parabolas near Sun become elliptical when the comet is farther out. Within the radius of Pluto, the various planets pull on a comet in different directions, but at larger distances, their pulls tend to become more parallel. This extra inward attraction may be just enough to change the orbit from parabolic to elliptical.

The reverse of expulsion from the solar system is possible. A comet in a parabolic or hyperbolic path, if it passes near a planet, can be deflected so that its path becomes an ellipse. Jupiter has been observed to cause a deflection in the orbit of a comet. The masses of the comets are very small even compared to the smaller planetary satellites, which are not perturbed measurably by the passage of a comet nearby. The reverse is easier, the perturbation of a comet by a satellite.

The nucleus of a comet reflects sunlight very much as a planet does, so it must be a solid body. The largest observed cometary nucleus had a diameter of about 100 km. The heads, in contrast, range from about 16,000 to over 200,000 km in diameter. The heads and tails are very diffuse. In 1910 when Halley's comet passed between Earth and Sun, it became invisible, with Sun shining thru with unnoticeably diminished vigor. The light from

Fig. 3.25 *Arend-Roland Comet in 1957. (Courtesy Lick Observatory.)*

the head has been analyzed spectroscopically and indicates the presence of carbon and nitrogen alone and in compounds with hydrogen and oxygen. In some cases, traces of metallic elements are also observed. Comets are believed to be composed of a nucleus of small particles bound together by frozen gas, some of which evaporates as the comet approaches Sun. A few comets reflect sunlight without altering it. These are believed to be almost entirely aggregates of dust. The tails of those comets which have them commonly show spectral lines for ionized carbon monoxide and nitrogen.

Several comets have broken to pieces as they passed near Sun, attesting to their low cohesiveness. Some comets travel in groups, and may be fragments of a once larger comet.

The heads and tails of comets are not visible except when the comet is near Sun, and they grow in size as the comet approaches that body. It is believed that they result from evaporation of gas from the nucleus due to solar radiant energy. Even some of the fine dust is spread outward by this evaporation. These particles, from atomic size on up, tend to be accelerated away from Sun by radiation pressure. The side of each particle facing Sun is irradiated more than the side away from it. The effect of this uneven illumination is to exert a small force outward from Sun on material in the head. The particles are not always moving away from Sun. As the comet

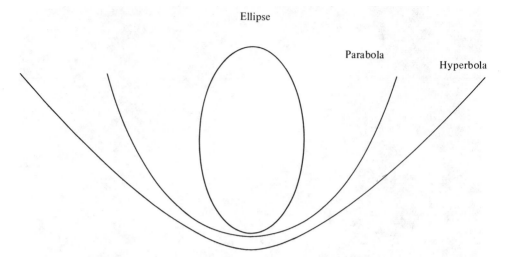

Ellipse

Parabola

Hyperbola

Fig. 3.26 *A parabola is less open than a hyberbola. Only the ellipse is a closed curve.*

approaches Sun, the tail lags behind, following it in. As the comet moves away, the tail precedes it. Each particle in the tail is actually in some kind of orbit around Sun. In this fashion a tail is formed from the diffuse head of the comet. Radiation pressure is aided by the solar wind, the stream of ions and electrons flowing out from Sun. This wind helps to spread cometary material outward. Variations in the brightness of comets have been correlated with variations in the solar wind and with fluctuations in the amount of ultra-violet radiation from Sun.

Surprisingly, the ultimate effect of radiation pressure is not to carry matter outwards from the solar system into space, but to draw it into Sun. Any particle in an orbit around Sun moves in a path whose radius is controlled by the sum of all the forces acting on it. The radial component of radiation pressure is opposed by the gravitative pull of Sun. There is in addition, however, a tangential component to radiation pressure, due to the fact that it takes radiation a small but finite time to pass any particle. The forward-moving side is in this manner illuminated more than the back side. This is most easily appreciated by thinking of a cube with one face parallel to the light rays being crossed (Figure 3.27). The effect of extra illumination on the forward face is to decelerate the tangential motion of the particle. Its path gradually becomes less eccentric and of smaller radius, until the particle finally falls into Sun. This effect, called the Poynting-Robertson effect after the men who first explained it, is negligibly small for any body large enough to see, but is strong enough that gas and fine dust in the solar system will be drawn into Sun in a few tens of millions of years.

From the evidence presented above, it is obvious that a comet is destroyed a little on each passage near Sun. The average life of a comet is estimated to be only 70 passages, although a few comets obviously survive

Direction of Light Rays

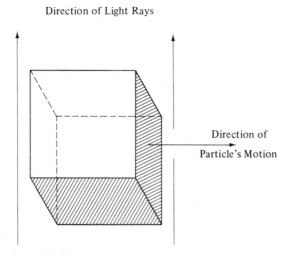

Direction of
Particle's Motion

Fig. 3.27 *The Poynting-Robertson effect. The forward side of the cube is illuminated because the body is moving into the path of the light rays.*

much longer than this. Halley's comet has already been seen 29 times without any reported change in its appearance. It should next appear in 1986. Other comets have been observed to get less bright with successive passages. Clearly there must be a supply of matter from which new comets are produced or a source from which other comets are attracted to our galaxy. One possibility is that there are many comets beyond Pluto with orbits about Sun which do not bring them close to Sun. Occasionally one of these comets is perturbed by some other star, or by a planet or even by another comet, so that it is diverted into the center of the system. Here its life is limited unless its orbit is again perturbed so much that it escapes. Whether comets are actually forming by the slow accretion of the matter scattered thru the galaxy beyond Pluto is unknown. If they are, then comets are a sample of the composition of this interstellar material.

3.9 PLANETOIDS AND METEOROIDS.

The presence of small planetoids with orbits more like those of planets than comets was pointed out in Section 3.2. Some of these are large enough for their orbits to be identified and the bodies themselves named. There must be a very large number which have never been seen because of their small size and their distance from Earth. Over 1600 have been found. A majority of these have orbits of average radius between that of Mars and Jupiter. Many have very eccentric orbits. Icarus passes nearer Sun than Mercury, although its orbit has an average radius greater than that of Earth. Others pass near Earth. Hermes comes within a million kilometers. Their orbits have a wide range of tilts relative to Earth's.

With so many bodies of this size known to be part of the solar system, one might ask if these are not the meteorites which strike Earth occasionally and have produced the craters on the surfaces of Moon and Mars. Many of the craters may have originated in this way, but probably not all of the meteorites striking Earth are of this sort. To see the reason for this, it is necessary to examine the features of meteors.

A *meteor* is a bright flash of light produced in Earth's atmosphere when a fast-moving body enters it. Very bright meteors are called *fireballs* or *bolides*. A *meteorite* is a piece of material which has fallen to Earth from outside. A *meteoroid* is a small piece of matter such as a planetoid moving thru the solar system. If a meteoroid comes close to Earth, it is drawn inward by Earth's gravity. It gains energy as it falls, like any other falling body. If nothing interfered with it on the way, a body dropped to Earth from an infinite distance would strike at a velocity of 11 km/sec. Over ninety per cent of this velocity is picked up in falling the last 30,000 km. Thus, any natural body falling to Earth will usually be moving with at least this speed.

If a body moves thru Earth's atmosphere at ten km/sec or more, the air does not have time to flow smoothly around it. The impact of the meteoroid on the atoms and molecules of the atmosphere is often large enough to ionize atmospheric atoms and molecules. This produces the meteor, which is a cloud of incandescent ions radiating light as the electrons return to their normal states. The impact also melts the surface of the meteoroid (Figure 3.28). There is not time for the heat to be conducted into the center of the body. Meteorites found within minutes of their fall are often so cold that ice condenses out of the air onto their surfaces. Nevertheless, the surfaces show distinct evidence of having been molten during the fall. The surface melts or vaporizes, and it may flow backwards along the edges of the body spewing hot droplets into its glowing tail. Thus large parts or all of the meteoroid are converted into fine dust high in the atmosphere.

A space vehicle entering Earth's atmosphere undergoes heating in the same way that a meteoroid does. This heat is a serious hazard to space travel, because it is generated faster than it can be radiated away. The heat generation is not all bad, however, because it uses up part of the kinetic energy of the object. The speed of a spaceship must be greatly reduced before it reaches Earth's solid surface or it will be destroyed by the impact.

Reduction in speed can be accomplished by two other means besides conversion of kinetic to thermal energy. One method is to use a retro-rocket. Such a device generates a back thrust by throwing mass forward, carrying with it part of the forward momentum of the space vehicle. Spaceships landing on Moon, where there is no atmosphere, use retro-rockets. On Earth it is more convenient to take advantage of the resistance of the air to the passage of the object.

If the space vehicle is blunt-nosed, it can be slowed also by the reaction of the shock wave produced in front of the body (Figure 3.29). When the spacecraft is moving thru the air faster than the speed of sound, individual

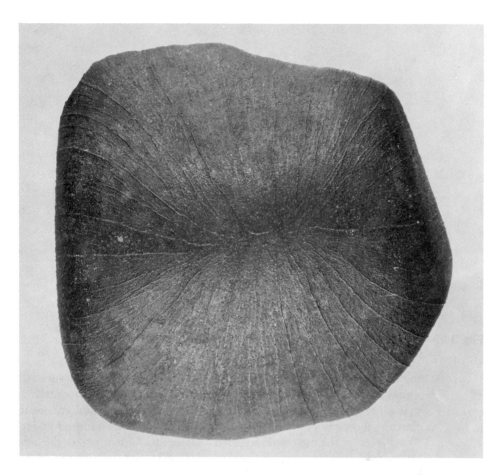

Fig. 3.28 *Streaked surface of a small meteorite, the result of fusion as it passed thru the atmosphere. (Courtesy Field Museum of Natural History.)*

gas molecules do not have time to flow past the body before more gas molecules are encountered. A cloud of compressed gas collects along the leading face of the moving object. This cloud of gas exerts a back pressure on the spacecraft, slowing it. Part of its energy of motion is imparted to the gas molecules, which flow away from the body violently. Because this wave front is travelling thru the air with the spacecraft faster than sound travels, there is a very sudden transition from undisturbed air in front of the shock wave to disturbed air behind it. This shock wave is of the nature of an explosion. The velocity of sound in the disturbed gas behind the shock wave front is greater then in the undisturbed gas in front of it, so that the tail of the disturbance keeps overtaking its front edge, transmitting a sudden change in pressure thru the atmosphere. Such sharp pressure changes are very irritating to the ear and may cause damage to sensitive objects. This is one of the problems confronting the air-transportation

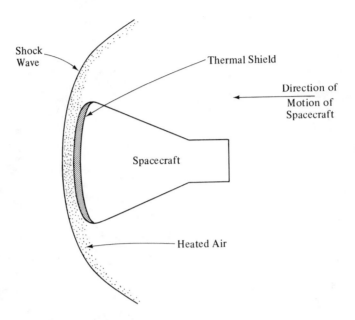

Shock Wave

Thermal Shield

Direction of
Motion of
Spacecraft

Spacecraft

Heated Air

Fig. 3.29 *Formation of a shock wave at the front of a spacecraft entering the atmosphere.*

industry in developing an aircraft that will travel faster than the speed of sound. Once this speed is exceeded, a shock wave is inevitably produced.

If a meteoroid is big enough it may penetrate thru the atmosphere to Earth's solid surface, where it is finally stopped, forming an impact crater. What is left of the meteoroid is called a meteorite. Meteorites are the only bodies in the universe whose total composition is known. Earth and Moon can only be sampled at or near their surfaces. Spectroscopic studies of Sun and other bodies give us evidence only of their near-surface materials. When a meteorite is found, it can be taken into the laboratory, sawed in half and studied in whatever detail is desired.

By photographing the paths of meteors from several places simultaneously, their velocities and trajectories can be calculated. Meteors can also be followed by radar tracking, as the ionized gas columns are good reflectors of radio waves. The paths of meteoroids show that almost all originated in the solar system. Less than three percent (if any) can have come from outside, because almost all velocities are below the escape velocity from Sun. The few whose membership in the solar system is uncertain may have obtained their peculiar orbits by being deflected by some solar-system body.

Many meteoroids arrive with orbits resembling those of comets, and probably were comets or parts of comets at one time. Others have orbits within the range of the planetoids and may be small planetoids. Meteoroids can approach Earth from any direction and at any time, but about five per cent of all meteoroids arrive at predictable times and directions.

The latter come in what are known as meteor showers, and must be members of clouds of small bodies travelling thru the solar system on regular orbits. There are at least a dozen meteoroid streams which Earth encounters once each year, and several hundred such streams have been cataloged. When Earth encounters such a stream many meteors are seen spreading from a single apparent direction in the sky.

If a group of meteoroids is travelling about Sun all with roughly the same orbit, and if they are spread evenly along the orbit, there will be a meteor shower annually each time Earth crosses the orbit. If the meteoroids are bunched so that they occur only at one or a few places along the orbit, then the repetition rate of showers will be less frequent. Such groupings of meteoroids are called *swarms*. They have been likened to a flying gravel pile. The gravel is rather thin, individual pebbles being typically about 150 km apart.

Some swarms of meteoroids appear to follow the orbits of known comets. Such a comet may be considered to be an unusually dense portion of a swarm of material. Other swarms may represent the remains of comets which have broken apart and become diffused. Some streams are narrow and Earth passes thru them in a few hours. Others last thru several successive nights. The broader streams are also the most diffuse ones. It is calculated that any stream in the solar system will tend to break up due to the perturbations produced by the gravitational pulls of the planets.

The non-shower-type meteors may be produced largely when Earth encounters meteoroids which have been deflected out of more normal orbits.

The total amount of material being swept up annually by Earth has been estimated as 4×10^9 kg. Even over 5×10^9 years this is only a very small fraction of Earth's total mass. At this rate, only 40,000 kg/m^2 of material would have been added to Earth since the oldest known rocks formed, which would be a layer 16 meters (52 feet) thick at a density of 2.5 gm/cc. The exact amount of material is hard to estimate as many of the particles swept up are so small they cause no observable meteor.

Very fine micrometeoroids will pit the surface of a polished metal plate in space. At one time it was feared that meteoroids of sufficient size to punch holes in space vehicles might be a serious hazard to space travel, but the best current estimates of their concentration suggest that the chances of such an encounter are very small.

Spectrographic study of meteors shows the presence of hydrogen, oxygen, nitrogen, silicon, sodium, calcium, magnesium, iron and nickel, materials also identified in comets. Individual meteorites which were seen to fall and were then found on the gound are extremely rare, although large numbers of meteorites which were not seen to fall have been identified. The meteorites which are found are probably all or nearly all of planetoidal origin. Observed cometary meteors have all been so small that they were destroyed in the atmosphere. (The Tunguska meteorite may be an exception to this.) Their compositions may therefore be quite different from planetoidal meteorites.

On the basis of their chemical composition, meteorites can be classified into four principal types (Table 3.4). Iron meteorites (*siderites*) consist nearly all of iron and nickel. Stony meteorites (*aerolites*) consist predominantly of silicates, especially iron and magnesium silicate. Stony-iron meteorites (*siderolites*) contain substantial proportions of both metal and silicates. *Tektites* consist largely of silica with a very low iron content. Some meteorites contain sulfur minerals, others are unusually rich in carbon compounds. All meteorites contain traces of many other elements. The range and variety of compositions even within each general class are quite great.

Among meteorites which are observed to fall, the stony meteorites are by far the commonest. Meteorite finds, however, are most often irons. This is almost certainly due to the fact that iron meteorites look distinctly different from terrestrial rocks, whereas stony meteorites do not. Tektites are also like terrestrial rocks in composition, and can only be recognized as meteorites by such features as evidence of surface melting in their transit thru the atmosphere (Figure 3.30). No tektite has ever been seen to fall as a meteor, leading some investigators to question whether tektites are not fragments of terrestrial rocks thrown about by the impacts of other types of meteorites.

The paths of the few meteors from which meteorites have been recovered are such as to suggest that they were planetoids before they fell to Earth. The fact that the planetoids occupy orbits whose average lies between Mars and Jupiter has led to the hypothesis that they may be either fragments of a former planet or pieces of the raw materials of a planet which never finished assembling. If either of these hypotheses is correct, then the composition of meteorites takes on special importance because they would then be a sample of the composition of a planet.

Further evidence for this comes from a peculiar relation between the distances from the planets to Sun known as Bode's rule:

$$R_n = (0.4 + 0.3 \times 2^n)R_e \tag{3.19}$$

where R_n is the average distance from Sun to the nth planet and R_e is the average radius of Earth's orbit. R_n is close to Venus' distance for $n = 0$, Earth's for $n = 1$, Mars's for $n = 2$ (Table 3.1). R_3 equals $(2.8)R_e$. This is the distance to the average radius of the planetoids. Jupiter and Saturn correspond to $n = 4$ and 5. The rule was already known before Uranus was discovered, and fits its orbit well for $n = 6$. Neptune and Mercury do not follow the rule, Pluto being found where $n = 7$. No theoretical reason why the rule should hold is known, but its prediction of the location of the planetoid belt lends credence to the possibility that the planetoids' origin is related to that of the larger planets.

Calculations of the effect of Jupiter's gravity on a planet in an asteroidal-belt orbit show that such a body would be less stable than the other

Fig. 3.30 *Natural (bottom row) and synthetic (top row) tektites, the latter made from molten glass in a wind tunnel. (N.A.S.A. Photo, Courtesy D. R. Chapman.)*

planets, and might well have broken up after a limited life. Under this view, different types of meteorites are postulated as having been formed from different layers in a differentiated planet. It is supposed that the interior of the planet was hot and that molten iron had settled to the center with silicates floating on top. Siderolites come from an intermediate layer where the separation was incomplete. This view is supported by the presence of crystalline structures in many meteorites indicating that they crystallized at pressures of 10^4 atmospheres or more. Such pressures are found on Earth only at depths of over 30 km.

On the other hand, a large portion of stony meteorites contain rounded bodies of silicate material called chondrules of a type never found in terrestrial rocks. Most theories of their origin require that they have crystallized in small bodies rather than in a large planet. Their prevalence would support the concept that the planetoids are an unassembled sample of planetary material.

Not all meteorites have necessarily come from the same source. The chondrule-bearing aerolites may be planetary raw material and the siderites and non-chondritic aerolites may be fragments of one or more than one assembled body. Neither theory answers the question: where did the material come from originally? Certainly it is different in composition from Sun. From the fact that there are both large low-density planets and

Table 3.4. *Composition of typical meteorites (mass percent). (After Wood, p. 348 of Middlehurst and Kuiper, "Moon, Meteorites and Comets" and Chapman and Scheiber,* Jour. Geophys. Res. v. 74, p. 6744.)

Composition	Siderites	Chondritic aerolites	Siderolites	Tektites
Fe	89.0	11.7	49.0	——
$FeO + Fe_2O_3$	——	12.0	6.7	4.9
FeS	1.8	5.9	0.5	——
SiO_2	——	38.3	17.1	70.9
Al_2O_3	——	2.7	0.4	13.5
MgO	——	23.9	19.8	2.5
CaO	——	1.9	0.3	3.1
Na_2O	——	0.9	0.1	1.4
K_2O	——	0.1	——	2.4
Ni	8.1	1.3	4.7	——
Co	0.6	0.1	0.3	——
TiO_2	——	0.1	——	0.8

small high-density planets in the same solar system, it is clear that bodies of different compositions can form close together. The pitted surface of Moon shows that bodies of a wide range of sizes have collided in the past. From studies of large impact craters such as Ries Kessel in Germany it is apparent that fragments from even a moderate (27 km wide) crater are thrown out for hundreds of kilometers. It is reasonable to suppose that much of the meteoroid gravel present in the solar system may be fragments generated in this way. This is especially true for the tektites.

Moon dust may provide a better average for meteorite composition than can be obtained from terrestrial meteorites. If Moon has no liquid core, as seems likely, it may also have had little alteration of its surface by actions from within. If this is the case, then dust from Moon's surface far from any recent crater should consist of a mixture of moon fragments and meteorite fragments. If Moon itself is built of accumulated fragments not subsequently differentiated, then the dust will be an average of meteoroid material. If, however, volcanic processes have occurred on Moon, the surface material will no longer represent an unbiased sample.

All the evidence shows that many changes are still occurring in the solar system. The fact that the Moon is gradually retreating from Earth and the possibility that Pluto may once have been a satellite of Neptune shows that large bodies can shift position. Perhaps the planetoids were formed by a collision between two bodies of differing characteristics. To choose between different theories of their origin, more data are needed. How do the chondrule-bearing meteorites differ in composition from the dust particles in comets? An answer to this would give evidence as to where each was most likely formed. Questions such as this may soon be answered by data gathered by space vehicles; and any day another meteorite may land some place on Earth giving new evidence.

BIBLIOGRAPHY

Baldwin, R. B., 1963, *The Measure of the Moon*. University of Chicago Press. 488 pp.

Baldwin, R. B., 1965, *A Fundamental Survey of the Moon*. McGraw-Hill Book Co., Inc. 194 pp.

Glasstone, S., 1965, *The Space Sciences*. D. Van Nostrand Co., Inc. 937 pp.

Hawkins, G. S., 1964, *Meteors, Comets and Meteorites*. McGraw-Hill Book Co., Inc. 134 pp.

Kaula, W. M., 1968, *An Introduction to Planetary Physics*. John Wiley and Sons, Inc. 490 pp.

Krogdahl, W. S., 1962, *The Astronomical Universe*. The Macmillan Co. 585 pp.

Kuiper, J. P.; B. M. Middlehurst, 1961, *Planets and Satellites*. University of Chicago Press. 601 pp.

Lundquist, C. A., 1966, *Space Science*. McGraw-Hill Co., Inc. 116 pp.

Many authors, 1970, *Science*, v. 167, no. 3918 (Special issue devoted to Apollo 11 results).

Mason, B., 1962, *Meteorites*. John Wiley and Sons, Inc. 274 pp.

Middlehurst, B. M.; G. P. Kuiper, 1963, *The Moon, Meteorites and Comets*. University of Chicago Press. 810 pp.

Ohring, G., 1966, *Weather on the Planets*. Doubleday and Co., Inc. 144 pp.

Wood, J.A., 1968, *Meteorites and the Origin of Planets*. McGraw-Hill Book Co., Inc. 117 pp.

4 DYNAMICS OF EARTH

4.1 EARTH'S ORBIT.

The center of gravity of the Earth-Moon system moves about Sun in an elliptical orbit of average radius 1.496×10^8km. At its greatest distance Earth is at 1.521×10^8 km. This point is called *aphelion* and occurs around July 4 (Figure 4.1). At its closest approach, *perihelion*, Earth is only 1.471×10^8 km from Sun. This occurs around January 3.

This difference in distance causes a variation in the amount of solar radiation received by Earth. The energy per square centimeter per minute is inversely proportional to the square of Earth's distance from Sun. At perihelion the flux is $\left(\dfrac{1.521}{1.471}\right)^2$ or 1.069 as great as at aphelion. On the whole, then, Earth is warmed more by Sun in January than in July. In the southern hemisphere this is true locally as well as for the whole Earth. In the northern hemisphere, the effect of distance from Sun is overshadowed by the effect of the tilt of Earth's axis.

The axis about which Earth rotates tilts with respect to the plane of its orbit by $23°26'59''$. As Earth moves about its orbit, this axis remains nearly constant in direction in space. Although the angle this axis makes with the plane of Earth's orbit remains constant, the angle it makes with a line from the center of Earth to Sun varies from $66.55°$ to $113.45°$ ($90 \pm 23.45°$). Thus the northern and southern hemispheres take turns being tilted toward and away from Sun.

The times when the axis has its maximum and minimum tilt toward Sun are called solstices. For the northern hemisphere, the summer solstice occurs on June 21 or 22 and the winter solstice on December 21 or 22 (Table 4.1). In between, when the angle passes thru $90°$, day and night are equally long. These times are called the equinoxes. The vernal equinox occurs on March 20 or 21 and the autumnal on September 22 or 23. The variation in date is caused by the fact that the calendar year varies in length, usually having 365 days, but in leap years, 366.

Solar heat flux at Earth's surface in summer is increased by two factors. First, the area over which the radiation from Sun is distributed depends on the tilt of Earth's surface with respect to the direction to Sun. In Figure 4.2, the radiation flowing in over the area B (perpendicular to Sun's rays) is

Table 4.1 *Time of solstices and equinoxes in Universal (Greenwich) Time. The times oscillate in a 4 year cycle. They will drift about 45 minutes every 4 years until the year 2100, when leap year will be skipped, moving the time forward again. (This data is published annually in the* American Ephemeris and Nautical Almanac.)

Year	Vernal	Summer	Autumnal	Winter
	March	June	September	December
1964	20, 2:10 P.M.	21, 8:57 A.M.	23, 0:17 A.M.	21, 7:50 P.M.
1965	20, 8:05 P.M.	21, 2:56 P.M.	23, 6:06 A.M.	22, 1:41 A.M.
1966	21, 1:53 A.M.	21, 8:33 P.M.	23, 11:43 A.M.	22, 7:29 A.M.
1967	21, 7:37 A.M.	22, 2:23 A.M.	23, 5:30 P.M.	22, 1:17 P.M.
1968	20, 1:22 P.M.	21, 8:13 A.M.	22, 11:26 P.M.	21, 7:00 P.M.
1969	20, 7:08 P.M.	21, 1:55 P.M.	23, 05:07 A.M.	22, 00:44 A.M.
1970	21, 00:57 A.M.	21, 7:43 P.M.	23, 10:59 A.M.	22, 06:36 A.M.
1971	21, 06:38 A.M.	22, 1:20 A.M.	23, 4:45 P.M.	22, 12:24 P.M.

spread out over the larger area A.° The larger J is, the more heat per unit area. At the equator, the decrease in the intensity varies from 0.914 to 1.000 so the heat flux changes less than 9 per cent seasonally. From the Arctic circle northward the intensity varies from zero to 0.403, so it is much warmer in summer than in winter.

The second factor controlling the amount of solar energy received per unit area is the number of hours of sunlight. Near the equator, day and night are always nearly the same in length; but as the Arctic and Antarctic circles are approached, the hours of daylight at the summer solstice increase to 24.

The amount of energy flux at Earth's surface per unit of time is called the *insolation*. It is less than the solar constant because of the effects of tilt of Earth's surface with respect to the incoming Sun's rays and because of the loss of energy in penetrating the atmosphere (Figure 4.3). This loss increases with latitude because the length of the path of Sun's rays thru the air increases as the angle the rays make with Earth's surface decreases. On the other hand, if insolation is measured per day, at high latitudes the effect of the increasing length of the day compensates for the effect of the tilt of Earth's surface, and insolation remains high in summer. The insolation in southern Greenland (latitude 65°N) on June 21, the summer solstice, is greater than at the equator. The long day is what makes it possible to raise crops as far north as the Arctic circle in spite of the short growing season.

Another factor is the difference in length of the seasons. As stated in Kepler's second law (Section 3.2), a line from Earth to Sun sweeps out a constant area per unit of time. This means that Earth must move more rapidly along its orbit near perihelion than near aphelion, making the northern winter pass more quickly than the summer. The northern summer is four days longer than winter. Consequently, total summer insolation is

---

°The ratio of B to A is called the sine of the angle J, abbreviated sin J.

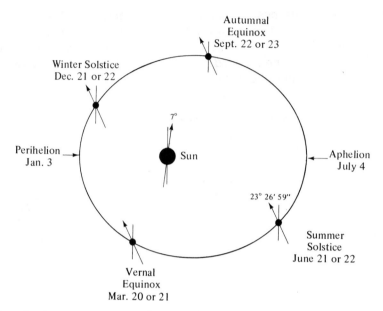

Fig. 4.1   *Earth moves in an elliptical orbit with Sun at one focus. (The ellipticity is exaggerated in this drawing.)*

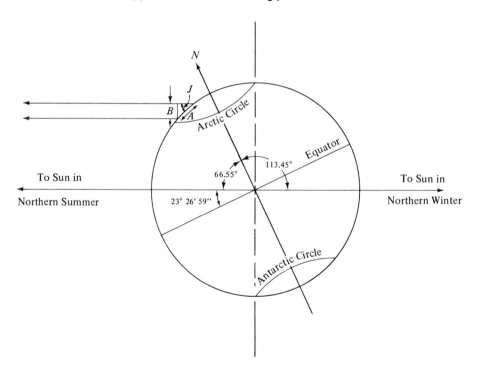

Fig. 4.2   *Insolation varies with the angle between Sun's rays and Earth's surface.*

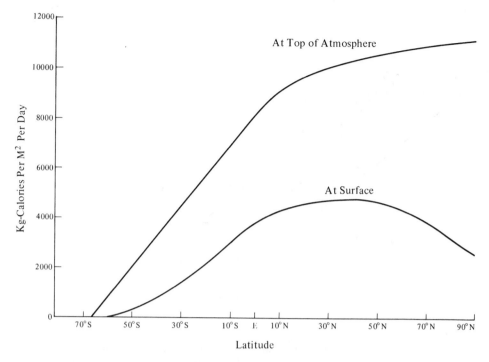

Fig. 4.3 *Solar energy distribution on Earth on day of northern-summer solstice.*

nearly the same in the northern and southern hemispheres. The increased distance of Earth from Sun is just about compensated for by summer being longer on the north half of Earth.

4.2 TIME UNITS.

The rate at which Earth turns on its axis gives us our basic unit of time, the day. The *solar day* is defined as the interval from one noon to the next. Noon occurs when Sun lies in the plane which includes the observer and Earth's axis of rotation. The solar day is a convenient unit for counting the passage of time because so many human activities are regulated by the alternation of night and day. It is not an accurate unit because successive days do not have the same length. To see why this is the case, it will be convenient to compare the solar day with other types of days.

Astronomers prefer to use the *sidereal day*, the time for Earth to rotate once using a line from Earth to any star except Sun for reference. This interval is more nearly a constant than the solar day. It is not a convenient unit for measuring the passage of time because there is one more sidereal day in a year than there are solar days. This is because Earth goes once around Sun in a year. Consider Figure 4.4. If Earth moves from A to B in one solar day, it rotates with respect to a distant star once plus the angle J_1,

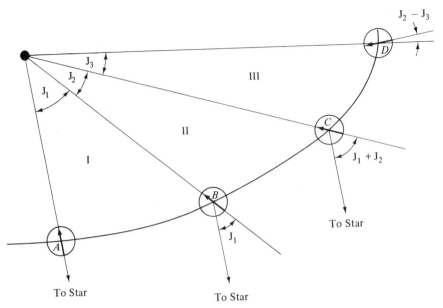

Fig. 4.4

the angle it has moved about Sun. The next solar day requires another sidereal day plus J_2 in addition. In the course of a year the J's add up to one whole extra rotation. Sidereal noon will occur at a different local time of day from one day to the next.

Now consider how far the Earth rotates in each of two successive time intervals of the length it took to get from B to C. From Kepler's second law (Section 3.2), area III must equal area II in Figure 4.4. For the areas to be equal, J_2 must be larger than J_3 since D is farther from Sun than B. In the time it takes Earth to get from C to D it will rotate once plus J_2, not once plus J_3. In other words, it takes slightly more than a solar day to go from C to D. Successive solar days are of different lengths because Earth does not progress around its orbit at a constant angular distance per solar day.

To avoid these differences, most people tell time using mean solar days. A *mean solar day* is the average length of time it takes Earth to rotate once on its axis with respect to Sun. It can be found by averaging the lengths of all the days in any year. The mean solar day is subdivided into 24 hours, the hour into 60 minutes, and the minute into 60 seconds. The sidereal day is 23.934 hours long.

Neither the mean solar day nor the sidereal day are of exactly constant length. The effect of the tides has already been discussed (Section 3.5). There are also small unsystemmatic fluctuations in the length of the day of as much as .005 seconds. These have, on occasion, set Earth as much as 40 seconds behind or ahead of schedule. Exactly what causes these changes is unknown. A part of the fluctuation appears to be seasonal, and may be due to changing patterns of circulation of the oceans and atmosphere. One of the fundamental laws of physics is that in any isolated system, momentum is constant. Considering only Earth's internal momentum (excluding that part involving its revolution about Sun and Moon), its angular momentum is largely in the form of its rotation on its axis. A small part, however, is in the form of motion of parts of Earth with respect to each other, for example, the air and ocean water with respect to the solid part of Earth. If the atmosphere were to develop a generally eastward drift at any one time, this would have to be balanced by a slowing down of Earth's rotation (a less rapid eastward rotation of the solid portion). Any such tendency of the air to drift tends to damp itself out by friction. Also, the atmosphere has so little mass compared to the rest of Earth that it would have to develop a strong drift to change the length of the day measurably. The observed changes in rate of rotation are greater than can be explained by variations in air and ocean currents.

There may, however, be one set of currents capable of explaining changes in the length of the day. Evidence will be presented later that Earth has a hot liquid core in which there are convection cells circulating. It seems probable that this core is not rotating at exactly the same speed as the rest of Earth. The evidence is that the magnetic field, which is believed to be generated in the core, is moving very slowly westward, suggesting that the core rotates faster than the rest of Earth. (Jupiter exhibits the same phenomenon. See Section 3.3.5.) A decrease in this drift would be compensated for by an increase in the rate of solid-earth rotation and vice versa. The amount of transfer of momentum required to explain the changes in length of the day by this process is well within the uncertainties in the amount of momentum in the core convection cells.

One other means is available for changing the rate of rotation. Angular momentum can be expressed by the equation

$$\Sigma \, MWR = \text{Constant} \qquad\qquad (4.1)$$

where M is mass, W is angular velocity, R is distance from Earth's axis of rotation and Σ means "sum up the effect of all" the masses everywhere in Earth. If R in Equation 4.1 is increased, W must be decreased. Thus if 10 millimeters of water are evaporated from the oceans and deposited as snow on the tops of high mountains, Earth has to slow its rotation. It will speed up again in spring when the snows melt and flow back to the sea. This effect is also too small to explain all changes in the length of the day, though seasonal mass transfers such as this may explain parts of the fluctuation.

Similarly, if rivers carry sediment and dissolved material to the ocean basins faster than new mountains are uplifted, then Earth must speed up. This uplift process is not a steady one, as earthquakes testify. No changes in the length of the day have as yet been correlated with individual earthquakes.

A change of .005 seconds in the length of the day will only be detected if it persists for enough days to accumulate a rotation of Earth's surface with respect to the stars by a measurable amount. The difficulty here is that the rotation of Earth with respect to the stars was how we measured time. In the United States, scientists at the U. S. Naval Observatory in Washington watch the position of Earth to give us Naval Observatory time with which to compare our clocks and watches. To see if Earth rotates fast or slow we must compare observatory time with some other standard. Imperfect though it is, the rate of rotation of Earth on its axis is one of the most constant quantities of which we know.

To get a more perfect unit of time we must find an oscillating system that is more invariant than Earth's rotation. One such system is the motion of an electron about the nucleus of an atom. To measure variations in the rate of rotation of Earth, the number of revolutions of an electron around the nucleus of a cesium atom are counted. A device which can do this is called an atomic clock. It is not that the period of revolution of one electron is necessarily any more constant than the period of rotation of Earth, but that we assume that by counting many revolutions, any small fluctuations will average out. There is the further advantage that, in the atomic clock, what is averaged is not merely the motions of one electron about one nucleus, but the average of many electrons about many nuclei. For the cesium atomic clock, 9,192,631,770 cycles correspond to one *ephemeris second*, which is by definition the mean solar second for the year 1900. This ephemeris second has been adopted as the international standard unit of time.

4.3 UNIVERSAL AND LOCAL TIME.

For convenience in keeping track of time intervals and comparing the times of events in different places, the world's nations have agreed to refer events to a uniform reference standard. This system, called *universal time*, is measured in mean solar days and fractions thereof (hours, minutes, seconds). Noon on the day of the vernal equinox at Kew Observatory at 0° longitude (Greenwich, England) is used for reference. For convenience, the hours are counted from twelve hours before noon to twelve hours past noon, i.e. from one "midnight" to the next. The vernal equinox was chosen as the reference day because of its relation to the year, as discussed below.

Because people are in the habit of regulating their activities by the position of Sun in the sky, almost all governments also define a *local time* which is displaced with reference to universal time by a fixed number of hours and minutes (Figure 4.5). Local times which cause local noon to approximate the time when Sun is at its zenith are called standard time. Many communities, especially in summer, use a local time which provides more daylight hours after noon than before noon. In a sort of verbal self-deception, this type of time is called "daylight savings."

In North America, there are eight time zones ranging from Atlantic standard time, used in the maritime provinces of Canada, which is four hours less than universal time, to Bering standard time, which is eleven hours less. Most countries use a local time which is displaced an exact number of hours with respect to universal time; but this is not always the case. In Afghanistan, local time is four hours and 26 minutes ahead of universal time.

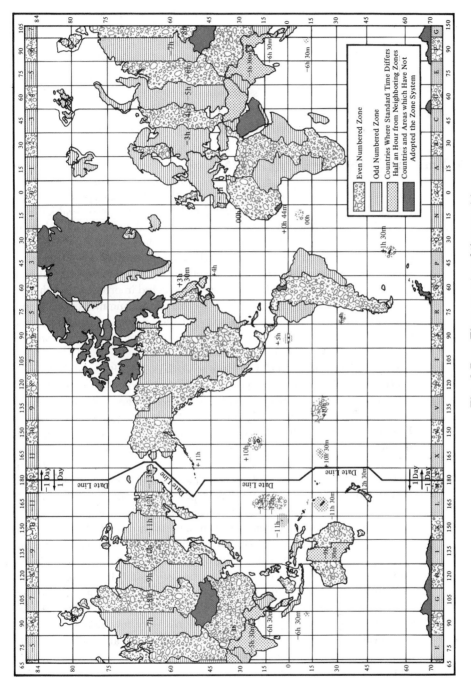

Fig. 4.5   *Time zones of the world.*

The use of local time can cause a peculiar situation when one is travelling. Bering time is eleven hours behind universal time, but places to the west of the zone using Bering time often use New Zealand time, which is twelve hours ahead of universal time. As one goes from Bering to New Zealand time, one shifts not one hour back but 23 hours forward in time. Going the opposite way, one changes 23 hours backward in time. Usually this also means changing the date. The line between time zones across which there is a whole day's difference is called the *International Date Line*.

4.4 THE YEAR AND THE CALENDAR.

For keeping track of long time intervals, the days are grouped into years. Just as there are several types of days, similarly there are several kinds of years. Our calendar year is based on the interval between vernal equinoxes. Either of the equinoxes could have been chosen equally well, or one of the solstices could have been used. These times are convenient for reference points because they can be relatively easily and accurately identified. A day related to the tilt of Earth's axis was selected in order that each year should contain exactly four seasons—winter, spring, summer and autumn. The interval between vernal equinoxes is called a *tropical year*. Years are numbered successively starting, in most countries of the world, from the year in which Christ is believed to have been born.

The tropical year is 365.2422 mean solar days long. Because there are a fractional number of days in a year, an extra day is added every fourth year at the end of February. Such years are called leap years. But one day every four years is too many to add; only .2422 of a day per year is needed. Therefore, every hundred years, leap year is skipped. This is done in years ending in 00 like 1900. In this fashion, 24 extra days are added per century. But this is .0022 days per year too few, so leap year is added every fourth century, even if the year ends in 00. The year 2000 will be a leap year. This is equivalent to adding .2425 days per year on the average, which is .0003 days too much. Every 3333 years one day too many is added.

Our present calendar system has been in use only since 1582, when Pope Gregory revised the calendar system established by the Roman Senate under Augustus Caesar in 8 B.C. Augustus's calendar did not provide for skipping three leap years every 400 years. As a result, the vernal equinox had moved twelve days ahead in the calendar since Augustus's time. No provision has been made for dropping the extra day every 3333 years. This task can safely be left to future generations. Very few calendar systems have lasted 3000 years without some modification, although our basic twelve-month-calendar concept dates back to 4236 B.C. when it was adopted in Egypt.

The tropical year is not the time it takes Earth to travel once around Sun with respect to the other stars. This interval is called the *sideral year*.

It is 365.2564 mean solar days long, or 1 1/25,800 tropical years. As a result, the stars visible in the sky on any particular date change slightly from year to year on a 25,800-year cycle. This is called the *precession of the equinoxes*. It involves a change of 50.2″ of arc per year in the direction that the axis of rotation of the earth points among the stars (Figure 4.6). At the present time, Earth's axis points very nearly at the star Polaris, known as the north star because of its usefulness in establishing that direction. But this will not always be the case in the future, nor has it always been so in the past.

The amount of precession is not constant but varies as much as 9.23″ from 50.2″ each year. This variation, called the *nutation*, of the direction of Earth's axis is a result of the gravitational pull of Sun and Moon on Earth's equatorial bulge, and is a measure of the distribution of Earth's mass in this bulge. If Earth were perfectly spherical and its mass symmetrically distributed with respect to its center, then there would be no nutation. The nearly constant rate of rotation of Earth on its axis is a result of the constancy of angular momentum. The direction and amount of rotation can only be changed by outside forces. Such a change is resisted by the moment of inertia of Earth.

The size of the nutation under the attraction of the known masses of Sun and Moon tells how evenly momentum is distributed in Earth. The moment of inertia of any body is

$$I = \Sigma\, M\, R^2 \qquad\qquad (4.2)$$

where M is the mass of any particle in the body and R is the distance from the axis to which the moment of inertia is referred. The complete moment is obtained by summing (Σ) the moments of every individual part. A body has a different moment about every different axis thru it. In particular, the moment is different for an axis thru the poles, I_p, and an axis lying in the plane of the equator, I_e. From the amount of the nutation, it can be calculated that for Earth

$$\frac{I_p - I_e}{I_p} = .003273 \qquad\qquad (4.3)$$

In other words the equatorial moment is .003273 smaller than the polar moment. This is a useful piece of information which will be needed later to find the distribution of mass in Earth's interior.

The third type of year is the period from one perihelion to the next. This is called the *anomalistic year*. Its length is 365.2596 mean solar days. Because the siderial and anomalistic years are of unequal lengths, the orbit of Earth must be rotating with respect to the stars. Since the two years differ by .0032 days, the ellipse rotates once every 365.2564/.0032 years, or once in 110,000 years.

Fig. 4.6   *The direction of Earth's axis in space precesses among the stars, making one rotation every 25,800 years.*

The difference between the anomalistic and the tropical years affects the intensities of the seasons. This difference, amounting to .0174 days per year, causes the relative positions of perihelion and the solstices to precess about Earth's orbit every 365.2422/.0174 or 21,000 years. As a result, perihelion and aphelion occur one day later every 57 years. At present in the northern hemisphere, the summer solstice occurs close to aphelion and the winter solstice close to perihelion. Because Earth is farther from Sun in summer than in winter, the differences in the seasons in the northern hemisphere are ameliorated and in the southern hemisphere intensified. In 10,500 years the situation will be reversed.

There is also a small variation in the shape of the orbit, which is believed to increase and decrease in ellipticity with a period of around 92,000 years. There is an additional variation in the tilt of Earth's axis,

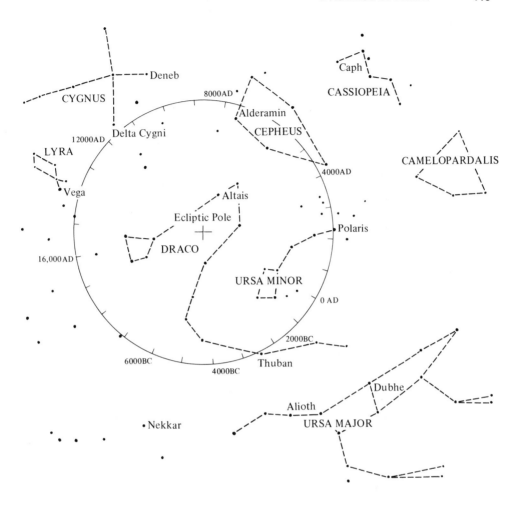

from 21½ to 24½° with a period of about 40,000 years. These variations also will have an effect on the seasons, which will go thru greater extremes when the ellipticity is large than when it is small. Differences in the total annual insolation at any given latitude are also introduced (Figure 4.7).

Such variations provide one possible mechanism which may be involved in the cause of glaciation. If it could be shown that glaciers are more likely to form under conditions of extreme seasonal variations (or the reverse of this), then it might be possible to correlate periods of glaciation on Earth with the orbital permutations. Unfortunately, neither the orbital permutations nor the intensities of glaciation on Earth are accurately known throughout Earth's history, so the problem of their relationship is insoluble until more data are available.

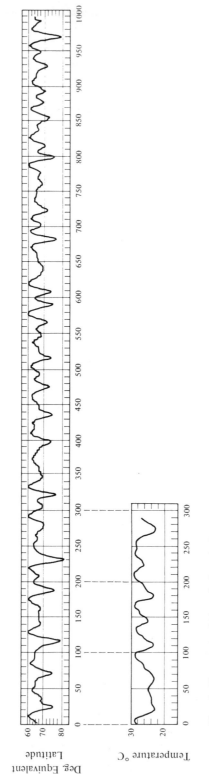

Time in Thousands of Years Before the Present

Fig. 4.7 *Comparison of climate based on geologic evidence (lower curve) with "equivalent" latitude (upper curve) of what is now 65°N as affected by orbital perturbations. (After Emiliani, Jour. Geol. v. 63 p. 117. Copyright 1955 and 1966 by the University of Chicago.)*

4.5 THE AXIS OF ROTATION.

From the preceding discussion, it is clear that nothing about the motion of Earth is constant: not the orbit, not the rate of rotation, not the direction of its axis. The same applies to the relation of the body of Earth to the axis of rotation. By observing the positions of the stars as a function of time with respect to the locations of astronomical observatories all over the world, it is found that the whole surface of Earth moves with respect to the axis of rotation (Figure 4.8). This is called the Chandler motion after the American astronomer who first measured it. The motion is at least in part periodic. The position of the axis of figure (fixed in Earth) moves about the axis of rotation in a roughly circular orbit with a period of 428 days (a little more than 14 months). Seen from the north, the axis of figure moves clockwise around the axis of rotation with a separation of around 0.4 seconds of arc, or 10 meters (30 feet) at Earth's surface.

The Chandler motion is at once both expectable and difficult to explain. Earth is not exactly a sphere, but bulges around the equator. If for any reason it is in rotation around any axis not an axis of maximum or minimum moment of inertia, it can be shown that internal forces will be generated which will cause the axis of maximum moment and the axis of rotation to move around one another. If the body is perfectly rigid (incapable of deformation), then this motion will continue indefinitely without change. If the body is deformable, and if this deformation involves friction between the parts, then the relative motion of the two axes will be damped out. In a deformable body, there will be a tendency for the body to adjust its shape so that the axis of maximum moment and the axis of rotation coincide. In other words Earth will try to change in shape so that the equatorial bulge lies exactly in a circle surrounding the axis of rotation. This process will tend gradually to reduce the size of the Chandler motion.

The motion should die out slowly. While this is happening, other processes, such as erosion and sedimentation, mountain uplift or ground displacements along faults, which occur during earthquakes, will continually introduce new asymmetries. This will prevent equilibrium from ever being achieved. The fall of a large meteorite would contribute to such asymmetry, both by the mass it adds and by the momentum it brings with it. The observed Chandler motion is reasonable for a body such as Earth.

The Chandler motion has never exactly repeated itself since measurements were begun. This leaves the possibility that there may be a gradual drift of the whole Earth (or of the whole crust of Earth) with respect to the axis of rotation. Such a shift is called *polar wandering*. Earth's poles do wander. The evidence comes from the study of variations in the direction of Earth's magnetic field (Figure 4.9). If the direction of magnetization of rocks formed many years ago in past geologic periods is examined, it is found to be different from the direction of the field today. But rocks forming today are magnetized in the direction of the present magnetic field. Clearly either the direction of Earth's magnetic field changes with time compared to the direction of the axis of rotation, or the position of points

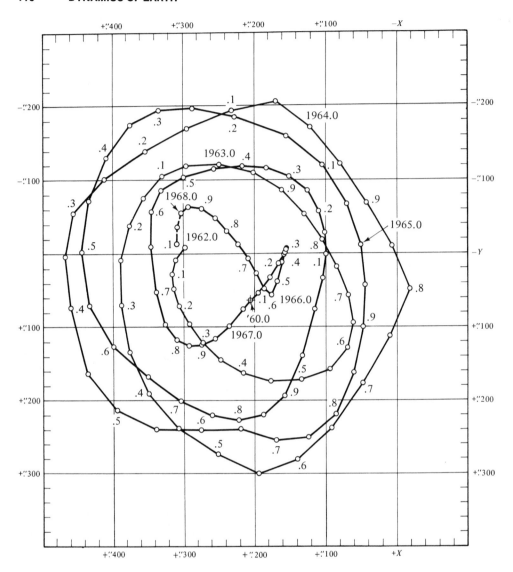

Fig. 4.8 *Chandler motion of Earth's axis of rotation 1962–1968. (Yumi, An-nual Report of Int. Polar Motion Service.)*

on the surface of Earth changes with respect to the direction of the axis of rotation, or both.

In Chapter 7 it will be shown that the most reasonable explanation of the cause of the magnetic field requires that it be so linked to the rotation that the directions of the magnetic axis and the axis of rotation will always be close, but not necessarily identical, to one another. Accepting this un-proven but probable hypothesis, then the surface of Earth has moved with respect to the axis of rotation. The pole has wandered. This does not prove that the body of Earth as a whole moves with respect to the axis of rota-

Fig. 4.9 *Past positions of the north pole of rotation on the basis of paleo-magnetic evidence gathered on different continents. Past times are shown in millions of years. (After Doell and Cox, 1961,* Advances in Geophysics *v. 8, p. 298. Copyright Academic Press.)*

tion, because all or parts of the surface can move with respect to the rest of Earth. This problem will be discussed in more detail later.

BIBLIOGRAPHY

Abel, G., 1966, *Exploration of the Universe*. Holt, Rinehart and Winston. 646 pp.

Holmes, A., 1965, *Principles of Physical Geology*. Ronald Press Co. 1288 pp.

Krogdahl, W. S., 1962, *The Astronomical Universe*. The Macmillan Co. 585 pp.

Strahler, A. N., 1963, *The Earth Sciences*. Harper and Row Publ. 681 pp.

5 MASS, MOMENT, SHAPE AND SIZE OF EARTH

5.1 WEIGHING EARTH.

In studying Sun, Mars and other extra-terrestrial bodies except Moon we had to content ourselves with such information as could be obtained by looking at them from our planet or by sending unmanned vehicles to make measurements. Our study of Earth can be more detailed because the subject is available for direct examination. On the other hand, Earth is so close to us that we can ordinarily regard only a part of it at any one time. As a result, measurement of its principal properties will continue to be indirect, involving a combination of observation and calculations based upon simple physical laws.

A good example of a property determined indirectly is the mass of Earth. This can be found, as was the mass of Sun, from knowledge of an artificial satellite's period of revolution, T, and distance from Earth. By analogy with Equation 3.4 we have

$$G\frac{M_e}{R^2} = \frac{4\pi^2}{T^2}R \tag{5.1}$$

where G is the universal constant of gravitation, M_e is Earth's mass, and R is the radius of the orbit of the satellite around Earth. R is hard to find precisely until Earth's radius is known, because distance to the satellite has to be measured from Earth's surface and appreciable error is introduced by neglecting the distance from Earth to the center of the Earth-Moon system.

An alternative method of finding Earth's mass is to use Newton's universal law of gravitation, Equation 3.18 (p. 60), more directly. To a first approximation, Earth can be treated as though all its mass were concentrated at its center. To the degree that Earth is a sphere whose density is a function only of distance from the center, the mass will attract as though it were concentrated at the center. If we measure the force acting on a unit mass anywhere above the surface, it is composed of several parts: the gravitational attractions of Earth, Moon, Sun, etc., and the centrifugal effects which result from Earth's rotation on its axis and its motion in space. By far the largest of these forces is the gravitational attraction of Earth. Neglecting all other components, anywhere above Earth's surface, from Equation 3.18

$$g = \frac{F}{M_1} = \frac{GM_e}{R_e^2} \qquad (5.2)$$

where g is gravity, defined as the force, F, per unit mass, M_1, resulting from the attraction of the mass, M_e, of Earth at a distance, R_e, from its center. This equation can be solved for M_e

$$M_e = \frac{gR_e^2}{G} \qquad (5.3)$$

To "weigh" the earth, we need only find the values of g, R_e and G.

5.2 RADIUS OF EARTH.

The radius of Earth is relatively easy to measure. Earth's roundness has been recognized by scholars since at least the time of Pythagoras, the Greek philosopher-mathematician who lived in the 6th century before Christ.

The method of measuring Earth's diameter was first described by another Greek, Eratosthenes, who lived in Alexandria, Egypt, around 200 B.C. The same method is used today. The instruments used to make the measurements, however, have been greatly improved. Eratosthenes happened to be in the town of Syene (now called Aswan) at noon on the day of the summer solstice, which is the time when Sun is at its maximum height above the horizon. He noticed that the bottom of a steep-walled, hand-dug well was, as nearly as he could see, completely illuminated, showing that Sun was directly overhead. He remembered that at Alexandria, Sun never rose that high above the horizon. He correctly reasoned that this was because of Earth's curvature. Alexandria is too far north for Sun ever to be overhead. A year later, at the next summer solstice, Eratosthenes measured the length of the shadow of a vertical pole at Alexandria. He knew that the angular width of the shadow of the pole (A_p in Figure 5.1) was the same as the angular distance from Syene to Alexandria (A_c in Figure 5.1) assuming that Sun's rays at the two cities were parallel. Noting that the width of the shadow was 1/50 of a circle, he reasoned that the circumference of Earth was 50 times the distance from Alexandria to Syene.

The accuracy of Eratosthenes' measurement is uncertain. He used the stadium as a unit of length. Unfortunately, there were three different units of length in use in Egypt all bearing that same name. The standard Olympic Stadium had a length of 185 meters; the Royal Egyptian Stadium, 210 meters; and the itinerary stadium, 157 meters. It is probable that Eratosthenes had the last of these in mind, as it is the one most commonly used to

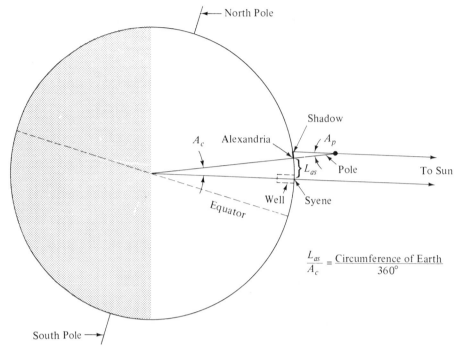

Fig. 5.1   *Eratosthenes' experiment to determine Earth's circumference.*

measure distances by travellers. He reported the distance from Syene to Alexandria to be 5000 stadia. Earth's circumference would, therefore, be 250,000 stadia, or 39,300 km. This is less than two per cent smaller than the true circumference of 40,000 kilometers. (The meter was deliberately defined in France with the intention that it would be 1/40,000,000 of Earth's circumference as nearly as the French revolutionists were able to measure this quantity.)

Another measure of Eratosthenes' accuracy is to compare his angular distance of 1/50 of a circle (7°12′) with the actual separation of Syene (24°5′ N. latitude) and Alexandria (31°12′ N. latitude), which is 7°7′. This also is an error of less than two percent.

It is often stated that men in the middle ages thought that Earth was flat. This may have been true for the uneducated masses of ordinary people, but Greek knowledge persisted among scholars, particularly thru the Moorish civilization of North Africa, with which Europeans had many contacts. Perhaps fortunately, Eratosthenes' calculation was replaced among these men by a later, less accurate measurement reported by another Greek, Claudius Ptolemy, in the second century A.D.. Ptolemy estimated the circumference to be only 33,000 km. Columbus set out to find Cathay with this figure in mind. If he had known Asia was 7000 km farther away, he might never have tried to reach it. Thus one sees Columbus, an Italian, sailing west with the support of a research grant from the Spanish government, having planned his experiment using data gathered 1200 years earlier by a Greek who lived in Egypt. He was looking for China and he discovered the island of San Salvador in the Bahamas. There is a lesson of some sort in this.

Significant improvements in man's ability to measure Earth's diameter followed Snell's introduction in 1617 of triangulation as a method for measuring distances. By 1735, Jaques Cassini had measured the length of 1° of arc in northern and in southern France. He found that a degree of arc was longer in the south than in the north. This meant that Earth is ellipsoidal rather than circular in shape.

Cassini's measurement suggested also that the radius was greater at the poles than at the equator. To see why this is so it is necessary to distinguish between radius and radius of curvature. In Figure 5.2, C is the center of the ellipse and P and Q are any two points on its edge 1° apart. If lines are drawn perpendicular to Earth's surface at P and Q, they will meet at a point O, which is the center of a circle which will pass thru P and Q and will closely approximate the ellipse from P to Q. The radius of this circle is called the *radius of curvature* of the arc PQ. Along the flattened part of the ellipse the radius of curvature is greater than the radius of the ellipse; and at the ends of the ellipse it is less than the true radius. For an ellipse, radius of curvature is large where radius is small and *vice versa*.

Cassini found that radius of curvature increased as one went southward in France. This meant that the radius of Earth was decreasing going from the north pole toward the equator. He concluded that Earth is a *prolate* ellipsoid (watermelon-shaped) rotating about its long axis (Figure 5.3).

When Isaac Newton in England heard this, he objected. Cassini's discovery did not seem reasonable to him. Because Earth is rotating, Newton felt it should bulge around the equator due to the centrifugal effect. He pointed out that it was to be expected from physical theory (as developed by himself) that Earth was an *oblate* ellipsoid (more nearly pumpkin than watermelon-shaped).

On checking Cassini's data, it was found that his calculated 1° arc lengths in northern and southern France differed by only 137 toises out of 57,000. (A *toise* is 6 French feet, or 6 feet 5 inches English measure. The French kings had bigger feet than the English kings.) This difference was less than the uncertainty of the measurements of either arc. Cassini's measurements suggested, but did not prove, that Earth was prolate.

To settle the question, expeditions were sent by the French National Academy to Lapland, as far north as it was possible to go in Europe in 1736, and to a point near the equator at the city of Quito, in what was then the Spanish colony of Peru but is now the capitol of Ecuador. The expeditions measured the length of a degree in each place, and proved clearly that Newton was right. Earth is an oblate ellipsoid.

Modern measurements indicate that the equatorial radius is 6,378.1 kilometers; the polar radius, 6,356.6 kilometers. The radius of a sphere with the same volume is 6,371.2 kilometers (3966, or about 4000, miles). There are also small variations in the radius with longitude around the equator.

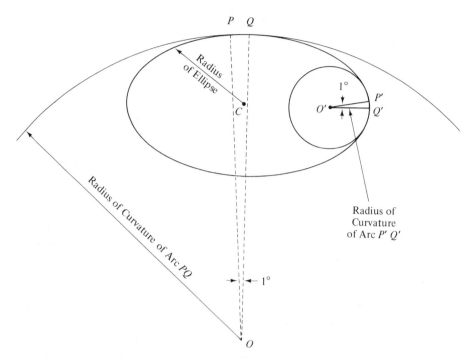

Fig. 5.2   *For an ellipse, radius of curvature is long where radius is short and short where radius is long.*

The sea-level radius at 15° east and 165° west longitude has been calculated to be over 200 meters longer than at right angles to this. This effect is much smaller than the variation of elevation represented by the continents and ocean basins. It is detected by measuring variations in the attraction of gravity, which varies inversely with distance from Earth's center, and from the perturbations of the orbits of artificial satellites, not by Eratosthenes' method using the length of one degree of arc.

When one speaks of the radius of Earth, what is meant is the distance from the center to the imaginary level surface which most nearly approximates the average elevation of the oceans. This surface is called the *geoid*. Its height under the continents is the height that sea level would follow thru the rocks if all were highly porous and if capillary forces were absent. The geoid is not a perfect ellipsoid, but rises under the continents and drops as it passes across the oceans as much as several tens of meters compared to this ideal shape. This means that the water surface is lower (at a lesser distance from Earth's center) in the middle of an ocean basin than along its edges.

The equatorial radius is $\dfrac{6{,}378.1 - 6{,}356.6}{3{,}378.1}$ or $\dfrac{1}{297}$ longer than the polar radius. This is less than one percent more bulge than would be expected if Earth were a fluid and had adjusted itself to the equilibrium shape under the influence of gravitational force and the centrifugal effect of rotation.

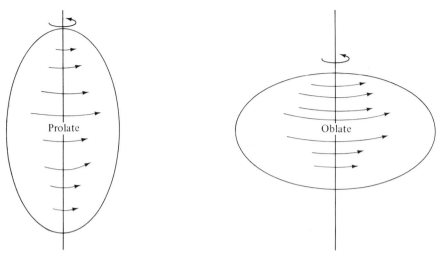

Fig. 5.3   *Rotating an ellipse about its longest axis generates a prolate ellipsoid; rotating it about its shortest axis generates an oblate ellipsoid.*

(Perturbations of artificial satellite orbits imply a slightly smaller value of ellipticity, $\dfrac{1}{298.26}$. For hydrostatic equilibrium the ellipticity would be $\dfrac{1}{299.7}$.) Considering the masses of rock moved about by erosion and the decrease in rate of Earth's rotation caused by the tides, the nearly hydrostatic condition leads to the conclusion that one of the following hypotheses is true. Either (1) for long-continued stresses Earth has no strength and gradually flows like a viscous liquid; (2) the rocks of Earth's interior are much weaker than the surface rocks, and allow Earth to adjust its shape to the equilibrium figure in spite of the stiffness of the crust; or (3) Earth was once either fluid or plastic and has only relatively recently solidified in the equilibrium shape. Hypothesis two is certainly true and perhaps one as well.

5.3 GRAVITY

Gravity is the second quantity which must be known to solve Equation (5.3) for Earth's mass. It can be measured by swinging a pendulum. The period (time of one complete swing) of a simple pendulum swinging under the influence of gravity, g, can be shown to be

$$T_p = 2\pi \sqrt{\frac{L}{g}} \tag{5.4}$$

where L is the length from the point of support to the center of mass of the pendulum. Squaring both sides of (5.4) and solving for g

$$T_p^2 = 4\pi^2 \frac{L}{g} \tag{5.5}$$

$$g = 4\pi^2 \frac{L}{T_p^2} \tag{5.6}$$

From the period of oscillation and the pendulum length, gravity can be calculated.

Gravity is found to be about 9.81 newtons/kg at sea level. Actually it varies from 9.780490 newtons/kg at the equator to 9.832213 newtons/kg at the poles. This variation is due to three factors we neglected in Equation 5.2. First, at the equator Earth's rotation produces a centrifugal acceleration which overcomes a part of the gravitational attraction and causes gravity to be lower than at the poles. The variation, however, is greater than would be expected due just to centrifugal effect alone. This is because the centrifugal effect causes Earth to bulge at the equator. This makes the equator farther from the center than the poles. Gravity measured at the equator will be less than gravity measured at a pole because of the $1/R_e^2$ term in Equation 5.2. Finally, the non-spherical distribution of Earth's mass affects the variation. The equatorial bulge has a slightly stronger attraction at the equator than at the poles, partially compensating for the effect of the difference in radius.

The other forces neglected in Equation 5.2 such as the attraction of Sun and Moon, add up to such a small fraction of gravity that they can barely be measured. They produce the forces which cause the tides. The tidal force has a maximum strength a little over one ten-millionth of gravity.

Gravity can also be studied using a very delicate spring balance. By comparing the stretch of a spring supporting a constant mass at two different locations, the difference in gravity between them can be found. The amount the spring stretches is proportional to the strength of gravity. Instruments which do this are called *gravimeters* (Figure 5.4). Differences in gravity as Earth's surface rises and falls due to the tidal forces of Sun and Moon are measured in this way. The amount of stretch of the spring which can be measured is less than a wavelength of visible light.

5.4 THE UNIVERSAL CONSTANT OF GRAVITATION

G, in Equations 5.1 and 3.18, is found by a laboratory experiment. The method most commonly used was developed by an Englishman, Henry Cavendish, in 1798 using an instrument called a torsion balance (Figure 5.5). Two round balls of known mass, M_2, are mounted on either end of a light rod, of length $2H$, which is suspended by a fiber which provides only a small resistance to being twisted. The rod and balls are free to rotate in a horizontal plane, except as this rotation is resisted by the torsion fiber. The reflection of a scale at a distance L from the rod is observed thru a telescope with a cross hair in it in such a manner that the cross hair appears to move relative to the scale when the rod turns. The strength of the fiber supporting the rod is such that if a twisting force of strength F is exerted at the distance H, the rod and balls will rotate thru an angle A given by

$$A = \frac{HF}{c_t} \qquad (5.7)$$

Fig. 5.4   *Gravimeter in use. (Courtesy Houston Technical Laboratories.)*

The quantity c_t is called the *coefficient of torsion* of the fiber. Rotation of the rod by an amount A will change the angle B by twice this amount. If the scale is at right angles to the line from the mirror, then a small deflection, X, of the image of the scale as seen thru the telescope is related to A by

$$2A = \frac{X}{L} \qquad (5.8)$$

If two additional balls of mass M_1 are placed at a distance D from the M_2 balls, the rod will be caused to rotate due to the attraction of these masses. According to the universal law of gravitation, Equation 3.18 (p. 60)

$$F = G\frac{M_1 M_2}{D^2} \qquad (5.9)$$

Substituting Equation 5.9 into Equation 5.7 and remembering that there are two forces exerted, one at each end of the rod

$$A = \frac{2HF}{c_t} = \frac{2HGM_1 M_2}{c_t D^2} \qquad (5.10)$$

(The gravitational pull of each M_1 ball on the more distant M_2 ball has been neglected. This force is small if the distance $2H$ is large.)

Equation 5.10 can be solved for the universal constant of gravitation obtaining

$$G = \frac{Ac_t D^2}{2HM_1 M_2} \qquad (5.11)$$

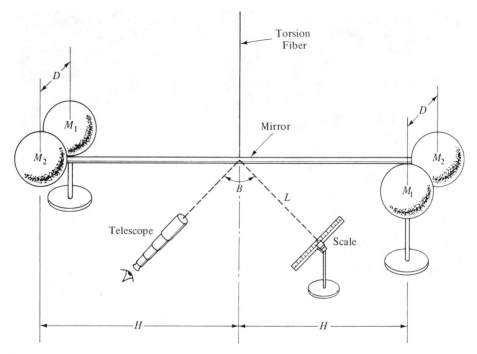

Fig. 5.5 *Principle of the Cavendish torsion balance for determining the universal constant of gravitation.*

All the quantities in this equation can be measured directly except c_t. To find c_t, the M_1 masses are taken away, and the horizontal bar is given a slight rotation. It will then oscillate in a horizontal plane as a torsion pendulum. Its natural period of oscillation will be found to be

$$T_p = 2\pi \sqrt{\frac{I}{c_t}} \qquad (5.12)$$

where I is the moment of inertia of the rod and balls. For a light-weight rod and heavy, round balls, the moment equals the square of the radius, H, times the mass M_2 of each ball. For two balls

$$I = 2M_2H^2 \qquad (5.13)$$

and (5.12) becomes

$$T_p = 2\pi \sqrt{\frac{2M_2H^2}{c_t}} \qquad (5.14)$$

Squaring both sides of (5.14) and solving for c_t

$$T_p^2 = 4\pi^2 \left(\frac{2M_2H^2}{c_t} \right) = \frac{8\pi^2 M_2H^2}{c_t} \qquad (5.15)$$

$$c_t = \frac{8\pi^2 M_2 H^2}{T_p^2} \tag{5.16}$$

Substituting this into Equation 5.11 and simplifying

$$G = \frac{8\pi^2 M_2 H^2}{T_p^2} \times \frac{AD^2}{2HM_1 M_2} = \frac{4\pi^2 HAD^2}{T_p^2 M_1} \tag{5.17}$$

All the quantities on the right in Equation 5.17 are measurable. The value of G has been determined with great accuracy at the U. S. Bureau of Standards in Washington and found to be $(6.673 \pm .003) \times 10^{-11}$ newton $-$ m^2/kg^2.

5.5 MASS AND DENSITY OF EARTH.

Now we have the three quantities needed to weigh Earth. Substituting into Equation 5.3 the value of gravity at 45° latitude, $g = 9.806$ newtons/kg the mean radius, $R_e = 6,371.2$ kilometers $(6.3712 \times 10^6$ meters) and $G = 6.673 \times 10^{-11}$ newtons $-$ m^2/kg^2 we obtain

$$M_e = \frac{gR_e^2}{G} = \frac{(9.806)(6.3712 \times 10^6)^2}{6.673 \times 10^{-11}}$$
$$= 5.965 \times 10^{24} \text{ kg} \tag{5.18}$$

More accurate allowance for the exact shape of Earth would give $(5.977 \pm .004) \times 10^{24}$ kg.

Dividing this by the volume of Earth gives a density of $5.517 \pm .003$ grams per cubic centimeter (5517 kg/m^3). This is about twice the density of most common rocks at Earth's surface and more than three times the density of many common soils. If the average density of Earth is twice the surface density, then density must increase going downward, and it must somewhere reach values greater than 5.517 grams per cubic centimeter.

5.6 MOMENT OF INERTIA.

One other important piece of information is available about the distribution of mass in Earth's interior. It was pointed out earlier that gravity varies on Earth almost exactly as would be expected for a rotating fluid. From considerations of physical theory, it can be shown that in this case, to a close approximation, gravity at the equator, g_{eq}, the rate of rotation, W, the mass, M_e, the polar and equatorial radii R_p and R_{eq} and the polar and equatorial moments of inertia, I_p and I_{eq} are related by the formula

$$I_p - I_{eq} \approx \left(\frac{2(R_{eq} - R_p)}{3R_{eq}} - \frac{W^2 R_{eq}}{3g_{eq}} \right) M_e R_{eq}^2 \tag{5.19}$$

All of these quantities except I_p and I_{eq} are known.

In Chapter 4 (Equation 4.3, p. 111) it was pointed out that the size of the nutation of the precession of Earth's axis required that

$$I_p = \frac{I_p - I_{eq}}{.003273} \tag{5.20}$$

Substituting (5.19) into (5.20)

$$I_p = \frac{M_e R_{eq}^2}{.009819} \left(\frac{2(R_{eq} - R_p)}{R_{eq}} - \frac{W^2 R_{eq}}{g_{eq}} \right) \tag{5.21}$$

Substituting in values, I_p is found to be 8.08×10^{37} kg-m². I_{eq} is 8.05×10^{37} kg-m².

The ratio of I_p to $M_e R_{eq}^2$ is a sensitive measure of the distribution of mass in Earth's interior. Rearranging Equation 5.21 and substituting values for known quantities, this quantity can be shown to be

$$J = \frac{I_p}{M_e R_{eq}^2} = 0.33078 \tag{5.22}$$

For a sphere of constant density throughout, the ratio would have been 0.4. For the ratio J to be so much less than 0.4, much of Earth's mass must be close to its center.

This and average density give two crucial clues to the composition of the rocks in Earth's interior. Density must increase with depth in such a manner that the average density is 5.517 gm/cc and J is 0.33078. These two requirements will be combined with other information in Chapter 10 to calculate the approximate variation of density with depth in Earth. Only a composition of rock whose density is close to the known value is possible at any depth.

BIBLIOGRAPHY

Anonymous, 1951, "Our Narrowing World." *Natl. Geol.* 100: 751–754.

Heiskanen, W. A.; F. A. Vening Meinesz, 1958, *Earth and Its Gravity Field.* McGraw-Hill Book Co., Inc. 470 pp.

Howell, B. F., Jr., 1959, *Introduction to Geophysics.* McGraw-Hill Book Co., Inc. 399 pp.

Hoyle, F., 1962, *Astronomy.* Doubleday and Co., Inc. 320 pp.

Kuiper, G. P. (ed.), 1954, *Earth as a Planet.* University of Chicago Press. 751 pp.

National Research Council, 1951, *Physics of the Earth II, Figure of the Earth.* 286 pp.

Stacey, F. D., 1969, *Physics of the Earth.* John Wiley and Sons, Inc. 324 pp.

Strahler, A. N., 1963, *The Earth Sciences.* Harper and Row, Publ. 681 pp.

6 THE ATMOSPHERE

6.1 COMPOSITION.

Seen from outside, Earth looks very different from any other planet (Figure 6.1 and Frontispiece). Compared to Mars, much thicker and more prevalent clouds obscure the surface. Mars's atmosphere is thin and its rare clouds are so tenuous as to be visible only thru large telescopes. Compared to Venus and the larger planets, Earth's atmosphere is thin, allowing parts of the solid surface of the planet to be seen. Although some areas are almost always cloud covered, a patient observer on Mars could eventually get a look at every part of Earth's surface. The unique characteristic of our atmosphere is thus its variability, which we call weather.

In terms of composition, the degree of this variability is very small. Table 6.1 lists the composition of the lower atmosphere. Up to 15 km, the principal variable is water vapor. The amount of water in the air can be expressed by stating the part of the air pressure which is exerted by the water vapor. The unit used is the bar, which is equal to 10^5 newtons per square meter. Normal sea-level atmospheric pressure is 0.987 bars (987 millibars). As a result, vapor pressure in bars is very nearly equal at sea level to the fraction by volume of the air which is water. (The mass fraction is 0.62 as great, because water vapor is lighter than air.) The maximum amount of water vapor which the atmosphere will hold depends upon temperature (Figure 6.2). Cold air has a small capacity for water and warm air a large capacity. On a warm day in summer, the air may contain several per cent water vapor. Below freezing, air is much dryer, containing at most only 0.6% moisture by volume (0.4% by mass), and usually much less. Thus, as air is warmed, its capacity to take on water increases rapidly. The family laundry, hung out in the morning when air tends to be warming, commonly dries quickly. After mid-afternoon, when the temperature starts to drop, the ability of the air to hold moisture declines, and its drying ability is likely to be less dependable.

The amount of water vapor can also be described by stating the ratio of the amount of water present to the amount that would be present if the air were saturated. This is called the *relative humidity*. A mass of air with a

Table 6.1 *Composition of the lower atmosphere.*

Element	Per Cent Volume	Per Cent Mass
Nitrogen	78.084°	75.51
Oxygen	20.946°	23.15
Argon	0.934°	1.28
Carbon Dioxide	0.033	0.046
Neon	0.00182	——
Helium	0.00052	——
Water	0.0 – 2.8	——
Ozone	0.0 – 0.001	——
Other gases	0.00033	——

° Dry composition

relative humidity of 80% generally feels damp because it evaporates water with which it comes in contact only slowly. Air with a relative humidity of 20% feels dry because it tends to evaporate moisture with which it comes in contact quickly. Relative humidity changes with temperature. The same body of air which has a relative humidity of 100% at 0°C would have a relative humidity of only 38% at 20°C. On cold winter days, outside air has a low vapor pressure, because vapor pressure cannot easily exceed saturation pressure. The same air brought indoors warms and drops to a low relative humidity. Thus, heated houses tend to be very dry in winter.

One of the results of the variation of the capacity of the air for water is that when moist air is cooled, water must condense out of it. The water may condense as either liquid droplets or solid ice particles. If these particles are small they stay suspended in the air. If there are enough of them, they form a cloud. Gaseous water allows sunlight to penetrate easily thru it, but liquid droplets and ice crystals tend to reflect and absorb light radiation. As a result, at Earth's surface a cloudy day is darker than a cloudless day; and the surface can be seen from above only in the absence of clouds.

Water droplets and ice crystals in clouds are denser than the surrounding gas, and are pulled downward by gravity. The rate of their fall depends on their size. Their motion is resisted by friction between the droplets and the air. Friction increases with the surface area, which is larger in proportion to the weight of the droplet for small droplets than for large. The velocity which a body will have when the frictional forces holding it back are exactly equal to the pull of gravity is called its terminal velocity. The terminal velocity of cloud droplets is commonly a few meters per hour or less. As a result most cloud droplets remain in suspension after formation until the temperature rises again and the cloud evaporates. If you watch clouds drifting overhead on a partially clear day, their shapes can often be seen to change as the edges evaporate.

Occasionally, however, the water forms droplets which are too large to stay in suspension. When this occurs, rain falls. The amount of rain which

Fig. 6.1 *Earth as seen from a Tiros satellite. The Arabian Peninsula and Mediterranean Sea are recognizable at the left. India is in the center of the picture. (N.A.S.A. photo.)*

can fall is strictly limited. If the vapor pressure is .01 bars, which is a common value near the ground, then roughly one hundredth of a column of air is composed of water. (Actually, the amount will normally be less than this, because vapor pressure tends to decrease with elevation.) This is equivalent to 6 centimeters (2.4 inches) of water. If much more rain than this falls on a given location in one storm, you may be sure that moisture is being brought in from outside the area during the storm. Fortunately, single storms rarely drop more than a fraction of this amount.

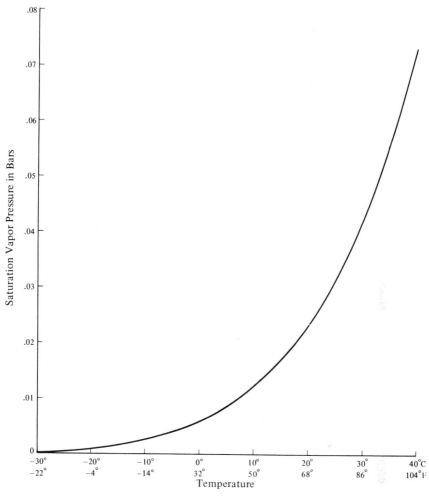

Fig. 6.2   *The amount of water vapor that the atmosphere can hold increases with temperature. Since normal sea-level atmospheric pressure is 0.987 bars, this curve very nearly represents the maximum volume fraction of the atmosphere which can be water.*

On the rare occasions when it does rain more than 6 centimeters in one storm, a flood is likely. To see why this is so, think of what fraction of an area is covered by lakes and streams. If 6 centimeters of rainfall must be carried away by streams covering only one percent of an area, then the average depth of the water in the stream will be 100 × 6 centimeters or 6 meters (20 feet) assuming that the rain accumulates much faster than the streams carry it away. This is an extreme supposition. Six-centimeter rainfalls are rare, and they occur usually in regions where there are very extensively developed stream systems. Also, the runoff rarely concentrates into the streams as fast as it falls. Most of it soaks into the ground and takes many days to work its way out of the area where it fell. But sometimes the factors which spread the peak stream load are not enough. Then there are floods.

The factors which control the size of droplets in a cloud are still imperfectly understood, but several are known to be important. First, temperature and pressure are not the only conditions which influence the condensation of moisture. For a droplet to form, several water molecules must join together. A single water molecule by itself, even well below the freezing temperature, still behaves as a gas. Its low temperature means only that it lacks the ability to leave a liquid droplet or a solid crystal if it is already a part of such a droplet or crystal. Thus, air which has cooled may contain much more gaseous water than is appropriate for its temperature. When it is in this condition it is said to be supersaturated.

Here the role of dust and other impurities in the air is important. Small solid grains picked up by wind, smoke particles, pollen, and salt crystals formed as ocean spray evaporates are always present to some degree in the lower atmosphere. These solid particles help water to condense and to crystallize as ice. A crystalline material has electrical charges on its edges due to the way the atoms are arranged in the crystal lattice. Water has the ability to polarize, and condenses more easily on such charged surfaces than on unpolarized material. Once water accumulates on the solid particles, all or part of the particle may become dissolved, providing ions in solution in the droplet. These ions are mobile, so that opposite sides of the droplet can become charged differently, making it easier for more water to condense. In this way a droplet forms.

The process will be aided by any tendency of the solid to react with water, thereby calling into play electrical forces which immobilize the water molecules. Where this tendency is strong, some droplets may form even when the air is not fully saturated with water.

Usually the number of solid particles available in the atmosphere to act as centers of condensation is large, and droplets are small. Droplets tend to remain separate either because they rarely touch one another or because most droplets are positively charged electrically, and as a result repel one another.

Occasionally, however, droplets grow so large that they fall. A large droplet falling thru a cloud of small droplets will sweep up the smaller droplets it strikes unless their electrical charges are large enough to move them out of the way. In this way the drop grows larger and larger, and as a result falls faster and faster. The maximum size is finally limited by a tendency of drops to be broken apart due to surface friction as they fall.

Electrical processes may either aid or oppose condensation. In a cloud, various processes tend to separate positive and negative charges. Two similarly charged droplets will repel one another (Figure 6.3). However, large electrical fields occasionally are present. When this is the case, each droplet becomes polarized. If two droplets approach one another with the positively charged side of one facing the negatively charged side of the other, these opposite charges may pull the two faces together more strongly than the two particles as a whole repel one another. This will cause them to coalesce. In this way, the formation of large droplets is accelerated.

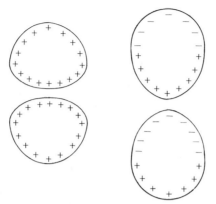

Fig. 6.3   *Positively charged droplets repel one another, but the unlike sides of polarized droplets attract.*

Lastly, if the average temperature is below freezing and both ice crystals and water droplets are present in a cloud, condensation will occur preferentially on the ice crystals, which will grow in size and sink thru the cloud picking up more water as they strike droplets.

All these processes of forming large droplets or snow crystals lead to precipitation. Water is removed from the air. Each time this occurs the composition of the air is changed, leaving it drier. Next time it is warmed it is then ready to take on more moisture, which evaporates from the sea or, on land, from lakes, streams, plants or the soil.

The temperature decreases going upward from Earth's surface, and the moisture separates out rapidly when the freezing temperature is reached (Figure 6.4). In middle latitudes, winds commonly carry moisture upward to about 12 km keeping it mixed in the atmosphere. Above 12 km the water content is uniformly low and clouds are rare. Jet aircraft fly above the clouds in this clear weather.

In addition to water and solid grains there are several other minor constituents of air whose composition is variable from place to place. The most obvious of these are the various products of burning, such as those produced by industrial processes, trash burning, heating fuels, forest fires, decay of vegetation and automobile engines. These include a wide variety of noxious gases such as carbon monoxide, sulfur dioxide, methane and ammonia. Some of these compounds are extremely irritating when breathed. Reactions between these gases and the oxygen and nitrogen of the atmosphere, stimulated sometimes by the ultraviolet rays of sunlight, create complex organic compounds which may turn the air into smog. If enough of these poisons accumulate in a body of air, it may become so rich in irritants that persons with heart or pulmonary weaknesses cannot breathe it safely, and they may even die. Normally, rain washes impurities out of the air and wind scatters them, keeping the level at any one place low. However, rain and wind come at irregular intervals, and long quiet periods can allow the level of irritants to build to dangerous levels (Figure 6.5). In deciding how much waste any one community can safely release into the at-

Fig. 6.4   *A thunderhead cloud is formed by condensation of moisture in a rising body of air. Rain often falls from such clouds. (Courtesy J. Dutton.)*

mosphere, man must plan for the rare occasions of minimum rate of renewal of the air. This will vary from one community to another depending on the local climate. Furthermore, some poisons are cumulative and long exposure to them even at a level which does not kill can gradually injure an individual.

One of the most important components of the atmosphere for life is carbon dioxide (CO_2). Plants use this in their growth metabolism, returning oxygen to the air with the help of the energy in sunlight. Oxygen is reconverted to carbon dioxide by burning or by the metabolic processes of animals. It might be expected that carbon dioxide would vary in quantity in the atmosphere depending on the rate of its production and removal. Measurable variations, however, are rare. This is largely because the oceans have dissolved in them many times the amount of CO_2 that is in the air. Any body of air which has a deficiency or excess of CO_2 and which is in contact with ocean water quickly receives or gives up enough CO_2 to keep the atmospheric concentration constant.

As one goes up in the atmosphere, there is a systematic variation in composition. Between 12 and 40 km, the atmosphere is unusually rich in ozone, a form of oxygen composed of molecules of three atoms in contrast to the usual two. Ozone is formed by ordinary two-atom oxygen combining with a free oxygen atom. Free oxygen atoms are formed by the breaking up of ordinary two-atom molecules by the ultraviolet radiation in sunlight. Ozone gradually breaks down reforming ordinary oxygen, so that

Fig. 6.5   *Pollutants are added to the atmosphere by many processes. (Courtesy H. Panofsky.)*

it must be continually replaced by new ozone. It is scarce below 12 km because the radiation which produces free oxygen atoms is almost all absorbed above this elevation.

Ozone has one property which is very important to life on Earth. It is highly absorptive to ultraviolet radiation, which it filters out of sunlight, thus preventing much of this powerful radiation from reaching the bottom of the atmosphere. Ultraviolet radiation is damaging to living tissue. Without the ozone layer or its equivalent, life of the sort which now exists would be impossible on Earth.

Above 100 km the degree of dissociation of molecules into atoms by the high-frequency radiation in sunlight becomes even more marked. From here outwards the composition of the atmosphere is largely controlled by diffusion of the light elements upward and the heavy elements downward under the pull of gravity. By 230 km, almost all oxygen molecules have been dissociated into free atoms. Nitrogen also dissociates, but much less strongly. As a result, above 230 km the proportion of oxygen to nitrogen in the atmosphere greatly increases (Figure 6.6). Two other elements, helium and hydrogen, become progressively more important constituents going upward, diffusing into layers on the basis of their atomic weights.

6.2 TEMPERATURE.

The influx of solar radiation is partly consumed in the atmosphere, partly reflected by clouds and surface water, rock or ground cover, and partly absorbed by water, vegetation or other surface material. Different parts of the spectrum are removed at different elevations (Figure 3.1, p. 51). The highest frequencies are removed first. Few X rays penetrate below 100 km. The ultraviolet is largely removed above 10 km. The atmosphere is most

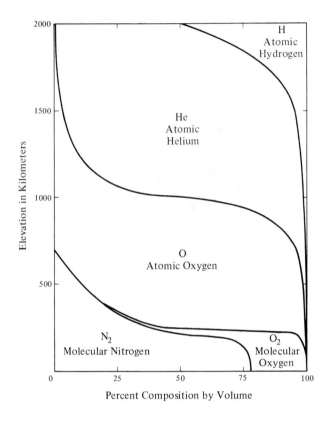

Fig. 6.6   *Variation of composition with elevation in the atmosphere.*

nearly transparent to visible-frequency radiations. Light rays, when not re-flected or absorbed by clouds, are largely absorbed by vegetation and by exposed rock, soil and water. Some infra-red radiation penetrates similarly to the ground, but much of it is absorbed in the lower atmosphere. The ground and the oceans tend to be warmer than the air due to this absorbed heat. At the surface, the air and ground easily exchange heat, holding the air, which is light and hence more easily affected, at the same temperature as the ground.

Water plays a particularly important role in the absorption of solar energy. At Earth's surface, energy is used in evaporating water into the air. This energy is later released, warming the air, by the condensation of the water to form liquid droplets and ice crystals.

All of the energy which has been stored in this fashion is eventually rera-diated. The rate of reradiation in relation to the rate of heating at different levels determines temperature in the atmosphere. Because of its low tem-perature, Earth reradiates electromagnetic radiation almost entirely at in-fra-red frequencies. Since infra-red is absorbed by carbon dioxide and wa-ter, including gaseous water, the lower part of the atmosphere is heated from below as well as from above. The original energy of sunlight may be absorbed and reradiated several times before finally escaping to space.

This produces what is known as the *greenhouse effect*. In the lower atmosphere, incoming light radiation penetrates more efficiently than outgoing infrared radiation, holding the temperature relatively high, just as in a greenhouse light enters thru the glass roof, but the infrared back radiation is reflected by the glass. The overall result of this is that the annual average temperature at Earth's surface at sea level exclusive of the polar areas generally lies between $-18°$ and $30°C$ ($0°$ and $86°F$), varying with latitude and other factors (Figure 6.7).

Heat storage is aided by daily cycles in the extent of the cloud cover. During the day the temperature of the atmosphere is raised on the sunlit half of Earth, and moisture tends to evaporate into the air both from the ground and from clouds. As a result, cloud cover generally is at a minimum in the morning or early afternoon, allowing Sun to heat the layer of air nearest the ground. At night, as the air cools by radiating heat away, clouds including haze and fog condense. Once they have formed, they slow further cooling by absorbing the energy being reradiated. This keeps the lower atmosphere warmer than it would be if heat were more freely reradiated.

As elevation above the surface increases, temperature normally drops (Figure 6.8). The rate varies from place to place and with time. The average is about $6°C$ per kilometer ($3.5°F/1000$ ft). This gradient results from three principal factors. First, the underlying solid and liquid Earth is the principal supplier of thermal energy to the air. Temperature decreases with distance from this source. Second, the concentration of water vapor decreases with elevation. Since moisture helps to absorb infra-red radiation, the lower air is warmed most.

Finally, vertical air circulation coupled with the pressure gradient decreases temperature with height. Pressure decreases with elevation because of the lessening weight of overlying air. As a mass of air rises, the decreased pressure allows it to expand. For the air to expand, energy is required. This energy is taken from the heat of the air. Therefore, expanding air cools. Sinking air correspondingly gets warmer. The rate of cooling which a body of air would experience if raised without gain or loss of energy is called the *adiabatic gradient*. The adiabatic gradient is about $10°C/km$. This is a more rapid drop in temperature than the average decrease in air temperature with elevation.

If the temperature gradient is less steep (fewer $°C$ per km) than the adiabatic gradient, and a parcel of warm air is moved upward adiabatically (without adding or taking away heat), the air will cool faster than the surrounding air temperature drops (Figure 6.8). It will then be colder and more dense than its surroundings; and it will tend to sink back to its previous position. On the other hand, if the surrounding air temperature falls faster than the adiabatic gradient, a raised air mass will be less dense than its surroundings, and will tend to continue to rise. As a result, when the temperature gradient is low (temperature falls slowly with elevation), air tends to be stable and not to mix vertically. When the temperature gradient exceeds the adiabatic gradient, on the other hand, active mixing occurs.

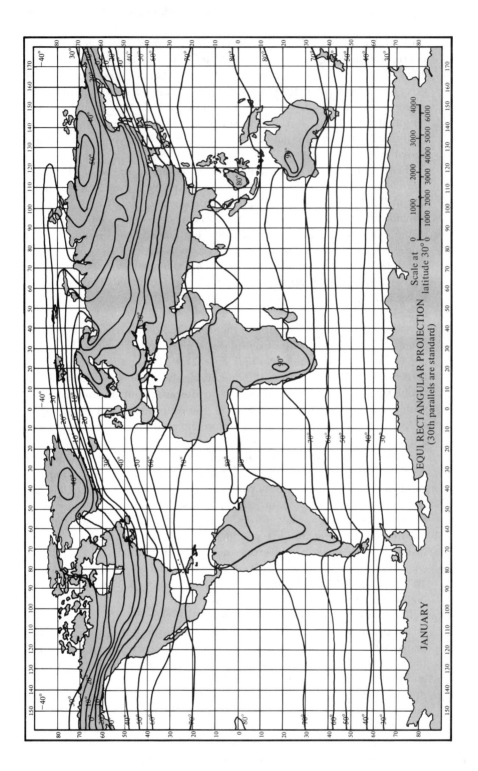

EQUI RECTANGULAR PROJECTION
(30th parallels are standard)

Scale at
latitude 30°

0   1000   2000   3000   4000
0   1000   2000   3000   4000   5000   6000

JANUARY

140

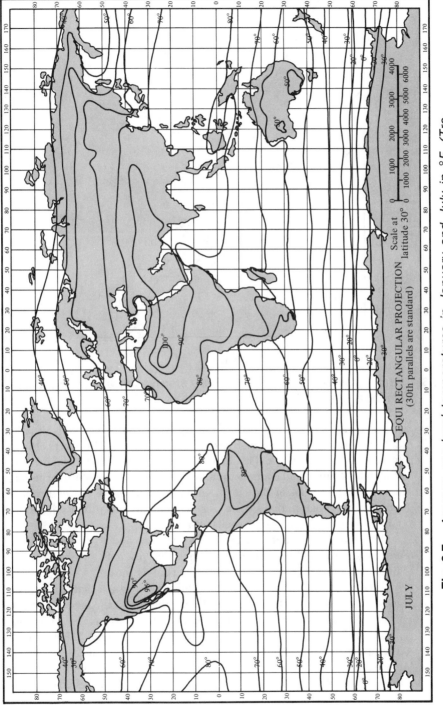

Fig. 6.7 Average sea-level temperatures in January and July in °F. (Trewartha, An Introduction to Climate, 4th ed., Courtesy of McGraw-Hill Book Co., Inc.)

141

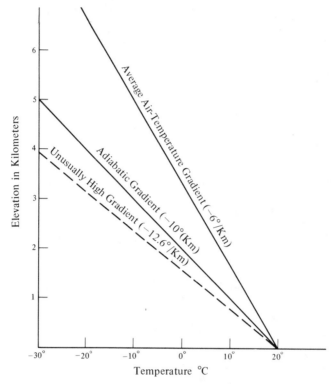

Fig. 6.8 *Usually, air temperature falls more slowly with elevation than the adiabatic gradient.*

The condensation of moisture from rising air complicates this situation. When moisture condenses or freezes, heat is released. This warms the air, making it more buoyant. This makes it want to rise more, which further stimulates cooling and condensation. If the condensed moisture forms large drops which fall out of the cloud, an unstable situation can be produced where rapid precipitation from a rising column of air is produced. Summer rain showers are produced by such cells.

Occasionally, a layer of warm air forms or is carried above cold air. Such a situation is called an *inversion*. An inversion presents a barrier to rising air masses, and tends to prevent vertical mixing of the air.

Usually, the gradient in the lower atmosphere is such that some mixing goes on all the time. Air is warmed near the ground, rises to elevations where it radiates heat away and descends again. This produces a regular, persistent pattern of currents in the atmosphere (Section 6.5, Figure 6.19). The upper limit of this zone of circulation occurs where heating of the atmosphere by the action of ozone in filtering out the ultraviolet slows the drop in temperature to below 2°C/km. This elevation is called the *tropopause*, and the layer of atmosphere below it, the *troposphere*. At the equator, the temperature reverses and starts to rise immediately above the tropopause. North and south of the equator, there is a layer of air of nearly constant temperature just above the tropopause. The thickness of this layer increases with latitude. The tropopause is higher at the equator, where it is around 17 km, than at the poles, where it is around 8 km. Its elevation rises in summer and falls in winter.

At the tropopause, temperature goes thru a minimum of around −55°C (polar regions) to −80°C (equator), then rises again (Figure 6.9). It is colder at the tropopause at the equator than at the poles because the troposphere is thicker at the equator. The layer of air above the tropopause is called the *stratosphere*. Because of the increase in temperature with elevation, there is relatively little actual mixing of the air here. This makes it a zone of generally quieter air than the troposphere. This makes travel in large airplanes which can fly above the tropopause more comfortable than travel in planes which fly in the often turbulent troposphere, which throws them about, sometimes causing nausea.

Around 50 km elevation the temperature reaches a maximum. This is the top of the ozone layer, whose absorption of ultraviolet causes its high temperature. The temperature here is about the same as at the ground, around 0°C or a few degrees higher. The height of the maximum temperature is called the *stratopause*.

From 50 to 85 kilometers, the temperature again falls, reaching a lower minimum than at the tropopause, around −95° to −140°C. This minimum is called the *mesopause* and the layer beneath it the *mesosphere*. Although this is again a region of falling temperature, the gradient is lower than in the troposphere so that there is less vertical mixing here than at low elevations. Airplanes cannot fly this high, for the air is too thin to support them. Observations of the upper layers of the atmosphere have been made with rocket and satellite-borne apparatus.

Above 90 km the temperature steadily rises again. This layer is known as the *thermosphere* because of its high temperature, which is the result of absorption of X-ray and ultraviolet energy. Atomic oxygen is an efficient absorber of these frequencies. The temperature rises rapidly at first and then more slowly. The maximum temperature reached varies from day to day due to small changes in solar radiation. It is highest at the peak of sunspot activity, and lowest at the minimum, the variation amounting to over 500°C.

There is also a daily cycle (Figure 6.10). Under the stimulus of sunlight, the whole atmosphere heats and expands. The upper atmosphere is particularly sensitive to this. A large bulge is produced over the sunlit half of Earth. The resulting rise and fall of the atmosphere is called the *thermal tide*. The thermal tide in the atmosphere is much larger than the gravity tide produced by the attraction of Moon and Sun. The latter has two peaks, one facing the attracting body and one on the far side of Earth. The atmosphere has a much lower viscosity than the oceans. As a result, unlike the ocean, the atmospheric tides exhibit no measurable phase delay in their positions with respect to the attracting bodies.

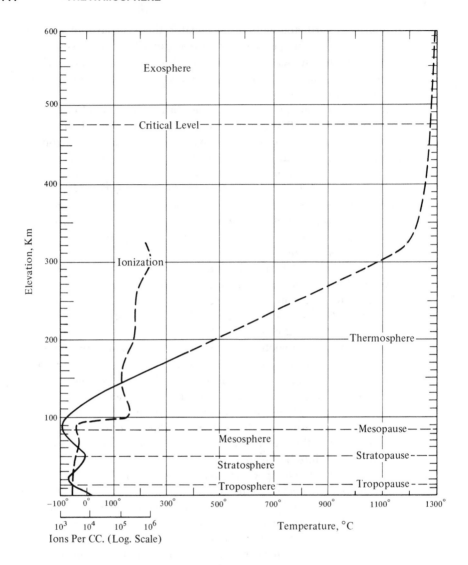

Fig. 6.9   *Temperature, ionization and layers of the atmosphere.*

The high temperature of the thermosphere is not a hazard to astronauts or unmanned rockets because the air is too thin to provide a large reservoir of heat. Any object at this elevation acquires a temperature controlled by the relative rates of absorption and reradiation of energy. The heat problem experienced by rockets moving thru the atmosphere occurs largely below the thermosphere. It is due to friction between the body and the air.

The upper limit of the thermosphere is hard to define. By 475 kilometers, individual atoms are so far apart that each atom travels a distance equal to its elevation above ground before colliding with another atom. This elevation is called the *critical level*. The layer of atmosphere above this is called the *exosphere*, because it is only loosely attached to Earth, and is much influenced by weather in the adjoining part of Sun's atmosphere. At an elevation of about 10,000 km (1.5 Earth radii) the air density is no greater than in the surrounding space.

Hydrogen and helium can escape to space from the exosphere (see Section 3.3.1, p. 63). Helium is replenished from Earth's interior, where it is formed by radioactive disintegration of thorium and uranium. Helium in the atmosphere at present amounts to about ten percent of that calculated to have escaped from the ground since Earth was formed. The rest presumably has escaped to space.

The rate at which the hydrogen content is changing on Earth is uncertain. Being lighter than helium, hydrogen escapes faster. However, there is also an influx from Sun's atmosphere, in which Earth moves. It is not certain whether influx or efflux is greater. Earth is unusual among the planets in the richness of its atmosphere in oxygen. This may be partly due to the inability of Earth to hold the more universally prevalent hydrogen. Both the atmosphere and the near-surface rocks of Earth are rich in oxygen. At the surface, hydrogen is almost entirely combined with oxygen to form water. Water is a common constituent of volcanic gases. It may be that Earth's deep interior is richer in hydrogen than its crust or atmosphere, the upper layers having lost hydrogen to space.

6.3 IONIZATION.

Earth's atmosphere gradually merges with Sun's atmosphere. The boundary occurs where individual atoms cease being carried along with Earth in its annual motion about Sun. In holding atoms, Earth's magnetic field plays an important role. A large percent of the atoms of the exosphere are ionized (electrically charged). A charged particle moving in a magnetic field is deflected by reaction with the field. Thus ions and free electrons in the exosphere tend to move in circles or spirals. The magnetic field holds in the atmosphere ions that would escape if the field were absent. The magnetic field can be visualized as dragging the outer fringe of the atmosphere along with the Earth.

Ions are produced by the energy of sunlight, especially by the X-ray and ultraviolet components, which are powerful enough to dissociate gas molecules and even to remove one electron from an atom. In the lower atmosphere, an electron or ion formed in this way soon encounters an oppositely charged particle with which it can recombine. As one rises thru the atmosphere, the distance a particle travels between collisions increases, and the life of the ion before recombination increases correspondingly.

In the lower atmosphere, there are only a few thousand ions present per cubic centimeter of gas. (At sea level, one cubic centimeter contains about 3×10^{20} atoms.) Going upward in the atmosphere the percentage of atoms ionized steadily increases. The total number of ions per unit volume also increases. There is a small maximum in ion density in the mesosphere at around 80 kilometers elevation. This is called the *D layer*. It is followed by a rapid increase starting at the bottom of the thermosphere (Figure 6.9). A second, sharper maximum occurs at a little over 100 km (the *E layer*). A third layer starts around 140 km and extends upward into the exosphere. This is called the *F layer*. The maximum concentration of ions per cubic centimeter usually is found between 200 and 300 km. Above this the density of atoms in the air decreases to so low a value that the ion density also falls off. At 1000 km about ten percent of the atoms present are ionized. The *D*, *E* and *F* layers together are sometimes referred to as the *ionosphere*.

The intensity of ionization fluctuates greatly, especially in the *F* layer. During the daytime the *F* layer divides into two parts. The main part, called the F_2 *region*, intensifies and expands upward as the atmosphere expands in the thermal tide (Figure 6.10). The intensity of ionization may vary by a factor of 5 or more depending on latitude. The elevation of the peak level of ionization commonly goes up and down over 100 km. This does not necessarily mean that the atmosphere swells this much, only that the locus of the most rapid formation of ions rises. During the daytime, a sharp rise in ionization is observed between 140 and 200 km, the F_1 *region*. This ionization fades away at night. The degree of ionization may fall below that of the more stable *E* region, in which, however, nighttime ionization also falls.

The *D* region also largely deionizes at night. This is the reason for the improvement in radio reception from distant broadcasting stations at night. The daytime *D*-layer tends to absorb radio signals but the *E*-layer is a good reflector of them (Figure 6.11). By day, only line-of-sight radio transmission is good. At night, during the weakening of the *D* layer, radio waves are reflected back to Earth by the bottom of the *E*-layer, permitting distant stations to be received.

It is the presence of large numbers of free electrical charges which causes the *E*-layer to reflect radio waves. These charges, especially loose electrons, are free to move about. This makes the ionosphere a good conductor of electricity. Electromagnetic waves are reflected by perfect conductors. Imperfect conductors reflect low-frequency waves and pass high-frequency waves. The maximum elevation to which a radio wave can penetrate into the ionosphere increases with its frequency. Electrical conductivity, and hence ionization, which determines conductivity, can be measured as a function of elevation by measuring the echo-time of radio pulses as a function of frequency. Echo-time increases with frequency. TV, FM, and radar-frequency waves normally pass right thru the ionosphere without being reflected. On rare occasions, TV signals are observed at larger than line-of-sight distances, indicating unusually high degrees of ionization in the upper atmosphere.

The *D*-layer also is subject to unusual increases in ionization. These result from events on the surface of Sun. The increased ionization of the *D*-layer may be strong enough to cause exceptionally large absorption of radio waves, but not strong enough to reflect them. The resultant *radio blackouts* are of several types. One type occurs only in the daytime and

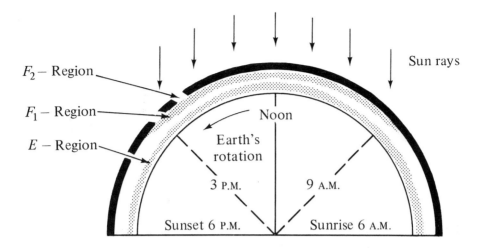

Fig. 6.10 *Sun's radiation warms, ionizes and expands the atmosphere daily.*

immediately follows solar flares, which emit bursts of X rays which ionize the atmosphere. A second type accompanies magnetic storms, in which clouds of electrons and positively charged hydrogen ions strike Earth. These are deflected by Earth's magnetic field over the whole of Earth's surface, so that unusual ionization occurs on the dark as well as the sunlit side. A third type, called *polar-cap blackouts*, results from penetration of bursts of low-level cosmic radiation into the areas near the magnetic poles where the magnetic field is least efficient in shielding Earth from high-intensity radiation.

Cosmic radiation has one useful effect. Neutrons are produced by the interaction of cosmic rays and atoms high in the atmosphere. These neutrons combine with nitrogen atoms, causing the nucleus to give off a proton (hydrogen nucleus) and converting the atom into a radioactive form of carbon. This carbon eventually decomposes back into nitrogen with the release of an electron. Half of the carbon decomposes every 5600 years. The radioactive carbon is circulated to the lower atmosphere, which contains a constant proportion of radioactive to ordinary carbon. Some of this carbon combines with oxygen to form carbon dioxide and becomes locked in various compounds in plants and animals. If the plant or animal dies, the radioactive carbon will gradually convert back to nitrogen. The proportion of radioactive to non-radioactive carbon in any such body decreases with the passing years and is thus a measure of how long it has been since the body ceased to react with the air. In this fashion the age of bones or fossil wood can be determined.

6.4 DENSITY AND PRESSURE.

At sea level the air pressure (weight of the overlying atmosphere) is equivalent to a column of water ten meters (34 feet) high. Its density is .0013

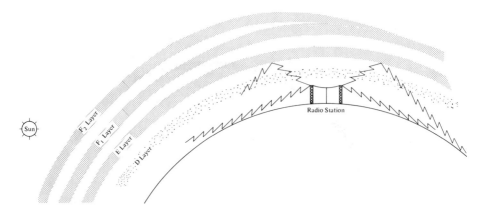

Fig. 6.11 *The E layer reflects radio waves. The D layer absorbs them, especially during the daytime.*

gm/cc. Both quantities decrease upward. The two do not decrease at exactly the same rate because gravity decreases with elevation and is less at the equator than elsewhere because of the centrifugal effect of Earth's rotation.

Half of the atmosphere lies below 5.5 km (18,000 feet) elevation. The troposphere contains nearly ninety percent of the atmosphere. Pressure and density decrease almost by a factor of ten for every 16 km (10 miles) of elevation up to the mesopause. Only about 1/100,000 is in the thermosphere and above. The filtering action which the upper layers perform on sunlight is remarkable in that such a diffuse body of matter is so efficient an absorber of energy.

The decrease of pressure, density and temperature with elevation has a peculiar effect on the transmission of sound thru the atmosphere. Sound travels at a velocity which depends on pressure, P, and density, d:

$$v = \sqrt{k \frac{P}{d}} \qquad \textbf{(6.1)}$$

where k is a constant. Density is directly proportional to pressure and inversely to temperature. As a result, in the atmosphere, neglecting horizontal pressure variations

$$v = v_0 \sqrt{\frac{t_k}{273}} = v_0 \sqrt{1 + \frac{t_c}{273}} \qquad \textbf{(6.2)}$$

where t_k is temperature in °Kelvin, t_c is temperature in °Celsius and v_o is the velocity at 0°C (273°K). As a result, sound velocity decreases with elevation. This has the effect of bending sound waves upward, so that sounds produced at the surface cannot be heard at large distances (Figure 6.12). This is why church bells are hung in towers. The higher the source, the farther away the sound can be heard. When there is a temperature inversion with warm air above cold air, sounds are bent down so that they may be heard at exceptionally great distances, sometimes being focused with high intensity at a distance of many kilometers.

Fig. 6.12 *Sound waves are bent upward in normal air, producing an acoustic shadow.*

6.5 GENERAL CIRCULATION.

One of the effects of the heating of the atmosphere by sunlight is that the air expands. When it expands it becomes less dense. When two adjoining columns of air differ in density, the heavier air tends to flow under the lighter air, which rises. This produces wind. At Earth's surface, winds blow in general from high-pressure areas to low-pressure areas. This flow is compensated higher up by winds in the opposite direction. Since more solar energy reaches the atmosphere near the equator than at the poles, one general feature of the atmospheric circulation is a rising trend at low latitudes and a sinking trend at high latitudes.

When air rises it cools, and its ability to carry moisture decreases. Hence low-pressure areas often have rain associated with them. Where air pressure is high, on the other hand, the air sinks and warms, increasing its moisture-carrying capacity. Hence, areas of high pressure are usually areas of clear weather.

Another example of the effect of high and low pressures on wind patterns and rain occurs along ocean shores. A given amount of sunshine will raise the surface temperature of rock or soil more than that of an equal area of water. There are four reasons for this. First, the light and heat radiation penetrate more deeply into the water than into the ground, so the energy is spread over a thicker layer. Second, wind and currents mix the water down to as much as several tens of meters depth, increasing the thickness of the layer heated. Third, part of the energy is used in evaporating moisture into the atmosphere from the water surface, leaving less energy to raise its temperature. Finally, water has a higher heat capacity than rock and soil, and, as a result, changes temperature less rapidly. As a result of these four factors, the sea surface temperature rises less rapidly by day and falls more slowly at night than does that of the land surface. The temperature of the layer of air near the surface tends to follow the surface temperature, so that a rising tendency develops in the air over land during the day and a falling tendency at night. This commonly produces sea-to-land breezes in late afternoon, bringing water-saturated air from which rain is easily produced. Just before dawn the wind direction is most often in the opposite direction. Just after dawn and just after sunset there are often periods of very calm air, as the heating and cooling patterns of day and night reverse.

Daily rain showers are produced in some areas by the sea-land breeze cycle. Tropical oceanic islands are particularly subject to this. The presence of an island in the tropics can sometimes be detected from distances where the island itself is below the horizon by the tendency of clouds to form over it because of a rising air current.

Similar sea-land cells develop over large continental areas, but the time-scale of the air currents tends to be of longer period. In summer, the continents are warmer than the adjoining oceans, and tend to draw in moisture-laden air. In areas like India and China this produces the *monsoons*, periods of strong sea-to-land winds which bring heavy rains. In winter the trend is the opposite. Winds are predominately from the interior toward the sea, and are dry.

The wind patterns which result from heating and cooling are made more complicated by the effect of Earth's rotation. An air mass at the equator has a west-to-east velocity of 1670 km/hr as a result of being carried along with Earth as it rotates. This decreases to 1200 km/hr at 45° latitude and to zero at the poles (Figure 6.13). The eastward velocity of quiet air thus decreases steadily with latitude. As a result, if a body of air in the northern hemisphere is moved one degree of latitude to the north without changing its eastward speed, it will acquire an eastward velocity relative to the underlying solid Earth. A body of air moving south is deflected in the opposite direction. This tendency increases in strength with distance from the equator. There is a similar tendency for rising air to drift to the west and for falling air to move eastward.

Even air moving eastward or westward is effected. This arises in the following way. The horizontal component of the centrifugal acceleration at a point P on Earth's surface is (Figure 6.14):

$$C = s \frac{V^2}{R}$$

(6.3)

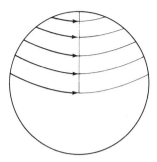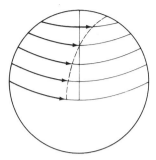

Fig. 6.13 *The eastward velocity of Earth's surface decreases from the equator toward the poles. As a result, a body of air shifted from the equator toward a pole develops an eastward velocity relative to the ground.*

where V is the eastward velocity at the point P, R is the distance from the axis of rotation, and s is a factor which increases with latitute from zero to 1.0.[*]

V is composed of two parts: WR, the rotational velocity of P about Earth's axis (W is the rate of rotation, once per day), and V_w, the wind velocity. Substituting in (6.3):

$$C = \frac{s\,(WR + V_w)^2}{R}$$

$$= s\left(W^2R + 2WV_w + \frac{V_w}{R}\right) \tag{6.4}$$

The first of these terms produces the equatorial bulge, and is exactly balanced by the slope of Earth's surface from the equator to the poles. The third term is much smaller than the other two, so that the force per unit mass deflecting any mass moving east or west is

$$F_c \approx 2sWV_w \tag{6.5}$$

F_c can also be shown to be the force on a mass moving in any other horizontal direction, although the derivation is more complicated in the general case. It may help to make this seem reasonable to realize that Equation 6.5 applies as the poles are approached, at which point the distinction between east-west and north-south ceases to have meaning.

This imaginary force is called the *coriolis force*. It acts on currents of water as well as air, causing circular ocean currents (see Section 8.3). In the northern hemisphere, coriolis force tends to deflect all horizontal motions

---

[*] s is simply the sine of the latitude angle, A. It is equal to the ratio of the length of the distance PQ in Figure 6.14 to Earth's radius, R_e.

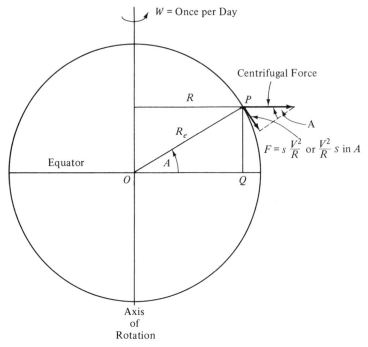

Fig. 6.14

to the right. In the southern hemisphere it tends to deflect motions to the left. It acts on wind, causing rising and falling cells to develop strong circular motions.

The effects of coriolis force can be seen on a weather map. In Figure 6.15, the contours represent lines of equal sea-level air pressure and the rods show the direction and strength of the wind. The air is flowing in the direction the rods point, and the markings show its strength. Around the low-pressure areas, marked L on the map, the circulation is predominantly counterclockwise because the winds are deflected to the right as they flow into the low-pressure area. Around the high-pressure centers, marked H, the circulation is predominantly clockwise because the winds are flowing outward. The shading represents areas where it is raining or snowing.

The movements of high- and low-pressure areas produce the changes in local atmospheric conditions called weather. These can be best understood if the atmosphere is visualized as composed of large air masses of different temperatures and humidities which are moved about by winds associated with the pressure centers. On a weather map, air masses are commonly indicated by drawing in boundaries between regions of contrasting temperatures. A *cold front* occurs where cold air is advancing, replacing warm air. The cold air is ordinarily heavier than the warm air and tends to flow under it, so that the boundary of the two masses is a surface which slopes back over the cold air mass (Figure 6.16). The leading edge of the front, however, may have a short steep section due to friction between the cold

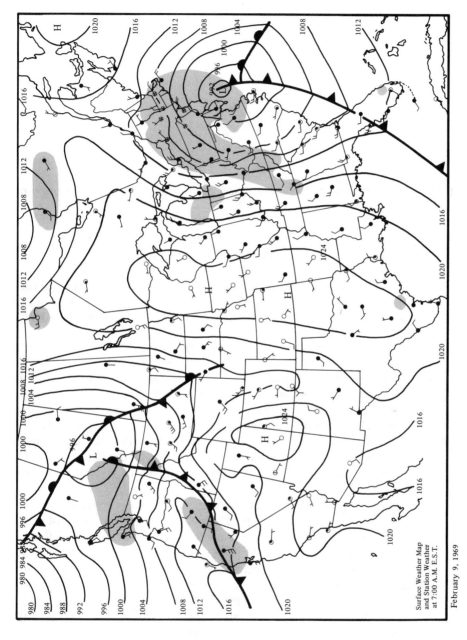

Surface Weather Map
and Station Weather
at 7:00 A.M. E.S.T.

February 9, 1969

Fig. 6.15 *Air circulates clockwise around a high-pressure center, counter-clockwise around a low-pressure center. Heavy lines are fronts.*

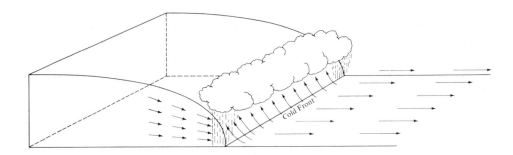

Fig. 6.16 *A cold front forms where a body of cold air flows under a body of warm air.*

air mass and the ground. If the warm air is moist, as it often is, a belt of showers commonly marks this steep edge of the front.

When warm air replaces cold air, the front tends to be much less sharp. The warm air mass spreads out over the cold air mass, which is held back by friction along the ground (Figure 6.17). Rains associated with warm fronts tend to be gentler and more prolonged than those associated with cold fronts.

Warm fronts tend to move more slowly than cold fronts. This happens because the advancing warm air usually has less density contrast with the cold air it is replacing than is the case with advancing cold fronts. As a result, cold fronts sometimes overtake warm fronts (Figure 6.18). The warm front may be driven up on the back of the preceding cold front and may be completely lifted off the ground, so that no warm front is observed at the surface. Since the raised warm air is cooled as it is lifted, this is a common source of gentle rainstorms. Even when rain does not result, the raised portions of warm air masses often form dense clouds, the familiar features of overcast days.

One general feature of atmospheric circulation is the equatorial rise and polar descent of air (Figure 6.19). In between, surface air must move southward and upper air must move northward. The coriolis deflection will produce high-level winds blowing west to east and low-level winds blowing east to west. This greatly complicates the general circulation. The warm equatorial air seeking to move northward tends to accumulate in a belt between 20° and 30° from the circle of latitude where Sun is directly overhead at noon. This circle shifts with the seasons as the direction of tilt of Earth's axis with respect to Sun changes. The accumulation of eastward-deflected air produces a belt of high pressures in which part of the air descends and joins with air working its way equatorward from the poles. At the surface, this produces the *trade-wind* belt, in which winds blow steadily from the northeast in the northern hemisphere, from the southeast in the southern hemisphere, except for a seasonal displacement north or south of the equator.

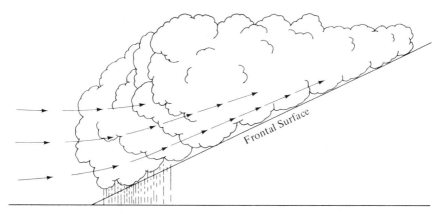

Fig. 6.17 *Cross sections of a warm front showing typical cloud deck formed in the rising warm air.*

Where the northeast and southeast trade winds meet is a narrow belt of low wind velocity, called the *doldrums*. Just beyond the trade-wind belt, in the zone of high pressures, winds are less regular. Here, another belt of frequently calm weather is found. This is known as the *horse latitudes*, because in the days of sailing vessels, many horses died and had to be thrown overboard when ships were becalmed here and ran out of water.

The area between 30° and 60° is called the *temperate zone*. Here the air motions are more variable. Warm air masses drift north and mix with cold air masses moving south from the polar regions. The circumference of a circle of latitude decreases going poleward. Therefore, there is a larger area of rising air in the equatorial belt than of descending air at the poles. This is one of the reasons for the accumulation of air around 30° which results in the high-pressure belt. It also results in air entering the temperate zone from the south as well as from the north. Obviously, air cannot continually enter this belt both from the north and the south or it would soon accumulate an excess of air. The continually changing weather patterns of this area result from the interference of these two trends with one another. Warm, moist, tropical air masses move into this belt from the south and meet cold, dry, polar air masses drifting down from the north. They are stirred together in the vortices surrounding high- and low-pressure centers, which drift eastward across the continents.

One of the effects of the descending-air belt around 20–30° latitude is that most of the world's deserts are found here (Figure 6.20). A *desert* is an area where rainfall is less than 25 cm (10 inches) per year. Falling air is warmed and as a result is undersaturated. Consequently, it does not easily produce rain. Lands in the horse latitudes are unusually dry except where monsoonal or onshore winds bring rain. Mexico, North and South Africa, Arabia and Australia are largely deserts as a result of this. India and southeast Asia are saved by the monsoonal effect. The polar regions are largely desert. Their icy condition results from low evaporation of what little moisture does get precipitated from the cold air.

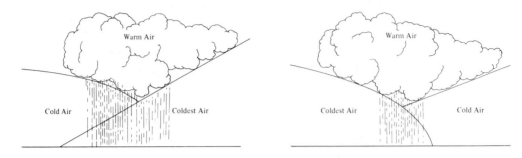

Fig. 6.18 *An occluded front occurs where a cold front overtakes a warm front raising a warm air mass off the ground.*

The equatorial regions, in contrast, benefit from rain produced by the rising air masses as they cool.

One other large-scale effect is important in controlling the distribution of rainfall. Because air cools as it rises, when an air mass must pass over a mountain range, or even a plateau of moderate height, there is a tendency for condensation to occur on the windward side of the high ground (Figure 6.21). As a result, clouds tend to form as the air is pushed upward by prevailing winds. If the air is near saturation before rising, rain will commonly be produced. The windward side of the mountains gets plentiful rainfall.

On the leeward side, the situation is reversed. On passing the high ground, the air descends and warms. Its moisture capacity is increased. Clouds and rain are rare. The leeward sides of high mountains lie in a dry, rainless *shadow zone*. The Great Basin of the western United States is such an area. Except for the western slopes of the larger mountain ranges, this area is a desert.

6.6 CYCLONES AND HURRICANES.

The cells of circularly moving air which characterize the middle latitudes are called *cyclones* and *anticyclones*. In the northern hemisphere, cyclones rotate counterclockwise due to the inflow of air toward the rising central area. They are accompanied by fronts where rain and high winds are common. Anticyclones rotate clockwise and are less intense as cold, dry air gently sinks.

Occasionally a cyclonic circulation becomes violently strong. This starts most commonly over the oceans in the tropics, where the maximum of solar energy is being supplied. When hot air rises at the center of a low-pressure area, the surrounding air spirals inward toward the central updraft, increasing in speed as the center is approached. Very strong disturbances of this type are called *hurricanes* in the Atlantic, *typhoons* in the Pacific

Fig. 6.19 *General circulation of the atmosphere including sea-level wind patterns.*

and Indian Oceans. In a hurricane or typhoon, wind often reaches peak velocities of over 200 km/hr (120 mi/hr) relative to the center of the cell. If the disturbance is moving at 50 km/hr (30 mi/hr), this speed will be added to the wind velocity on one side of the cell and subtracted from it on the other side. In the northern hemisphere, the high-velocity side is on the right of the path of travel.

At the center of the hurricane is an area of calm, the *eye* of the storm. Since air is moving predominantly upward here, the only surface motion results from the drift of the low-pressure area as a whole, and even this will be cancelled out by the circulation at one edge of the eye.

Hurricanes develop frequently in the Caribbean Sea and nearby parts of the Atlantic Ocean in late summer and early fall. At first they drift west-

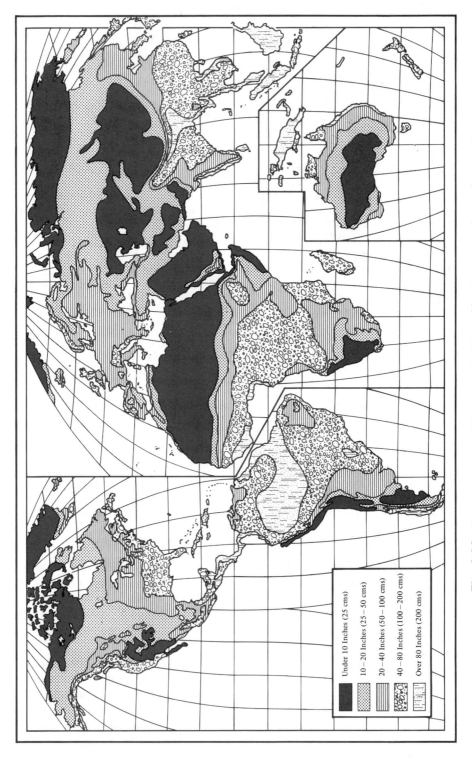

Fig. 6.20 *Desert areas of the world. (After Yearbook of Agriculture, 1941, p. 668.)*

Under 10 Inches (25 cms)

10 — 20 Inches (25 — 50 cms)

20 — 40 Inches (50 — 100 cms)

40 — 80 Inches (100 — 200 cms)

Over 80 Inches (200 cms)

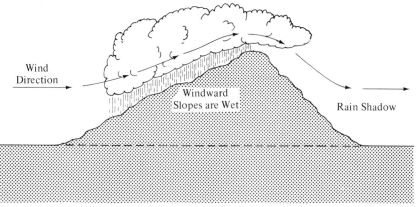

Fig. 6.21 *As an air mass rises to pass over a mountain range, cooling causes rain on the windward side.*

ward with the trade winds (Figure 6.22). On rare occasions they cross Central America and blow themselves out over the Pacific Ocean. More often they swing north into the Gulf of Mexico or the North Atlantic. Here they can be picked up by the westerly-wind system and carried northeastward. Usually they die out at sea, but sometimes they cross onto the land where they do extreme damage.

The powerful winds of a hurricane build up huge waves on the ocean surface. They may also raise the sea surface in a broad ridge many kilometers across and meters in height which inundates low-lying shorelines especially at high tide, and may completely flood small islands (Figure 6.23). Such long waves are known as *storm surges*. They are also often miscalled "tidal" waves. They are produced by wind, not by gravitational forces.

One of the most famous hurricanes of all time was the New England hurricane of 1938. It began on September 4 as an ordinary low-pressure center over the central Sahara Desert in western Africa. The trade-wind system carried it westward, gathering strength, past the Cape Verde Islands toward the West Indies. As it approached Puerto Rico it passed out of the trade-wind belt. It swung northward, the usual path for such storms, where it came under the influence of the mid-latitude belt of westerlies. Storm warnings were issued in Florida on September 18. At this time, a broad high-pressure area was spread across the whole of the central United States, very slowly pressing eastward. This thrust the low-pressure center of the hurricane to the north, and it turned past the Bahama Islands and up the U.S. Coast, the path taken by hurricanes which usually die out over the North Atlantic Ocean. But on this occasion a large, summer high-pressure area had developed over much of the North Atlantic, and it was sitting still rather than drifting eastward. Between these two highs lay a long trough of low pressure. Repelled by the steep pressure gradients to the west and unable to climb the high to the east, the hurricane moved up this low-pressure channel at around 40 km/hr.

With the hurricane 500 km (300 miles) off the coast and shipping warned of its presence, the Weather Bureau thought its job was done. Hurricanes usually break up without crossing the Atlantic Coast once they get

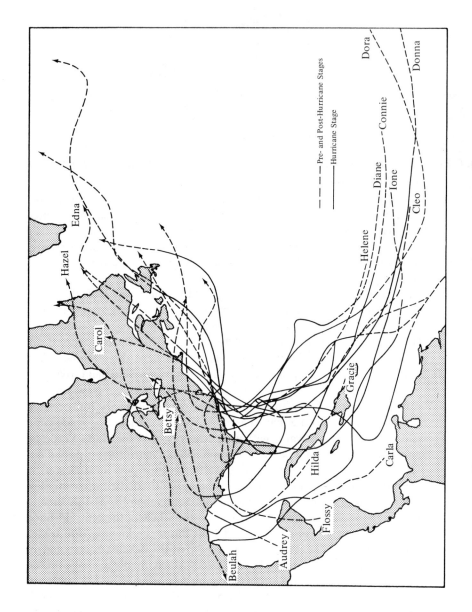

Fig. 6.22 *Typical hurricane paths. (Courtesy E.S.S.A.)*

Fig. 6.23 *A hurricane-generated storm surge floods the Rhode Island coast. (Courtesy Providence, R. I. Journal Co.)*

North of Florida. Not since 1821 had such a major tropical hurricane gone inland fully developed, and there was no one left alive to recall that storm, nor were there any systematic records from that time to forewarn that such an event might be reoccurring.

The first warning that this hurricane was not going to stay at sea was issued at 2 P.M. on September 21. At this point the center of the storm was 125 km (75 miles) off of Atlantic City, New Jersey, and headed straight for Long Island and Connecticut. Forty-five minutes later the center passed onto land, with waves which at places were 9 to 12 meters (30 to 40 feet) high. The storm coincided with a spring (very high) tide. In Providence, Rhode Island, the water rose 5.4 meters (17.6 feet) above the normal high-tide mark. Downtown streets were flooded above the tops of automobiles and street cars. Peak wind velocities were over 300 km/hr (187 mi/hr) at Harvard Observatory.

Generally unwarned and almost completely unprepared, 494 people were killed and $300,000,000 damage was done in New England. Thousands of centuries-old elm trees were knocked down which blocked streets and ripped out telephone and power lines. Communications throughout New England were paralyzed for several days.

West of the storm center, the event was no more intense than a violent thunderstorm. The author was moving into his undergraduate dormitory room preparatory to the start of classes the following Monday. Rain squalls with strong winds littered the streets with branches, but life went on as usual for a late summer day in New Jersey. Only on hearing the evening radio news did people realize that a major disaster has passed them by and struck their neighbors in nearby states. Classmates were unable to get transportation out of New England by train or bus or airplane, and arrived days late for the start of classes.

Such a disaster could not occur today. The typical cloud patterns of a hurricane are easily recognized on the photographs radioed to Earth by Tiros weather satellites, which circle the earth once every 100 minutes (Figure 6.24). Hurricanes can also be followed by radar, which receives characteristic reflections from the cloud banks which surround the eye. Each hurricane is watched carefully as it develops and progresses. Police and civil defense workers evacuate populations from threatened areas before the storm strikes. With ample advance warning, damage is kept to a minimum. In 1938 the hurricane had to be followed by reports from ships at sea, which naturally avoided it in so far as possible, from the few islands in its path, and from land measurements made far from the center of the storm. Modern instrumentation has made man no longer dependent on the crude tools of direct observation.

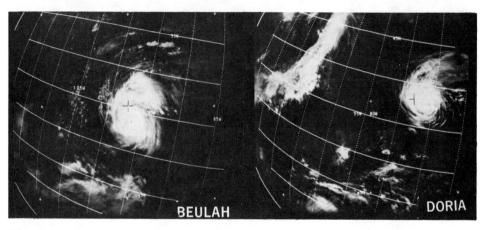

Fig. 6.24 *Two 1967 hurricanes as seen from a Tiros satellite. (E.S.S.A. photograph)*

6.7 TORNADOS.

In cumulative destructiveness, a much smaller type of storm provides as much hazard to the United States as hurricanes. This is the tornado. There are on the average 236 tornados in the United States annually. They can occur elsewhere, but the climatic conditions in the central United States make this area the most prone to tornados of any place on Earth.

Tornados develop under conditions similar to those which produce thunderstorms. They occur where there are large masses of warm moist air being carried upward. This updraft is accompanied by condensation of rain or hail. In a tornado this updraft becomes so violent that an intense

Fig. 6.25 *A tornado. (E.S.S.A. photograph)*

circular eddy develops with winds as high as 800 km/hr (500 mi/hr). A tapering funnel composed of moisture rapidly condensing out of the air extends from near the ground to great heights (Figure 6.25). This funnel is usually only a few tens or hundreds of meters in diameter. Within it barometric pressure may be so low that buildings literally explode as the eye of the tornado passes over them due to the expansion of their contained air.

The tip of the tornado funnel may rise above ground level, but where it touches the ground it typically leaves a path of devastation in which all except the sturdiest structures are completely destroyed or badly damaged

Fig. 6.26 *Path of the April 19, 1968 tornado across Greenwood, Arkansas. (E.S.S.A. photograph)*

and in which trees and crops are levelled (Figure 6.26). Tornados typically travel at around 40–60 km/hr (25–40 mi/hr). Wise midwestern farmers build tornado shelters, underground cellars with heavy doors and flat roofs, to which they and their families can retreat to safety when they see a tornado coming. Even a steel automobile is not a safe place to sit out a tornado, as cars have been known to be picked up and carried away. Caught in the open, the safest thing to do is to lie flat in a ditch or other narrow depression and let the tornado pass overhead. The danger in the open is not just the risk of being picked up and carried away, but the hazard óf being struck by an object carried by the violent winds. Pieces of straw have been embedded in boards by tornado winds (Figure 6.27).

Waterspouts at sea are miniature versions of the same general sort of storm as a tornado. Most water spouts are relatively small vortices which quickly break up. Dust devils are similar small cells on land.

Fig. 6.27 *Flying debris can become a lethal weapon during a hurricane or tornado. (E.S.S.A. photograph)*

BIBLIOGRAPHY

Donn, W. L., 1965, *Meteorology*, McGraw-Hill Book Co., Inc. 484 pp.

Glasstone, S., 1965, *Sourcebook on the Space Sciences*. D. Van Nostrand Co., Inc. 937 pp.

Johnson, F. S. (Ed.), 1965, *Satellite Environment Handbook*. Stanford University Press. 193 pp.

Kuiper, G. P. (Ed.), 1954, *Earth as a Planet*. Chicago University Press. 751 pp.

Leet, L. D., 1948, *Causes of Catastrophe*. McGraw-Hill Book Co., Inc. 232 pp.

Spar, J., 1962, *Earth, Sea and Air*. Addison-Wesley Publ. Co. 152 pp.

Strahler, A. N., 1963, *The Earth Sciences*. Harper and Row Publ. 681 pp.

Trewartha, C. T., 1968, *An Introduction to Climate*. McGraw-Hill Book Co., Inc. 408 pp.

7 GEOMAGNETISM

In preceding chapters it was pointed out that Earth, Jupiter, Sun and many other stars have magnetic fields. These fields influence the motion of electrically charged particles approaching the bodies from outside or moving within their component layers. Like other features of the environment, the magnetic field is undergoing constant changes. The detailed nature of these changes provides important evidence on invisible processes in the interiors of planets and stars. To understand these processes, it is necessary to understand what magnetism is and how it affects and is affected by rocks, molecules, atoms and the component parts of atoms.

7.1 NATURE OF MAGNETISM.

Consider a hydrogen atom. It consists of a positively charged nucleus and an electron circling the nucleus somewhat like a planet about Sun. The electron is negatively charged. Its motion about the nucleus constitutes a loop of electrical current. Where an electrical current flows, a magnetic field is produced. A current loop produces a field thru the center of the loop and returning around the outside (Figure 7.1). Each hydrogen atom is, therefore, a tiny magnet similar to an iron bar magnet or the needle of a compass.

The degree to which an atom behaves as a magnet depends on the pattern of the electron orbits about the nucleus. If several electron-current loops tend to cancel one another out, then the atom has only weak polarity. If there is more current in one sense than in the reverse, then there will be a strong magnetic polarity.

When atoms combine to form molecules or crystals, they may pack together so that the effects of adjoining atoms add or they may pack so that the effects oppose. In the absence of a magnetic field, atoms usually orient themselves nearly randomly with respect to their magnetic polarity, so that a crystal of the material has no magnetic polarization. This is the case also for fluids, where individual molecules generally have no systematic orientation with respect to one another, and hence no resultant polarity.

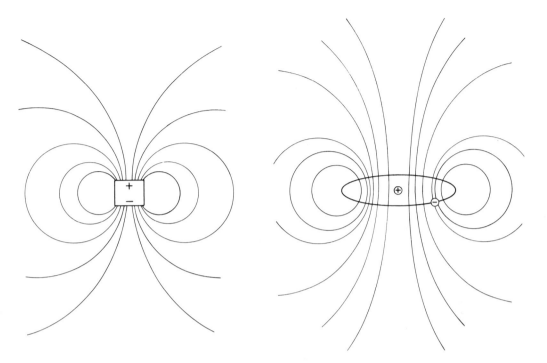

Fig. 7.1   *Electrical current flowing in a loop produces a current like that of a bar magnet.*

In both solids and fluids, in the presence of a magnetic field atoms have a tendency to align themselves so that their magnetic axes are parallel to the field. Where the alignment enhances the field, the material is said to be *paramagnetic* (Figure 7.2). Where it tends to oppose the field, the material is said to be *diamagnetic*. A few materials, notably iron and certain compounds containing iron, are very strongly paramagnetic and are called *ferromagnetic*. The property which represents the degree of alignment is called the susceptibility of the material. *Susceptibility* is the ratio of the intensity of magnetization to the magnetic stress causing it. Diamagnetic bodies have susceptibilities less than that of empty space; paramagnetic and ferromagnetic bodies have susceptibilities greater than that of space.

Ferromagnetic materials have an additional characteristic property. If a ferromagnetic body is subjected to a magnetic field, causing its atoms to take on a systematic alignment, and then the field is taken away, some of the atoms will remain aligned. In this condition the body is said to be *permanently magnetized*.

Permanent magnetization of this type may be acquired when a solid originally crystallizes from a gas or liquid, or on exposure to a magnetic field later in its life. The polarization it acquires when it first solidifies is often very tenaciously retained, even though the solid is later exposed to

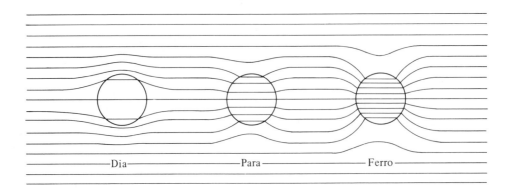

Fig. 7.2 *Effect of dia-, para- and ferromagnetic materials on a magnetic field.*

magnetic fields of different orientations from those under which it originally formed.

The one thing which can most easily wipe out permanent magnetization is heat. For every material there is a temperature above which the component atoms become so agitated in their places in the crystal lattice that they lose any tendency to retain a set magnetic orientation on removal of the field. This temperature is called the *Curie point*. The Curie point for iron is 780°C. For magnetite, the commonest easily magnetized mineral, it is 580°C. The Curie point changes only very slowly with pressure, decreasing as pressure increases when it changes at all.

A magnetizable material cooled thru the Curie point retains permanently a part of the polarization it had at the time of its cooling. If the material is part of a body of rock in Earth, the direction of this polarization will be that of Earth's magnetic field. The direction of polarization thus shows the direction of Earth's field at the time the rock formed. Such polarization is called *fossil magnetism*.

7.2 EARTH'S MAGNETIC FIELD.

Earth's magnetic field varies from place to place. It can be described by stating its strength and direction (Figures 7.3 and 7.10). The unit of strength is the *oersted*, which is the field which will exert a force of 1 dyne (10^{-5} newtons) on a unit magnetic pole. (A *unit magnetic pole* is one which will exert a force of one dyne on another unit pole one centimeter away.) Earth's field varies from about 0.25 to 0.7 oersted at the surface, with rare anomalies of twice that strength. The magnetic field is much more variable than the gravitational field, which varies by less than six tenths of a per cent from place to place at Earth's surface.

The strength of the magnetic field can be measured in many ways. One of the simplest is to rotate a coil of wire in the field and measure the resulting electrical voltage generated in the coil. This voltage will be propor-

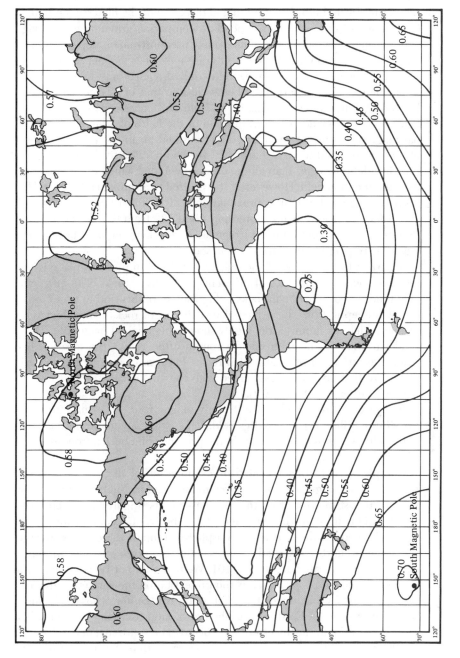

Fig. 7.3 Strength of Earth's magnetic field at the surface in 1965. (After U. S. Navy Hydrographic Office Chart 1703).

tional to the field strength and to the rate of rotation of the coil. By spinning three coils at right angles to one another in known orientation with respect to Earth, both the strength and direction of the field can be found. Other measuring instruments depend on comparing gravitative and magnetic torques on a bar magnet. One widely used instrument depends on the effect of the field on the degree of saturation of a bar magnet (Figure 7.4). Another depends on the magnetizability of hydrogen atoms (Figure 7.5).

The field is often described as being close to that of a large bar magnet located at Earth's center. This is far from an accurate description. The field can be better described as the sum of the fields of a large number of bar magnets of various strengths, directions and positions in Earth. If one of these is assumed to be at Earth's center and parallel to the axis of rotation, it will be stronger than all the others combined which are required to describe the total field. However, if it is desired to approximate the field by a single magnet, the best approximation will be for a magnet displaced about 436 km from Earth's center and tilted 12° from the axis of rotation (Figure 7.6). The field is highly irregular. A very large number of magnets would be required to produce the field shown in Figure 7.3.

A part of the variation of the magnetic field from place to place is the result of the susceptibilities of different near-surface rocks. The magnetic field tends to concentrate in rocks of high susceptibility and to avoid rocks of low susceptibility. The principal constituent of rocks controlling such magnetic field variations is the highly susceptible mineral magnetite. Thus, local variations in magnetic field strength often reflect the location beneath the surface of magnetite-rich rocks. Patterns of magnetic-field variations are thus commonly associated with the geological structure of buried rock formations (Figure 7.7).

In rare cases, rock bodies are strongly magnetized, sometimes in directions quite different from Earth's field. Very strong anomalies in the magnetic field may thus be indicators of the presence of a buried iron-ore body (Figure 7.8).

Complexity is further revealed by examination of the direction of the field. Direction is given by stating declination and inclination. *Declination* is how far east or west of north a compass points (Figure 7.9). A compass needle aligns itself in the direction of the horizontal component of the magnetic field. In the northern hemisphere, compass directions converge on a point in northern Canada. This is called the *north magnetic dip pole*. In 1960 it was located at 74.9°N, 101°W. It should not be confused with the *north magnetic pole*, which is the projection on the surface of the axis of the imaginary bar magnet placed at Earth's center whose field most nearly approximates Earth's field. The north magnetic pole was located at 78.6°N, 70.1°W in northwestern Greenland. The south magnetic dip pole in 1960 was on the Adelie coast of Antarctica south of Australia at 67.1°S, 142.7°E, while the south magnetic pole was at 78.6°S, 109.9°E in Wilkesland.

Inclination (dip) is the angle a magnetized needle balanced on a horizontal axis makes with the horizontal (Figure 7.10). At the magnetic dip

Fig. 7.4 *Airborne magnetometer in use. The sensitive element is towed be-hind the plane to keep it away from magnetic fields produced by the aircraft. (Courtesy U. S. Geological Survey. Photo by E. F. Pat-terson.)*

poles the needle points straight down and up, and the field is vertical. An ordinary compass needle must be weighted on one end so it will sit horizontally. The amount of torque needed depends on latitude. Some compasses have a weight whose distance from the balance point can be adjusted to allow for the effect of magnetic latitude on going from one area to another.

7.3 VARIABILITY OF THE FIELD.

Description of the field is further complicated by the fact that it is continually changing. At the present time, the average strength of the field is gradually decreasing, and the pattern is drifting westward at a rate of about 0.18° of longitude per year. The past history of the field is incompletely known, but such data as are available show that it varies in a complicated, non-cyclic manner. Detailed measurements have been widely made for less than 100 years. Systematic, nationwide observations in the United States began in 1899. Isolated series of measurements extend back to 1540, at which time they were begun in England by William Gilbert, who was Queen Elizabeth's physician (Figure 7.11). Gilbert was the first person to point out that Earth's field approximates the field of a large bar magnet at Earth's center.

Fig. 7.5   *A proton precession magnetometer in use. A change in the magnetic field causes the magnetic axes of hydrogen ions to oscillate at a measurable rate dependent on the strength of the remaining field. (Courtesy Varian.)*

Short-term cyclic variations with daily and monthly components result from electrical currents in the ionized layers of the atmosphere (Figure 7.12). The atmospheric tides move electrical charges in Earth's main magnetic field. This induces current loops which flow in regular patterns (Figure 7.13). These move relative to the surface as Earth rotates, the center of the pattern closely following the position of Sun.

The daily magnetic field fluctuations induce electrical currents which flow in the solid part of Earth and in the oceans. Ocean water is a good conductor because many ions are present. In contrast, the rocks of the continents are poor conductors except where they are water saturated (near the surface). Nevertheless, a recognizable pattern of electrical currents can be traced by placing electrodes in the ground and measuring the daily fluctuation in voltage (Figure 7.14).

In addition to these regular variations, a wide variety of transient oscillations are observed. These can be used to probe Earth to study its struc-

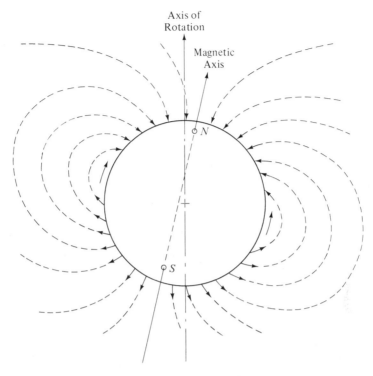

Fig. 7.6   *Earth behaves like a giant magnet.*

ture, as the currents are deflected by poorly conducting bodies and concentrated by good conductors. They show that at a depth of several tens of kilometers, conductivity increases greatly.

The effects of sun-spot disturbances on atmospheric ionization were discussed in Sections 6.3 and 3.1. The atmospheric electrical currents which are produced cause characteristic disturbances in the magnetic field (Figure 7.15).

Many human activities produce electrical currents which are accompanied by magnetic fields. This is particularly true around mining operations, where direct current is often used rather than the commoner alternating current. Both types of current produce magnetic disturbances, but alternating-current effects are easily distinguished from natural fields by their frequency.

7.4 HISTORY OF EARTH'S FIELD.

Figure 7.11 shows how the direction of Earth's magnetic field has varied for the past four centuries at two places. To extend these measurements to earlier times requires that the direction of the field be determined from the magnetization of objects of known age. Studies of the magnetization of bricks in hearths where fire has allowed magnetizable material to cool thru the Curie point show that the variation is irregular rather than cyclic, and

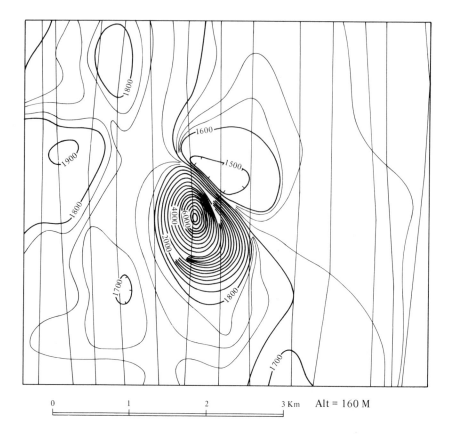

Fig. 7.7   *Magnetic field strength variations in gammas (10⁻⁵ oersteds) due to the presence of magnetite-rich rocks near Chibougamau, Quebec, Canada. (After Gaucher, Geophysics v. 30 p. 766.)*

that the magnetic axis may depart substantially from the axis of rotation. When all data for the past 7000 years are used, however, the average direction is close to the present axis of rotation.

By studying the permanent magnetization of igneous rocks, knowledge of the direction of magnetization has been extended back for about 3.5 million years (Figure 7.16). During this interval the direction of the field has reversed at least nine times.

The history of reversals has been extended back to seventy million years by studying the magnetization of the ocean floor. Detailed regional surveys of the magnetic field over parts of the ocean have been made. If the field of a dipole approximating Earth's field is subtracted from the observed field, what is left is typically a sequence of bands of alternating magnetic polarity (Figure 7.17). The bands occur in correlatable pairs on either side of certain oceanic ridges. The same sequence of bands is found in both the Atlantic and Pacific Oceans. It is just as though material were rising to the surface under these ridges and spreading horizontally from there in both directions. As newly solidified material at the surface cools

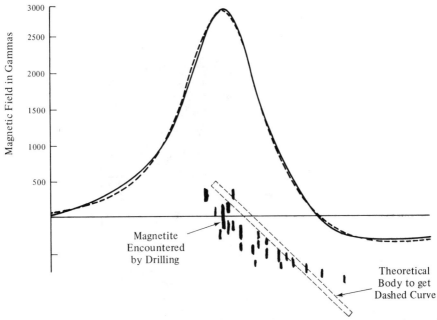

Fig. 7.8   *Vertical component magnetic anomaly over a magnetite deposit in southeastern Pennsylvania. Solid curve is observed anomaly. Dashed curve is calculated anomaly for a buried magnetized rod. (After H. P. Ross, Penn. State Univ. thesis.)*

thru the Curie point, it becomes magnetized in the then-current direction of the field. Successive bands of alternating polarity result. The central bands can be correlated with the known field reversals of the past 3.5 million years. They indicate an ocean-floor drift of up to 6.0 cm/year. Assuming the rate in any one region has been constant for many millions of years, the belts lying further from the central ridge can be used to date the reversals farther back in time. Figure 7.18 shows the resulting pattern of reversals for two areas projected to eleven million years. By using data from many areas, the history of field reversals has been extended to over 70 million years.

It is much more difficult to determine what the strength of the field was formerly. The direction of permanent magnetization of rocks is not easily changed, but the strength of this magnetism is subject to many types of interference. Only in obviously undisturbed cases can the strength of rock magnetism be used to measure the strength of the field which produced it. The best estimates available indicate that Earth's field has decreased by roughly one third in the last 2500 years. There is some evidence that previously it was increasing.

If the present rate of decrease in field strength continues for another 5000 years, the main field will disappear. Using its past history as a guide, it may then be expected to build up again. The rejuvenated field may be in the same sense as at present or reversed.

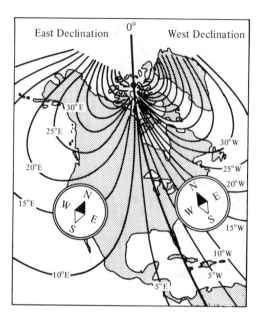

Fig. 7.9 *Magnetic declination is the difference between north and the direction a compass needle points. (From* Investigating the Earth, *by the Earth Science Curriculum Project,* © *1967 by the American Geological Institute. Reprinted by permission of Houghton Mifflin Company.)*

If one examines the direction of permanent magnetization of rocks more than eleven million years in age, it is found that the north magnetic pole was not always close to the present geographic pole. If the position of a pole is plotted as a function of time for any one area, it is found to shift slowly away from the present pole position (Figure 4.9, p. 117). Pole position must be plotted without regard to polarity, because of field reversals. Although the sequence of positions found from the study of successive rock samples exhibits a great deal of scatter, there appears to be a gradual drift of the pole position. The scatter is what might be expected if there is a certain amount of fluctuation of the magnetic poles about the geographic axis, as seems to have been occurring for the past 2100 years (Figure 7.11). This short-term oscillation is superposed on a slower, gradual drift. The polar wandering is most easily explained by postulating that the surface layers of Earth have shifted with respect to the axis of rotation.

The sequences of positions found for the poles using data from different regions, however, differ from one area to another. For any one continent, the long-term component of the path of polar wandering seems to be the same. It is only when the paths for different continents are compared that a discrepancy appears. This difference can be accounted for by supposing that the continents have moved with respect to one another. It is as though, like an iceberg in the ocean, each continent is a rockberg floating

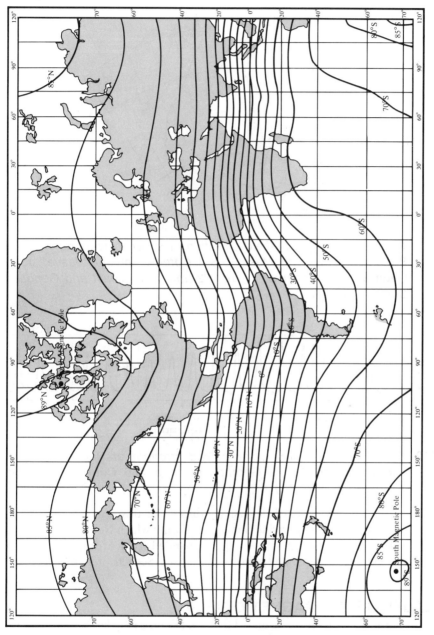

Fig. 7.10 *Magnetic inclination (dip) in 1965. (After U. S. Navy Hydrographic Office Chart 1700.)*

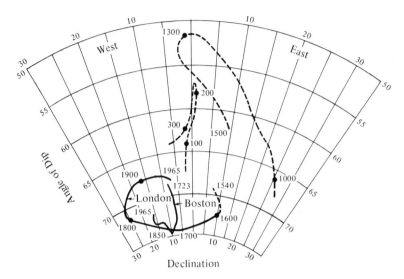

Fig. 7.11 *Variation of magnetic-field direction at several locations in Great Britain and at Boston, Mass., U.S.A. (After Aitken and Weaver,* Jour. Geomag. and Geoel. *v. 17, p. 392 and U. S. Coast and Geodetic Survey reports.)*

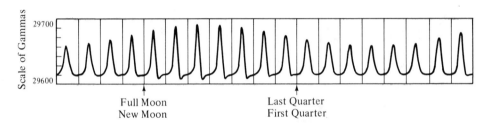

Fig. 7.12 *Cyclic variations in magnetic field strength at Huancayo, Peru. (Chapman and Bartels,* Geomagnetism, *p. 266. Courtesy The Clarendon Press, Oxford.)*

in the underlying Earth layers. The history of this drift has been traced back for several hundred million years, during which North America separated from Europe, South America separated from Africa, and Australia and Antarctica separated from southern Asia. (See Chapter 15.)

7.5 CAUSE OF THE MAGNETIC FIELD.

For many years it was assumed that Earth's magnetic field was caused primarily by the permanent magnetization of the rocks of which it is composed. When it was found that rocks will not retain magnetization at high temperatures, it was immediately realized that the layer of cool magnetizable rock near the surface was too thin to be the source of a field as strong as that of Earth. By this time the existence of atmospheric electric

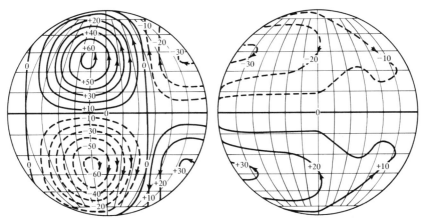

Fig. 7.13 *Atmospheric electrical current system responsible for the daily magnetic field variations. Units are in thousands of amperes. (Chapman and Bartels,* Geomagnetism *p. 696. Courtesy The Clarendon Press, Oxford.)*

currents and the field variations resulting from them were known, so it was natural to postulate a system of currents within Earth which would cause the main part of the magnetic field.

The problem in explaining the field by an electrical current loop in Earth is finding some mechanism which will cause such a current to flow. To cause a current to flow, the resistance of the conducting medium must be overcome. This is a dissipative process, and there is a transformation of energy from electrical to thermal form. The near-surface rocks are highly resistive. The short-period (one year or less) current patterns which are observed at the surface are associated with the external field, and no measurable component related to the main part of the field is observed. Resistance decreases at a depth of a few tens of kilometers, and is undetermined at great depths. It is reasonable to postulate that it is much less in the central parts of Earth, where the material is probably very different in composition from the near-surface material (see Chapter 10). It is here that the electrical currents responsible for the field are believed to be generated.

The most reasonable source for the energy needed to keep the currents flowing is Earth's internal heat. The annual heat loss thru the surface represents more energy than that needed to maintain the magnetic field. How much heat is available in the interior is unknown, but the amount required to maintain the magnetic field is a tiny fraction of what is there (see Chapter 12). Furthermore, if the electrical currents dissipate their heat in the region where they are themselves generated, there is no overall energy transfer out of this region.

To transform thermal energy into electrical current, it is postulated that a system of convection cells exist in the central part of Earth. These convection cells move conducting material in Earth's magnetic field. If a conductor is moved in a magnetic field, an electrical voltage is generated,

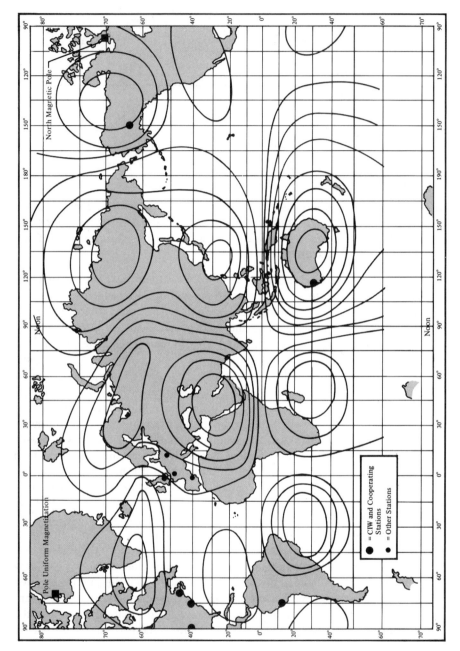

Fig. 7.14 Earth current system at 6 hours, universal time. (Chapman and Bartels, Geomagnetism, p. 442. Courtesy The Clarendon Press, Oxford.)

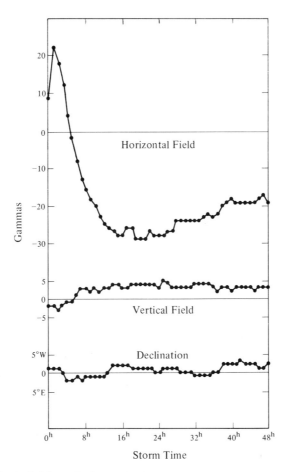

Fig. 7.15 *Typical field variations during a magnetic storm. (Chapman and Bartels,* Geomagnetism, *p. 276. Courtesy The Clarendon Press, Oxford.)*

causing a current to flow. A convecting core thus constitutes an electrical generator. The resulting current patterns are very complicated and difficult to predict exactly. Patterns are possible, however, in which the electrical currents create magnetic fields which stimulate further the generation of current, and hence an even stronger field. In this way, a small field, once created, can gradually be built up into a field such as that observed. This is known as the *dynamo theory.*

This concept is particularly appealing because it explains not only the main part of the field but also its variations. The convection cells can reasonably be expected to undergo changes in their detailed pattern with time, just as the magnetic field does. Eddys in the convection cells easily account for much of the variation of field strength from place to place. The rates of change of the field are of the order of magnitude of what

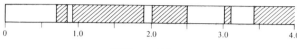

Millions of Years

Fig. 7.16 *Polarity of Earth's magnetism as a function of past time. Shaded areas represent periods of reversed polarity. (After Vine,* Science *v. 154, p. 1407. Copyright 1966, American Association for the Advancement of Science.)*

would be expected for eddys hundreds of kilometers in cross section. The slow westward drift of the magnetic field is explained by a slow westward drift of the convection-cell system. The sudden small changes in the length of the day (see Section 4.2, p. 105) are explained by changes in the momentum distribution between Earth's rotation and the rate of overturn of matter in the convection cells.

The convection pattern does not have to be of such a nature as to build up a magnetic field, and there is no necessity that it be maintained once it exists. Therefore, it is not surprising that the present field should be weakening. The theory requires that the main part of the field roughly parallel the axis of Earth's rotation, but permits it to depart moderately from this direction, as is observed. The field can reasonably be expected to die out occasionally, as it has done repeatedly in the past. When it reforms, it may do so with the present polarity or the opposite polarity, explaining the known reversals of the field.

Why the convection cells almost always tend to arrange themselves in a pattern which does produce a strong magnetic field is not yet fully understood. It might seem more reasonable that only occasionally would the field be strongly developed. This, however, is not the case. Almost all magnetizable rocks appear to have some permanent magnetization, presumably induced in them by the field at the time of their formation. This means that Earth has had a magnetic field almost all the time since the oldest rocks formed.

Several other theories to explain the main part of the magnetic field have been proposed, and abandoned as unlikely. The gradual diminishing of the field at present led to the supposition that Earth was formerly subjected to an externally applied field. The external field was then removed. Removal of the field induced a system of electrical currents in Earth's conducting interior. These currents and the magnetic field they produced are postulated to have been dying out ever since. Their long persistence results from the tendency of any change in magnetic field strength to induce electrical currents which oppose the change, and hence maintain the field.

The discovery that Earth has had a field throughout most of geologic time makes an external cause unlikely. The external field strength needed to produce a field such as Earth's by this method is much greater than has been observed in any stellar system at the distance of Earth from Sun.

The fact that most stars have magnetic fields suggested the possibility that a magnetic field is a fundamental property of all rotating bodies. If

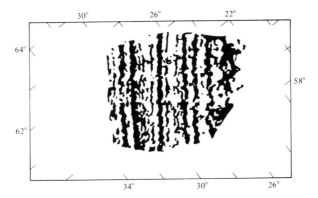

Fig. 7.17 *Magnetic anomalies over the Reykjanes section of the Mid-Atlantic Ridge. (Reprinted with permission from J. R. Heirtzler,* Deep Sea Research *v. 13, p. 428. Copyright 1966 by Pergamon Press.)*

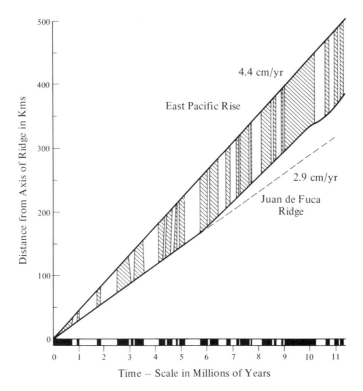

Fig. 7.18 *Magnetic-field reversal predicted by the ocean-floor spreading hypothesis. Upper scale is for East Pacific Rise. Lower scale is for the Juan de Fuca Ridge west of Vancouver, Canada. (After Vine,* Science *v. 154, p. 1411. Copyright 16 Dec. 1966, American Association for the Advancement of Science.)*

this were the case, then the magnetic field should decrease downward into Earth in mines and bore holes. Although the expectable amount of this decrease is so small that it cannot easily be observed in the presence of magnetic noise due to various causes, there appears to be no tendency for the field to decrease with depth.

Deep-seated electrical currents remain the only known reasonable cause of the magnetic field.

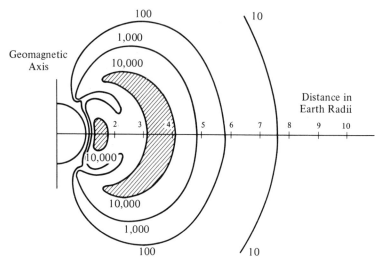

Fig. 7.19 *The Van Allen belts. Scale is measured radiation counts per second using a geiger counter. (After Van Allen,* Jour. Geophys. Res. v. 64, *p. 1684.)*

7.6 SOME EFFECTS OF THE MAGNETIC FIELD.

The magnetic field performs a useful function in protecting Earth from high-energy charged particles and in dragging the outer fringes of the atmosphere along with Earth thru space. It reaches far out in space, dying off with the fourth power of distance. Charged particles crossing the magnetic field are deflected at right angles to both the field and their direction of motion. Many such particles, especially electrons and hydrogen ions, acquire spiral paths which concentrate them in certain regions of Earth's atmosphere known as the *Van Allen belts* after their discoverer (Figure 7.19). The particles may originate in the upper atmosphere or outside of Earth. Hydrogen ions rising upward in the exosphere tend to be deflected by the field, making it more difficult for them to escape from Earth. Electrons and excited nuclei in the solar wind tend to be deflected by the field, preventing them from falling directly into the lower layers of the atmosphere.

Charged particles moving parallel to the magnetic field do not undergo this deflection. Thus the immediate vicinity of the magnetic poles is less

Fig. 7.20 *An aurora. (E.S.S.A. Photo.)*

protected than other areas from charged particles from Sun or elsewhere in space. This is believed to be the cause of polar radio blackouts (see Section 6.3).

If there were no magnetic field, there would be a steady flux of such particles everywhere on Earth. Most of these would be intercepted by gas molecules in the atmosphere, but a certain percentage would reach Earth's surface. Particles of this sort are very penetrating, far more so than X rays or gamma radiation. They constitute, essentially, bursts of cosmic rays, the most powerful electromagnetic radiation of all. Cosmic rays are thought to be one of the principal causes of changes in the genes. Since most such genetic changes are bad, any increase in the cosmic ray flux would increase the percentage of defective off-spring of every species, but it would also accelerate the appearance of new species.

During the periods when the magnetic field is reversing there must be intervals of increased cosmic ray flux. The genetic damage which would result has been suggested as a possible cause of the extinction of some species of animals such as the dinosaurs and the rise of large numbers of new species. How important the increase in cosmic ray flux is in controlling the rate of evolutionary changes in species is difficult to assess. A small amount of genetic damage must be produced by the current flux of high-intensity radiation. The expectable increase during periods of field reversal is not great enough that it would be expected to be a serious danger to living organisms. The population of any species is sensitive to many factors, however, and even a small effect on the birth rate could have a large influence

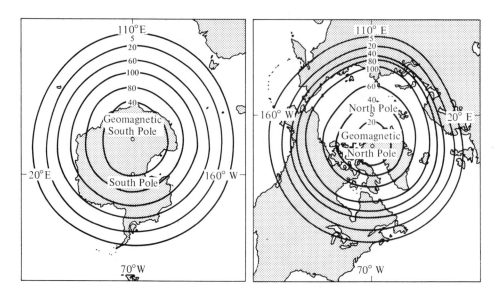

Fig. 7.21 *Average annual frequency of auroras. (After Feldstein et al.,* Jour. Phys. Soc. Japan *v. 17, Suppl. A1, p. 250.)*

on the survival of species which are having a hard time competing for living space for other reasons.

7.7 AURORAS.

One of the most striking consequences of the magnetic field is the *aurora borealis* (northern lights) in the northern hemisphere and the *aurora australensis* in the southern (Figure 7.20). The auroras appear to be due to the influx into Earth's upper atmosphere of electrons trapped in the Van Allen belts. Their original source is sunspots and solar flares. Clouds of hydrogen ions and free electrons striking Earth accumulate in the outer of the two main Van Allen belts and flow out of the ends, the flux being strongest in two circles surrounding the magnetic poles at a distance of about $22\frac{1}{2}°$ (Figure 7.21). As the charged particles flow earthward thru the atmosphere, they excite atoms in the mesosphere and thermosphere, which then glow in characteristic colors, most commonly a pale green due to atomic oxygen, but sometimes with reddish or violet tints. Distinctive patterns are observed, most commonly described as "curtains" or "bundles of rays." These patterns are constantly changing. They are visible only at night, but are presumably present by day also.

Some light of the same type as the aurora can be detected every night and at all latitudes. This is called the *night airglow,* and is believed to result from the excitation of the upper atmosphere by ordinary solar radiation.

BIBLIOGRAPHY

Chapman, S., 1951, *The Earth's Magnetism*. Methuen and Co., Ltd. 127 pp.

Chapman, S.; J. Bartels, 1940, *Geomagnetism*. Oxford University Press. 1049 pp.

Cox, A.; G. B. Dalrymple; R. R. Doell, 1967, "Reversals of the Earth's Magnetic Field." *Scientific American* 216(2):44–54.

Doell, R. R.; A. Cox, 1962, "Paleomagnetism." *Advances in Geophysics* 8:221–313. Academic Press.

Hines, C. O.; J. Paghis; T. R. Hartz; J. A. Fejer, 1965, *Physics of the Earth's Upper Atmosphere*. Prentice-Hall, Inc. 434 pp.

Howell, B. F., Jr., 1959, *Introduction to Geophysics*. McGraw-Hill Book Co., Inc. 399 pp.

Jacobs, J. A., 1963, *The Earth's Core and Geomagnetism*. Pergamon Press. 137 pp.

Johnson, F. S., 1965, *Satellite Environment Handbook*. Stanford University Press. 193 pp.

Kuiper, G. P. (Ed.), 1954, *Earth as a Planet*. Chicago University Press. 751 pp.

Nagata, T., 1961, *Rock Magnetism*. Maruzen Co. Ltd. 350 pp.

Spar, J., 1965, *Earth, Sea, and Air*. Addison-Wesley Publ. Co. 156 pp.

Stacey, J., 1965, *Physics of the Earth*. John Wiley and Sons, Inc. 324 pp.

Strahler, A. N., 1963, *The Earth Sciences*. Harper and Row, Publ. 681 pp.

Svendsen, K. L., 1962, *United States Magnetic Tables for 1960*. U. S. Govt. Printing Office. 87 pp.

Vine, F. J., 1966, "Spreading of the Ocean Floor: New Evidence." *Science* 154: 1405–1415.

8 THE OCEANS

Earth is possibly unique among the members of the solar system in that it has the right range of temperature and pressure conditions for its surface to be partially covered by large bodies of liquid. The smaller members up to Mercury and Mars are too small to hold enough atmosphere to produce a surface pressure allowing condensation in liquid form. Although it has been suggested that Venus may have seas of liquid hydrocarbons, the atmospheric temperature is too high for this to be probable. The larger planets have thick atmospheres. The depth of these layers is probably so great that their lower portions are undistinguishable from liquids. A gas, if compressed to the density of the equivalent liquid, merges indistinguishably into this form. If this is the case, then the solid surfaces of these planets are everywhere submerged in an ocean of liquid methane.

Earth has just enough ocean to cover 70.8% of its surface. The layer of water which fills the low areas on its surface is called the *hydrosphere*. Almost all of the hydrosphere is ocean. Only one part in 2700 of all surface water is locked in lakes, rivers and groundwater. There are, however, about 45 grams of water in the form of ice for every gram of fresh water on land.

8.1 COMPOSITION OF OCEAN WATER.

On the average, 3.45 grams of other material are dissolved in every 100 grams of ocean water. The amount of dissolved solid material per 100 grams of water is called the *salinity*. In addition, small amounts of all the atmospheric gases are dissolved in seawater. The most important dissolved materials are listed in Table 8.1. Traces of almost all of the elements found on Earth are found in the oceans.

Water is evaporated from the sea surface, the rate varying with time of day, with the season, with latitude, with water and air temperature, with sea surface roughness, with wind and with other weather conditions. Estimates of the average amount of water evaporated per year range from 76 to 106 cm per cm^2 of surface area, or about 1/4000 of the ocean annually. This evaporation loss is about 334,000 cubic kilometers annually (3.34 $\times$ 10^{20} cc). Since the area of Earth is 5.101 $\times$ 10^{18} square kilometers, this is

Table 8.1   *Principal materials dissolved in seawater.*

Element or ion	Grams per gram concentration
Chlorine	.01898
Sulfate (SO_4)	.00265
Sodium	.01056
Magnesium	.00127
Calcium	.00040
Potassium	.00038
Bicarbonate (HCO_3)	.00014
Carbon dioxide gas	$6.7 - 11.0 \times 10^{-5}$
Oxygen gas	$0 \ - 1.3 \times 10^{-5}$
Nitrogen gas	$1.0 - 1.8 \times 10^{-5}$
Nitrogen compounds	$6 - 700 \times 10^{-9}$
Phosphorus	$0 - 90 \times 10^{-9}$

equivalent to a layer of water a little over 65 cm (26 inches) thick. This is several times the amount of water in the air at any one time, so the water evaporated from the oceans must be precipitated as rain relatively quickly. About 89% of the evaporated water returns directly to the oceans as rain. The remaining 11% is carried onto the land, where it is precipitated, often evaporated and reprecipitated, and eventually returns to the sea as rain, thru streams and in underground seepage.

The returning water most often reaches the sea far from where it originally evaporated. As a result, the salinity of ocean water varies from place to place. The highest salinities are found in the sub-tropical high-pressure zone around 30° latitude, where dry descending air currents have a large capacity for moisture, which is carried out of the region by the prevailing winds. In the open ocean the salinity is prevented from rising much above 3.7 percent by the tendency of exceptionally saline waters to sink, with less saline water rising to the surface or flowing in from the sides. In shallow places, especially in seas which have only limited openings to the oceans, the salinity can rise to higher values. In the Red Sea, which is surrounded by deserts with almost no streams returning fresh water, the salinity is at places over 4.1 percent.

The lowest values of salinity are found at the mouths of rivers where fresh water flows into the ocean. Being lighter than salt water, the river water floats on the sea surface until the action of wind and waves mixes it with the underlying layers. The flow of the Amazon River is so great that its proximity can be detected by the low salinity of the ocean water off its mouth long before land is in sight.

Ocean water is rich in carbon dioxide, both in the form of gas and as carbonate and bicarbonate ions. There is everywhere an abundant supply for plant growth. Plants, however, also require sunlight. They flourish only near the surface; and they live, therefore, almost entirely in the first few tens of meters; although in very clear water, such as the Mediterranean, plant life may persist to over 100 meters depth.

Oxygen is dissolved from the air at the air-water interface. Animals, particularly micro-animals, gradually use up the oxygen, so that the concentration varies greatly, especially with depth. Where the oxygen supply has been exhausted, the waters are barren of animal life.

Nitrogen is present both as a gas and in the form of compounds. Phosphorus is present as phosphates. Both elements are important in the life cycles of marine organisms. When a plant or animal dies, unless it is eaten by some other creature, its remains sink to the ocean floor where they either decompose or become buried in sediment. Nitrogen and phosphates are brought to the surface life zone by upwelling of water from the deeper parts of the ocean and by rivers. Rivers draining populous areas often carry large concentrations of plant nutrients in the form of sewage.

The water at the ocean bottom is often low in oxygen and supports little life. Under these conditions, the water becomes enriched in nitrogen, phosphorus and other compounds derived from overlying zones of more abundant life. Ocean currents rising from great depths bring with them these nutrients; and, on picking up atmospheric oxygen at the surface, can produce regions of great variety and richness in living creatures (Figure 8.1).

Streams add material to the oceans in several ways. In the weathering of rocks on land, some of the minerals are dissolved. The resulting salts are carried by streams to the sea. In addition, streams carry considerable solid material. The larger particles are dragged along the bottom by currents and are dropped as soon as the velocity decreases. As a result, they tend to be deposited near the mouths of the rivers which bring them to the sea.

The smaller particles may be held in suspension for a long time. Near the shore, where wave action keeps the water stirred, such suspended matter represents an important additional component in water composition. In the open ocean, where there has been ample time for all except the smallest particles to settle out, it is unimportant.

Along the ocean bottom there is a constant interchange of matter between the solids exposed to the sea and the materials in solution. Living organisms play an important role in adding or removing elements. Many organisms take calcium and magnesium carbonate from water to form their shells and skeletons. On death, these materials collect on the ocean floor or as a part of coral reefs, where they may become locked in sedimentary rocks. Unlike most compounds, the solubility of carbonates in water is greater at low than at high temperatures. In the deepest parts of the oceans, where the temperature is below average, carbonates tend to be dissolved. As a result, only silicate skeletons and shells are deposited here.

These variations in composition of ocean water from place to place, small though most of them are, are like climatic factors on land. The varieties of life which flourish depend on the presence or absence of conditions needed to support each species. Thus the fauna and flora vary from place to place with the concentration of the nutrient elements and the degree of freedom from pollutants such as mud in suspension. They depend also on conditions such as the amount of sunlight and the temperature.

Fig. 8.1 *Major areas of upwelling of oceanic waters.*

8.2 TEMPERATURE.

The temperature of ocean water ranges from a high of about 28°C (82°F) to a low of –1.9°C (28°F). In the Arctic and Antarctic, where air temperatures drop far below this, ice freezes on the surface and provides a transition zone of high temperature gradient.

Heat is added to ocean water at its upper surface by solar radiation. Heat is added from Earth's interior along the ocean floor. Heat may also be brought to the ocean by the inflow of warm streams or rain, although precipitation is more often than not cooler than the water into which it falls. Solar heating is by far the most important heat source.

Heat is lost from the oceans by conduction or radiation from the sea surface and by the energy used to evaporate seawater.

Over three quarters of the solar energy penetrating the surface is absorbed in the upper ten meters of the ocean (Figure 8.2). Visible light penetrates more deeply in clear water than does infra-red heat radiation. Below ten meters the spectrum of the light becomes progressively richer in green. The depth of penetration is reduced by the presence of bubbles or solid material in suspension.

Water has a large heat capacity compared to most other substances. As a result, it changes temperature relatively slowly. The seasonal and daily temperature variations are, therefore, much less than for air, rarely exceeding 8°C (14°F). Such changes are limited largely to the first few tens of meters depth. Water tends to warm cool air and to cool warm air which is in contact with it, greatly ameliorating climatic extremes in the vicinity of the oceans. Fog is often formed where the overlying air has a different temperature from the underlying water. The most common occurrence of fog is when warm, moist air has moved over colder water. Here the drop in air temperature produces fog. But fog can also be produced when cool air moves over warm water. In this case, the air is heated near the water surface, and its capacity for moisture increased. As a result, the layer of air just over the water becomes wetter than the immediately overlying air. Vertical mixing can then produce irregular patches of fog called sea smoke.

Temperature normally decreases with depth in the oceans. Unlike fresh water, the density of seawater increases with a decrease in temperature at all temperatures, so that cold water has a tendency to sink. Near the surface there is a layer from a few tens to a few hundreds of meters thick in which the temperature is kept nearly constant by the mixing action of waves. Beneath this, except under polar conditions, temperature decreases at first rapidly and then more slowly (Figure 8.3). The layer of rapidly decreasing temperature just below the mixed layer is called the *thermocline.*

A drop in temperature produces an increase in density. The thermocline is thus also a layer of rapidly increasing density across which water does not easily mix. This is easily noticed in a submarine, which must either adjust its buoyancy as it rises or sink thru the thermocline, or must use power to counteract its effect.

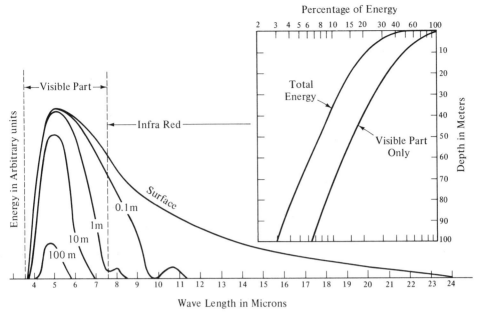

Fig. 8.2 *Relative amounts of solar energy penetrating to different depths in clear water. (After Sverdrup et al.* The Oceans, *p. 105. Courtesy Prentice-Hall, Inc.)*

In the deepest parts of the oceans, the gradient sometimes reverses and there is a slight rise in temperature as the bottom is approached. The rate of increase is always less than the adiabatic gradient (see Section 6.2, p. 137). If this were not the case, vertical convection would develop.

8.3 OCEAN CURRENTS.

Ocean currents are produced largely by two causes, density difference and wind. Vertical currents form where heavy water sinks and light water rises, with consequent horizontal flow between such areas. Wind is the principal cause of horizontal currents. Wind can also cause vertical currents. Where horizontally moving water approaches a shoreline, it has no place to go but down or to one side; and where it is blown offshore, it is often replaced by water welling up from below. Currents are also produced by the tides (see Section 8.6 below). Near the shore, the inflow and outflow of tidal water may produce currents which are much stronger than density-and wind-produced currents.

The density of seawater is controlled by three factors: temperature, pressure and composition. The most important of these is composition. Seawater of 3.485 percent salinity is 2.8 percent denser than fresh water at the same temperature. Salinity varies due to surface evaporation, with corresponding density variations. Around latitude 25° where evaporation exceeds rainfall, salinity is higher than anywhere else at the ocean surface, and there is a tendency for the water to overturn, with the saline water sinking.

Density may be even more greatly increased by the presence of suspended solids in the water. Very large storm waves stir sand and mud from the bottom into shallow seawater. Density currents composed of this weighted water may flow down the floor of the ocean from the near-shore shallows, where rivers dump most of their sediment load, to the deepest parts of the oceans, carving channels on the sloping bed of the ocean just as a river does on land. Such submarine channels are commonest off the mouths of large rivers where sediment to feed the density currents is plentiful. A large canyon of this sort occurs off the mouth of the Hudson River (Figure 8.4).

Pressure is the least important of the three factors. Water shrinks by a factor of only about .0045 per kilometer of depth, or about 2.7% from the surface to a depth of six kilometers.

The effect of temperature is more complicated. For salinities of 24.7 parts per thousand or more, water increases in density with decreasing temperature at all temperatures. Cold seawater, therefore, tends to sink. Less saline water expands at temperatures close to freezing. For fresh water, density decreases below 4°C. Furthermore, fresh-water ice floats in a liquid of the same composition.

When seawater freezes, the ice crystals which are formed are less saline than the water. As a result, they float. The water which is left behind is more saline than before. In the polar areas, water tends to be heavy and to sink both because it is cold and because it is enriched in salinity by the ice which forms at the surface. The temperature factor is the more important. Some of the ice melts seasonably returning low-salinity water to the oceans. Lateral differences are developed, however, because of the drift of the ice from its place of formation. The two main areas of sinking of water are the Antarctic Ocean and the North Atlantic east of Greenland.

Wind produces ocean currents by frictional drag of the air moving over the surface of the water. This exerts a force on the water surface in the direction that the wind is moving. The water motion is influenced by the coriolis effect, which, in the northern hemisphere, causes it to drift to the right of the force causing the motion. Since each layer of water drags the underlying layer with it, there is a progressive rotation of the motion with depth (Figure 8.5). The expected deviation of the surface-water current from the wind direction is 45°. At a depth which depends on latitude and wind velocity, the water flows in the opposite direction from the wind. Its velocity decreases with depth, and is small beneath the depth where direction is reversed. The overall effect of the coriolis force is that shallow ocean currents tend to flow at right angles to the winds that produce them.

In the open ocean, the world-wide patterns of wind and the climatic patterns of evaporation and precipitation, with consequent variations in water salinity and density, cause corresponding persistent ocean currents (Figure 8.6). The principal cells of this current system are centered on the

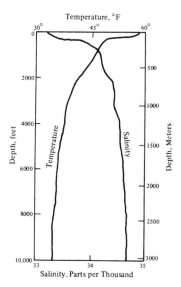

Fig. 8.3   *Typical profiles of temperature and salinity versus depth. (After Bowditch, American Practical Navigator p. 744. U. S. Navy Hydro. Office.)*

subtropical high-pressure areas, from which the trade winds blow equatorward and to the west, and middle-latitude winds blow poleward and to the east. The oceanic cells are called *gyres*, and particularly in the northern hemisphere their pattern is modified by the shape of the ocean basins.

The north Atlantic circulation is made up of two gyres. The larger is a subtropical gyre. Its southern portion is a broad westward drift from the bulge of Africa past the north coast of South America into the Caribbean Sea and the Gulf of Mexico. From there, the accumulated water flows thru the narrow opening between Cuba and Florida at a velocity (at its center) of 6 km/hr (4 mi/hr). Here it is joined by the Antilles current, which sweeps northwest from the central Atlantic past the Bahama Islands, forming what is at first called the *Florida Current* and, farther north, the *Gulf Stream*. In the open Atlantic it tends to preserve its identity as a water mass, floating on the ocean surface because its temperature is higher than that of deeper waters. As far north as Cape Hatteras it hugs the coast, providing warm water for swimming along the beaches of the southeastern United States. From Cape Hatteras north it shifts gradually seaward and broadens. As it approaches Europe, it is renamed the *North Atlantic Current*. Part spreads southward as the Canary Current, and rejoins the North Equatorial Current to complete the subtropical gyre. The remainder bathes the coast of Europe from Spain to Norway in warm surface waters. Part extends on into the Arctic Ocean enabling shipping lanes to be kept open all year to Murmansk on the Arctic coast of Russia.

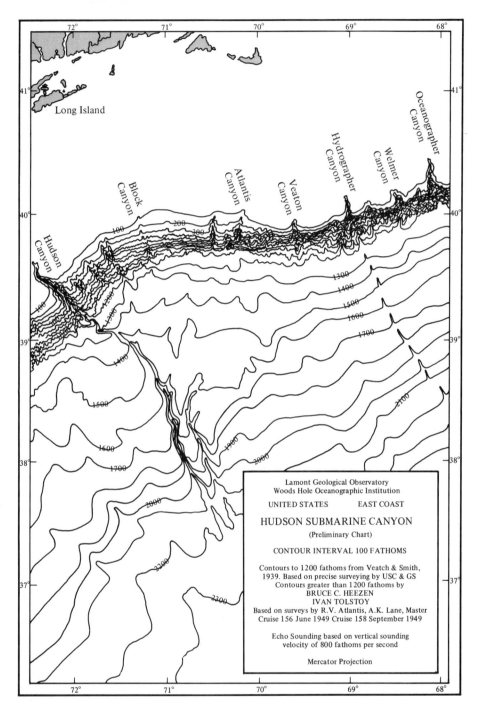

Fig. 8.4 *Topographic contour map showing the location of submarine canyons off the northeast coast of the United States. (After Heezen et al.,* Geol. Soc. Am. Sp. Paper 65, *pl. 2.)*

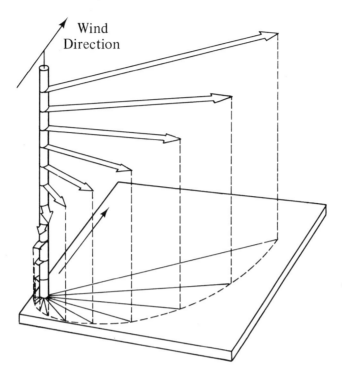

Fig. 8.5   *The direction of an ocean current tends to rotate clockwise with depth in the northern hemisphere. This is called the Ekman spiral. (After Sverdrup et al.,* The Oceans, *p. 493. Courtesy of Prentice-Hall, Inc.)*

The remainder swings east past Iceland to join the East Greenland and Labrador currents, which flow south out of the Arctic Ocean. Together this flow forms a small subpolar gyre rotating counterclockwise in the northeast Atlantic Ocean, with its southern edge a part of the North Atlantic Current. This subpolar gyre brings cold water to the beaches of eastern Canada and the northeastern United States as far south as Maryland. It also includes some of the most prolific fishing grounds of the world, from which a large portion of the fish harvest of the Atlantic is obtained.

Similar gyres exist in other oceans. The large subtropical gyre of the North Pacific consists of two cells, one east and one west of Hawaii, with a poorly developed zone of opposing currents separating them. The California current, a part of the North Pacific gyre, brings cold water south along the coast of North America.

A small subtropical gyre exists in the Bay of Bengal east of India, but the expectable corresponding current in the Arabian Sea to the west is not clearly developed. Large counterclockwise gyres occur in the South Atlantic, South Pacific and Indian Oceans. These are bordered on the south by the belt of uninterrupted ocean at 50°S latitude surrounding Antarctica. Here a persistent eastward circumpolar current is found, driven by the

Fig. 8.6 *Principal surface ocean currents. (After Bowditch, Am. Practical Navigator, p. 720. U. S. Navy Hydro. Office.)*

powerful westerly (northwest-to-southeast) winds of the roaring forties (Figure 6.20, p.158). Small clockwise gyres are developed south of each of the main oceans, where space is available, between the Antarctic continent and the circumpolar current.

The North and South Equatorial Currents are driven by the trade winds, which cause an accumulation of water on the west sides of both the Atlantic and Pacific Oceans. As a result, sea level is higher on the west than on the east side of each ocean. Between the trade-wind belts, in the low-wind zone of the doldrums, a weak downhill flow of water moves from west to east. This is called the *equatorial countercurrent*.

Ocean vessels often take advantage of the current patterns to speed their travel. The center of the Gulf Stream and North Atlantic current is sought by eastbound ships and avoided by westbound ones.

The centers of the subtropical gyres are regions of slow water circulation. Because evaporation exceeds precipitation, there is a tendency for water to sink. These areas become depleted in nutrients, and as a result they support a relatively low concentration of plant and animal life.

Measurements of ocean-current velocities are made less easily than wind velocities because there is no convenient fixed surface on which to stand and make the measurements. Surface currents can be traced by drifting with the current and marking successive positions with respect to nearby land or by star positions. Electronic position-finding devices or satellite sightings are also used. Such methods, however, are slow and tedious. Currents can also be measured by noting the effect of Earth's magnetic field on a moving conductor. If an electrical conductor is moved in a magnetic field, a voltage is developed at right angles to the field and to the motion of the conductor. Since salt water is a good conductor, two electrodes immersed in a moving ocean current will have a voltage generated between them by the movement of the ocean. The electrodes can be towed behind a ship and a continuous recording made indicating the velocity at right angles to the line joining the electrodes. By laying out a grid of such measurements along intersecting lines, the current velocities can be mapped. The instrument used to make this measurement is called a *geomagnetic electrokinetograph*, or GEK.

Theoretically the GEK can also be used for measuring currents at any depth. In practice, deep currents are usually measured by comparing their velocity with the surface velocity. This can be done by lowering a measuring device on a cable or by releasing a pressure-sensitive device which will float at a set depth beneath the surface and which will send out sound signals by which its location can be followed from the surface. All of these devices are used primarily for surveying horizontal currents. Vertical currents in the oceans are much less thoroughly studied than horizontal currents.

The general pattern of deep circulation is still not thoroughly mapped. The polar sinking of cold water causes deep water to drift equatorward. Estimates of the rate of overturn of oceanic waters along this path, based on the rate of decay of radioactive carbon, which was originally derived from the atmosphere, indicate that it takes from 200 to 1800 years for bottom water to return to the surface.

There are also currents at intermediate levels which play an important role in returning wind-driven surface waters to areas of surface divergence. An important current of this type, the Gulf Stream Countercurrent, flows southward along the west edge of the north Atlantic basin. The Cromwell Current in the South Pacific flows eastward beneath the South Equatorial current.

One of the most remarkable features of the ocean circulation is the persistence of thin layers of water flowing over large horizontal distances. The main area of descending water in the northern hemisphere is near the center of the small gyre south of Greenland. Water sinking in this area can be traced flowing southward nearly to Antarctica, where it is overlain and underlain by northward flowing currents (Figure 8.7).

8.4 WAVES.

In addition to being the principal cause of currents in the oceans, winds produce most of the waves on the sea surface. The simplest type of waves is gravity waves, so called because the energy and propagation of the waves result from the pull of gravity on the raised ridge of water in the wave. Waves of this sort are formed when wind drags patches of water along, piling them on top of adjoining portions. This process is aided by the force of the wind blowing against one side of the waves. The size of the resulting waves depends upon the speed of the wind, the area over which it is acting and how long it has been blowing. When the wind has been blowing for many hours over a large expanse of water, the maximum height of the waves will be proportional to the square of wind velocity, and is given by the formula

$$H = .0023 \ V^2 \tag{8.1}$$

where height is in meters and velocity is in kilometers/hr. A 20 km/hr (12.4 miles/hr) wind will produce waves 0.9 meters (3 ft) high. A 100 km/ hr (62 miles/hr) wind will produce waves 23 meters (75 ft) high.

Although the crest of a wave seems to progress forward, there is actually little or no forward transport of water in a wave in the open ocean. At the surface, a water molecule moves in a nearly circular path: up, forward, down and back (Figure 8.8). In a wind, the crest of the wave may slip forward over the underlying water or be carried forward as spray resulting in a drift of the water. In the absence of wind, there is no tendency for the water to progress. Once a wave is formed, however, the pull of gravity on the raised ridge makes the wave propagate. Waves in the area of their formation are rough and irregular, with small waves superposed on large ones. After a wave has travelled far from its source area, it becomes smoother and simpler in shape, and is called a *swell*. Short-wavelength waves are damped out relatively rapidly, but long-wavelength waves die out slowly. As a result, long swell from a storm thousands of kilometers

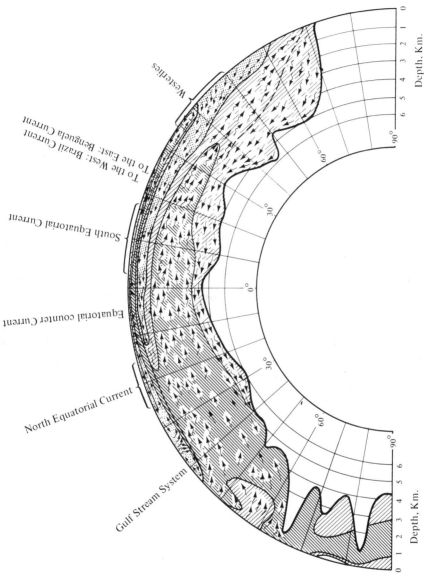

Depth, Km.

Westerlies

To the West: Brazil Current
To the East: Benguela Current

South Equatorial Current

Equatorial counter Current

North Equatorial Current

Gulf Stream System

Depth, Km.

Fig. 8.7 Simplified pattern of north-south water circulation in the Atlantic Ocean. (Reprinted with permission from A. Defant, Physical Oceanography, p. 695, copyright 1961 Pergamon Press Ltd.)

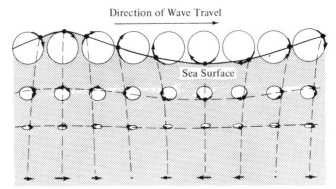

Fig. 8.8   *Particle motion in a gravity wave in deep water.*

away may reach a shoreline far from any storm. Gravity waves propagate with a velocity which depends on the wavelength and the water depth.° When the wave is travelling in deep water, waves travel with the approximate velocity

$$V_{deep} = \sqrt{\frac{gL}{2\pi}} \qquad (8.6)$$

where g is gravity and L is the wavelength. Thus, in the open ocean the long-wavelength waves travel faster than the shorter waves.

In shallow water, the velocity approximates

$$V_{shallow} = \sqrt{gH} \qquad (8.7)$$

where H is water depth. Thus, the velocity decreases as the shoreline is approached.

°The complete equation for the velocity of gravity waves on the surface of a liquid layer is

$$V = \sqrt{\frac{gL}{2} \tanh \frac{2\pi H}{L}} \qquad (8.2)$$

where g is gravity, L is the wavelength, and H is water depth. Tan$h$ is a trigonometric function such that

$$\tanh\left(\frac{2\pi H}{L}\right) = \frac{e^{\left(\frac{2\pi H}{L}\right)} - e^{-\left(\frac{2\pi H}{L}\right)}}{e^{\left(\frac{2\pi H}{L}\right)} + e^{\left(\frac{2\pi H}{L}\right)}} \qquad (8.3)$$

where $e = 2.7183\ldots$. When $2\pi H$ is less than L, this can be found from the convergent series:

$$\tanh\left(\frac{2\pi H}{L}\right) = \frac{2\pi H}{L} - \frac{1}{3}\left(\frac{2\pi H}{L}\right) + \frac{2}{15}\left(\frac{2\pi H}{L}\right)\cdots \qquad (8.4)$$

When $\dfrac{2\pi H}{L}$ is small, the case for shallow water, this reduces to

$$\tanh\left(\frac{2\pi H}{L}\right) \approx \frac{2\pi H}{L} \qquad (8.5)$$

For large values of $\left(\dfrac{2\pi H}{L}\right)$, the case for deep water, tan$h \left(\dfrac{2\pi H}{L}\right)$ approaches unity.

In deep water the amplitude of wave motion dies off rapidly with depth, being less than one twentieth of the surface amplitude at one half a wavelength. Therefore, in the deep ocean, there is no measurable disturbance at the bottom no matter how great the storm at the sea's surface.

As a wave approaches a shoreline, on the other hand, sooner or later it reaches a point where the effect of the bottom becomes important. This happens first for the longest wavelengths. These are slowed in their rate of travel. This means that their wavelength is decreased. The distance from one crust to the next becomes less. But the energy involved in the wave motion must be conserved. One result is that the wave height increases.

At the same time, the water along the ocean bottom begins to be involved in the wave motion. The lowermost layer cannot move up and down, but is dragged forward and back, stirring up loose sediment along the bottom and holding it in suspension due to turbulent motion of the bottom water. The depth to which material is stirred in this fashion extends to much deeper water when wave heights are large and when wavelengths are long than under normal conditions. Thus, the largest storms disturb bottom mud that is unaffected by smaller storms.

Where the wave motion is constrained by the bottom, the water molecule orbits become flattened ellipses instead of circles (Figure 8.9). As each wave approaches the beach, the water surges landward under the crest of the wave, and withdraws under the trough. This stirs the bottom sediment, allowing only large grains to accumulate along the shoreline.

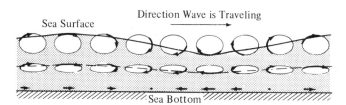

Fig. 8.9   *Wave orbits in shallow water.*

Another effect of the decrease in velocity of propagation with decrease in water depth is that the direction the wave advances is turned toward the shore. Along gently sloping beaches, the waves always approach the shore steeply, even though farther out they may be travelling nearly parallel to it (Figure 8.10). Also, the top of the wave tends to override the bottom, and a breaker is formed (Figure 8.11).

Where shorelines are irregular in shape, waves may be concentrated at some points, spread out at others. Waves tend to converge on headlands and to be spread out along embayments (Figure 8.12).

Local currents are set up along the shoreline by this focusing of wave action. Where waves are concentrated, there is an excess of water, which must flow away near the surface or along the bottom. Bottom currents of this sort are called *undertow* and usually are limited to brief pulses following each wave. Surface currents are more often continuous or last for longer periods. Between areas of wave concentration, *rip currents* may develop. These are narrow surface currents which flow rapidly for a few tens

Fig. 8.10 *Waves turn toward the shore as the water becomes shallower. (Photo by Robert Brigham. Courtesy Woods Hole Oceanographic Inst.)*

or hundreds of meters, then die out when deeper water is reached. They are a serious hazard to inexperienced swimmers, since they often flow out faster than a man can swim. However, since they are narrow, escape is easy if one swims parallel to the shoreline until free of the rip current, then to shore.

8.5 STORM SURGES AND TSUNAMIS.

In addition to the ordinary waves which have periods of a few seconds or tens of seconds, there are also longer-period waves. The crests of these waves are far apart, and their heights are usually less than the heights of shorter-period waves. As a result, they rarely break as they reach a shoreline, and so are not readily seen as waves by an observer. If the height to which successive waves flow onto a beach is watched, it is often easy to recognize a periodic variation in wave size. Sometimes it is every seventh wave which comes in most strongly, sometimes the interval is much longer. This is an effect of long-period components of the wave motion.

Fig. 8.11 Waves appear to rear up out of the water as they prepare to break on a beach.

Fig. 8.12 *Shallowing water tends to concentrate wave action on headlands, to spread it out in embayments. (U. S. Navy Photo. From Turekian, Oceans, p. 87. Courtesy Prentice-Hall, Inc.)*

The tides are waves of this sort, with periods as long as a day. Tide gauges provide efficient means of studying the long-period waves in the presence of shorter-period oscillations. A typical *tide gauge* consists of a float which follows water height attached to a recording device of some kind which makes a record of water level (Figure 8.13). By placing the float in a chamber which is connected to the ocean by so small an opening that very little water will enter the chamber during the duration of one short-period wave, rapid wave motions can be filtered out, leaving a clear record of long-period water-level changes.

Winds and air-pressure changes affect water level. High air pressure depresses the ocean surface, and low air pressure allows it to rise. A sea-level change of 25 cm (10 inches) will be produced by an air-pressure change of 2.5 percent, as might occur during the passage of a middle-latitude cyclone. Such water-level changes, however, are small compared to the effect of strong, persistent winds. Such winds can pile water against a coast. Hurricane winds often raise the water level by several meters. When a large storm approaches a coast at the same speed that waves travel, exceptionally large amplitudes can be achieved, in extreme cases over ten meters (33 ft), with consequent great loss of life and property damage. In this circumstance the wave may arrive relatively suddenly, being greatly

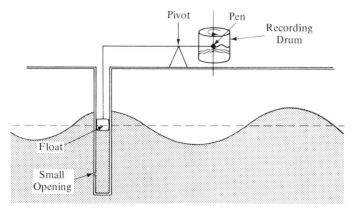

Fig. 8.13 *Principle of the tide-gauge.*

steepened as it slows down on reaching shallow water. Such waves are called *storm surges*. A 12-meter (40-foot) storm surge in 1737 in the Bay of Bengal is estimated to have drowned 300,000 persons.

Equally devastating waves called *tsunamis* sometimes accompany earthquakes. (They are also often miscalled tidal waves, even though they have no relation to the tides.) The principal wavelength of such waves is commonly 100 km (60 miles) or greater in the open ocean, and the periods commonly range from a few minutes to three quarters of an hour (Figure 8.14). With such long wavelengths, the velocity is always proportional to the square root of water depth (see Equation 8.7). The heights of tsunamis in the open ocean are so small that they are invisible there; but as the wave approaches shallow water, the wavelength decreases and the height increases. The amplitudes are also greatly affected by the shape of the ocean-bottom. Depressions and ridges in the sea bottom focus tsunamis along some stretches of coast and attenuate them along others.

Tapering embayments in the coastline may further increase wave height. A long, narrowing estuary can cause wave height to build upwards by progressively constricting the landward surge of water. In shallow water the wave sometimes breaks like an ordinary sea wave and rolls inland with a nearly vertical front a few meters high. Tsunamis have been known to reach heights of 60 meters (200 feet).

The details of the mechanism by which tsunamis are produced are uncertain. Changes in sea-floor level of up to 15 meters (50 feet) occurred during the 1964 Alaskan earthquake, whose tsunami was observed all around the Pacific basin and caused extensive damage as far away as California. Where there is a raising or lowering of the sea floor accompanying an earthquake, it is to be expected that a corresponding wave will be generated on the sea surface. Consider the case of a displacement of the ocean floor as one block is raised and another lowered (Figure 8.15). Assuming that the displacement occurs in much less time than it would take a gravity wave to cross the affected area, portions of the sea surface are raised and lowered corresponding to the bottom changes. These surface-elevation changes then propagate outward as ordinary gravity waves. The energy is

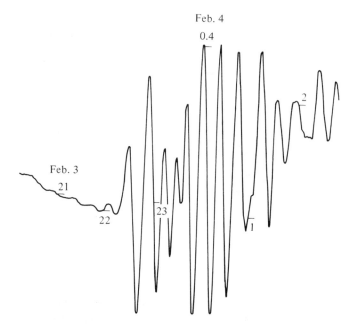

Fig. 8.14 *Honolulu, Hawaii, tide-gauge record showing the 3–4 February 1923 tsunami caused by an earthquake near Kamchatka, U. S. S. R. Time in hours is shown. (U. S. Coast and Geodetic Survey, Pub. 30–3, p. 32.)*

absorbed only very slowly in the open ocean, so that the wave reaches coasts attenuated only by the effect of spreading with distance. The larger the area of sea bottom affected, the larger the energy in the wave.

Tsunamis commonly consist of a sequence of waves, not just a single large wave. The first observed motion of the tsunami is often a retreat of the sea. A sure warning of an impending inundation is the sudden withdrawal of water beyond normal low-tide level. The water can be expected to return to a height above its previous level equal to the amount of its withdrawal below normal. The five to twenty minutes between retreat and advance allow time for persons to escape drowning if they act quickly enough. In Japan and Chile, where tsunamis occur often enough that many persons have experienced them, such warnings rarely go unheeded. When an earthquake occurs, persons living along the coast have learned to watch the water level and to flee if it drops excessively.

Tsunami waves travel large distances and may arrive many minutes after the earthquake which caused them. The 1868 tsunami at Arica, Chile began a half hour after the earthquake, and the largest wave did not occur until two hours after the shock. It was great enough to carry a U.S. gunboat, the *Wateree*, three kilometers (two miles) inland, where it left it high and dry on land. Here it had to be abandoned, to the embarrassment of its captain and the consternation of authorities in distant Washington, who found it difficult to understand this unauthorized invasion by a naval vessel so far inland into a friendly territory.

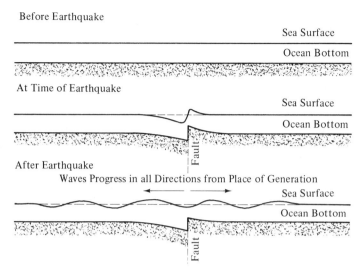

Fig. 8.15 *Generation of a tsunami by a fault in the ocean floor.*

Unfortunately, there is not always a clear warning, and the tsunami sometimes arrives first as an unanticipated inundation. Since it also affects areas far from where the earthquake causing it was felt, the countries facing the Pacific Ocean, where most large tsunamis occur, have set up a warning system under the leadership of the United States Coast and Geodetic Survey. A network of seismic observatories records ground motions at twenty places around the Pacific. When a large earthquake occurs, these stations record characteristic ground vibrations from which the distance to the place where the earthquake occurred can be determined (see Chapter 13). These arrivals are reported by telegraph to a central observatory at Honolulu, Hawaii. From the reported times of arrival, the location of the earthquake is determined.

If the shock occurred along the edge of the Pacific in any of the areas known from past experience to be potential sources of tsunamis, a check is made whether the tide gauge nearest the earthquake area recorded a tsunami and how large a wave was observed. If a large tsunami was recorded, all countries bordering the Pacific are warned.

The operation of the system is illustrated by what happened following the 1964 Earthquake in Alaska. At 03:36:13 universal time on March 28 (13 seconds after 05:36 P.M. local time on Good Friday afternoon March 27) one of the largest earthquakes of the twentieth century occurred under Prince William Sound southeast of Anchorage, Alaska. In a few seconds millions of dollars of damage was caused by the shaking in Anchorage, Valdez and Seward. A large section of the city of Anchorage began to slip downhill as a giant landslide, carrying with it homes, roads and people (Figures 8.16 and 13.29, p. 361). Within less than five minutes the main shaking had ceased. Within twenty minutes, seismographs all over Earth would record the shock. But as the shaking began to die off in Alaska, most of the world was still undisturbed by what had happened; and even in Alaska the worst was yet to come.

Fig. 8.16 *Part of the Turnagain Heights Area of Anchorage, Alaska, after the 1964 Earthquake. (U. S. Coast and Geodetic Survey photo.)*

The vibrations spread thru Earth at velocities ranging from 6 to over 13 km/sec. They reached Hawaii at 03:44, eight minutes after the shaking began in Alaska. At the Tsunami Warning Center, the ground began to move back and forth. To a person in the observatory, the motion was too small to feel; but sensitive instruments responded to the movements, which were substantially stronger than the normal, small vibrations due to wind, traffic, surf and other causes. An alarm bell began to ring, and the seismologist on duty turned to look at a recording drum on which an amplified tracing of the ground motion was being written (Figures 8.17 and 8.18). From the size of the ground motion, it was immediately apparent that this was a big earthquake. A teletype message was sent to other observatories asking for the times of arrival of the earthquake motion on their instruments. As the answers came in, it quickly became clear that an earthquake had occurred somewhere near Seward, Alaska. At 05:02, the following message was sent out via Federal Aviation Administration and Department of Defense communications systems to government agencies all around the Pacific: "This is a tidal wave advisory. A severe earthquake has occurred at Lat. 61N, Long. 147.5W, vicinity of Seward, Alaska at 0336Z (3:36 universal time), 28 March. It is not known, repeat not known, at this time that a sea wave

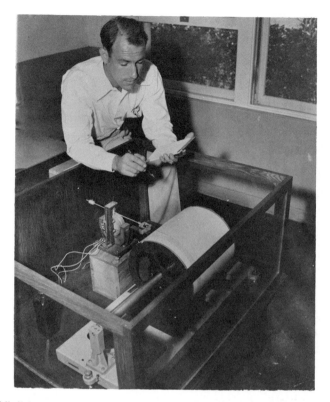

Fig. 8.17 *Visibly recording seismograph at Honolulu, Hawaii Tsunami Warning Center. (Honolulu Advertiser Photo.)*

has been generated. You will be kept informed as further information is available. . . . Its ETA (estimated time of arrival) for Hawaiian Islands (Honolulu) is 0900Z (9:00 universal time)°. . . ."

By this time the tsunami had already begun its deadly work. Waves sometimes over 52 meters (170 feet) high had struck near Valdez, killing 31 persons, and had washed the Middle Rock Light in Valdez Narrows off its pedestal 11 meters (35 feet) above sea level. At Cordova, which is located on Prince William Sound, tsunami waves were nine meters (30 feet) high. Around Seward and Kodiak, roughly 100 and 300 km to the southwest on the Gulf of Alaska, the waves were equally great. At Seward, most of the business district and ninety percent of the industry of the town was destroyed. One railroad locomotive was carried 150 meters (500 feet) inland (Figure 8.19). The collapse of oil-storage tanks spread oil throughout much of the inundated area, where it caught fire, causing additional damage even after the waves receded.

°This and subsequent quotations are from "Preliminary Report, Prince William Sound, Alaskan Earthquakes March–April 1964," U. S. Dept. of Commerce, Coast and Geodetic Survey, April 17, 1964, 98 pp.

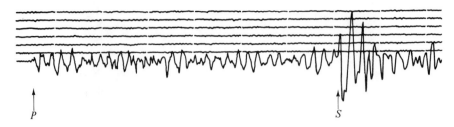

P   S

Fig. 8.18 *A part of the Honolulu, Hawaii, seismogram of 7 September 1948,
showing the arrival of a Tonga earthquake. This arrival led to the
first tsunami warning issued by the U. S. Coast and Geodetic Survey. (Courtesy R. A. Eppley.)*

In the central area of earthquake damage, normal communications facil-
ities were knocked out of operation, and no information of exactly what
had happened there was as yet available. At Kodiak, the tide gauge was
damaged by the earthquake and remained out of operation throughout the
tsunami. Communications from there to Honolulu, however, remained
open; and at 05:55 Kodiak sent the first definite word that a tsunami was
spreading: "Experience seismic sea wave at 0435Z. Water level 10–12 feet
(3–4 meters) above M.S.L. [mean sea level]." The first wave rose to a maximum height of about 5 meters (16 feet), then fell equally far below normal. With this information in hand, at 6:37, Honolulu issued a definite
warning: "This is a tidal wave warning. A severe earthquake has occurred
at Lat. 61N, Long. 147.5W vicinity of Seward, Alaska at 0336Z 28 March.
A sea wave has been generated which is spreading over the Pacific Ocean.
The ETA of the first wave at Oahu is 0900Z, 28 March. The intensity cannot, repeat, cannot be predicted. However, this wave could cause great
damage in the Hawaiian Islands and elsewhere in the Pacific Area. . . ." Estimated arrival times for several dozen points around the Pacific area were
given.

The tsunami by this time had already begun to arrive at Sitka, where the
nearest tide gauge to survive the earthquake was located. Four surges rose
at Sitka, the first starting at 5:08 and rising 1.9 meters (6.1 feet), the largest
cresting at 4.5 meters (16.6 feet) at about 10:50, after which the oscillations gradually died off.

The Honolulu message was received at the California Disaster Office in
Sacramento, which alerted sheriffs, police chiefs and civil defense directors
in coastal communities. In Crescent City, California it was already 10:37
P.M., local time, when the message was sent out. The first wave was to arrive in 62 minutes. Police and civil defense workers immediately began the
task of sending people from their homes and other places near the beach to
the safety of higher ground. The first wave, at 7:39 universal time (11:39
P.M. local time), was only one meter (three feet) high. The second and third
crests broke on the beach without causing serious damage. Some of the
evacuees thought that the danger was past and began moving back. The
fourth crest arrived at 9:20 (1:20 A.M. local time). It and the fifth wave
were 3.7 meters (12 feet) high, and they substantially destroyed 27 blocks

Fig. 8.19 *Tsunami damage at Seward, Alaska. (E.S.S.A. photo.)*

of waterfront property, drowning eleven persons who had returned to this area. Elsewhere on Oregon, Washington, California and Hawaii there was damage; but generally, warned of the possible danger, people were able to climb out of the way.

The problem of evacuating threatened areas involves psychological as well as technical problems. When the warning system was first tried in Hawaii, the first two tsunamis for which warnings were issued were too small to cause appreciable damage. Many persons failed to recognize the seriousness of the danger, and were lulled into a false sense of security by the failure of significant danger to materialize following the announcements. On May 23, 1960, a ten-meter (35-foot) high tsunami inundated Hilo, Hawaii following a large earthquake in Chile. Although nearly everyone in the flooded area heard the warnings, less than one-third of the people evacuated the area. There were even persons who drove to the coast to see the wave come in and were caught. Many who received the warnings did no more than go indoors, assuming incorrectly that their houses offered sufficient security. Sixty-one persons were drowned and 282 injured.

Displacement of the sea floor on faults is not the only mechanism by which tsunamis are generated. At Seward and Valdez in Alaska the ocean bottom is unusually steep near the shore, at places dropping off as steeply as 35°. In 1964, large blocks of alluvial deposits along the shore line were loosened by the earthquake vibrations and slid down these slopes. At Seward, the water surface was seen to drop along the shore as the waterfront sank. Farther out, the surface was seen to have a "boillike" agitation. Apparently, the sinking of alluvium near shore raised a corresponding crest farther out (Figure 8.20). At Seward and Valdez, the waves resulting from this type of source were the largest waves of the tsunami.

It has been proposed that submarine landslides on a large scale provide the complete mechanism by which many tsunamis are produced. The degree to which observed tsunamis are generated by this process in contrast to other types of sea-floor displacement (e.g. faulting) is unknown. The 1922 Chilean earthquake had an epicenter (source area) well inland, yet it produced a tsunami. The tsunami progressed southward along the coast arriving at successively later times at points approaching the epicenter. However, it is known that the ground displacements accompanying big earthquakes extend to large distances from the locus of maximum shaking, so that the sea floor may be depressed or raised even when the center of the shaken area is inland.

Regardless of how important landsliding is in causing tsunamis, the tallest sea wave ever reported was produced by a rockslide triggered by an earthquake. The 1958 Lituya Bay earthquake in southeastern Alaska caused a huge 900 meter (3000 foot) high rockslide to slip down the 40° slope of the Fairweather Mountains into the head of long, narrow Lituya Bay. An estimated 90 million tons of rock fell into the water. The resulting wave surged across a spur jutting into the bay, washing away trees and soil to a height of 520 meters (1720 feet). The wave proceeded down the bay with diminishing amplitude. Two fishermen who survived in small boats described the wave as about 50–75 feet (20 meters) high at a distance of six miles (10 kilometers) from the head of the bay. A third vessel sank, and its crew was never found.

A similar rockslide in northern Italy in 1964 produced a wave 240 meters (800 feet) high which overtopped Vajout Dam and drowned 2000 people.

Large sea waves can also be produced by volcanic action. The volcano Krakatoa, which is an island in the Sunda Strait between Java and Sumatra, erupted in a series of violent explosions in 1883. The largest explosion destroyed two-thirds of the part of the volcano above water, and produced a wave which drowned an estimated 36,000 people in Java.

8.6 THE TIDES.

Tides are waves. If Earth were a smooth, completely water-covered ellipsoid, the tides would progress around its surface as two systems of waves, one following the position of Moon and the other, Sun (see Section 3.5). The tidal bulges produced by the attractions of these bodies cannot travel around Earth because land bodies and sub-oceanic ridges interfere. As a result, the actual tides in each ocean, sea, strait or bay are determined as

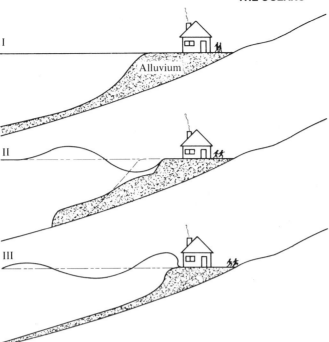

Fig. 8.20 *Submarine landslides have been proposed as a cause of some tsunamis.*

much by the reaction of the sea bottom to the movement of the ocean water as by the tidal forces.

As Sun and Moon pull the ocean waters west and east, they are deflected to the side by the coriolis effect. This tends to change the tidal oscillation into a rotary motion, counterclockwise in the northern hemisphere and clockwise in the southern hemisphere. In the middle of each such rotary circulation is a point where the rise and fall of the tides are minimal. Such places are called *amphidromic points*. Lines where the tides pass thru their maximum at the same time are called *cotidal lines* (Figure 8.21).

Where narrow openings restrict the flow of water, as in bays and seas, the tidal motions become further complicated. In the North Sea between England and the continent of Europe, the Atlantic Ocean tide reaches its peak at the west end of the English Channel earlier than at the Shetland Islands north of Scotland. As a result, two different tidal waves feed in water, and a complicated system of waves with three amphidromic points results (Figure 8.22).

Where a tide enters a channel which becomes narrower along its length, the tidal wave may increase in height as its width is constricted. As a result of this and other factors, a great variation in the heights of the tides is observed from one nearby point to another. In extreme cases, where a tide must proceed up a long narrow estuary, a *bore* is developed. This is a steep-fronted wave like a breaker on a beach which progresses as a wall of water. It results from the greater velocity of the wave in deep than in shal-

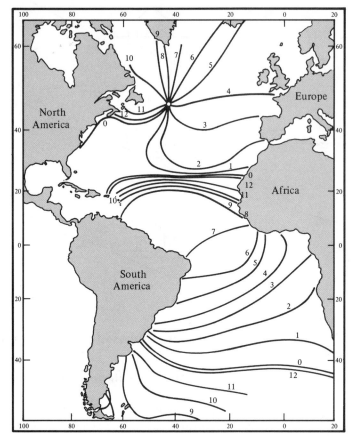

Fig. 8.21 *Cotidal map of the semidiurnal tides in the Atlantic Ocean. (After Defant,* Physical Oceanography *v. 2, p. 502. Copyright 1961, Pergamon Press, Ltd.)*

low water, so that the rear of the wave continually catches up with the leading edge. This is the same phenomenon which occurs in a shock wave produced by an aircraft travelling above the speed of sound in air. A well-known example of a bore occurs in the Bay of Fundy in eastern Canada (Figure 8.23).

Another factor which affects the tides is the tendency for bodies of water to oscillate more easily at certain frequencies than at others. Anyone who has tried to carry a broad shallow bowl of soup across a room knows that each little jog tends to set the soup oscillating back and forth in the bowl. The bigger the bowl, the longer is the natural free period of oscillation of the pool of liquid in it. It is much easier to set a body of liquid in oscillation at one of its natural free periods than at any other frequency. For a basin of length, L, and depth, D, the principal natural free period in seconds is

$$T = \frac{2L}{\sqrt{gD}} \qquad \textbf{(8.8)}$$

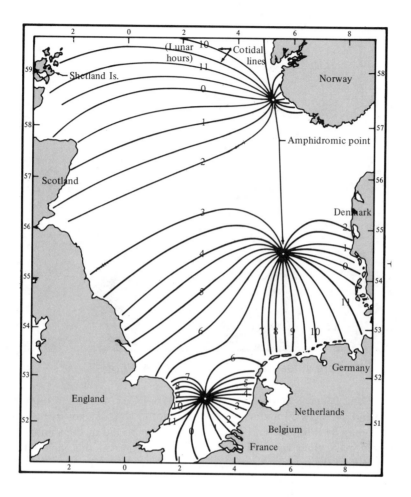

Fig. 8.22 *Cotidal map of the North Sea. (After Doodson and Warburg, 1941, Admiralty Manual of Tides, Admiralty Chart 301.)*

where g is the force of gravity. The free period of a basin 5000 kilometers wide and six kilometers deep (such as the North Atlantic Ocean) is 69 minutes. This is too short a period to be noticeably stimulated by the tides. Shallower seas, however, can be resonated, contributing to the complexity of the tidal motion.

In the case of an estuary open at one end, the effective length is doubled since the open end acts like the middle of an enclosed body. The natural free period then becomes

$$T = \frac{4L}{\sqrt{gD}}$$ **(8.9)**

Fig. 8.23   *Tidal bore in the Bay of Fundy. (E.S.S.A. Photo.)*

Long Island Sound, just east of New York City, is 145 km (90 miles) long and averages 20 meters (65 feet) in depth, which makes its free period 11 hours 34 minutes. This is near enough the period of the semidiurnal tides to make their rise and fall three times greater at the enclosed west end than at the open east end (2.3 meters compared to 77 cm).

Other forces whose frequency of application coincides with one of the natural frequencies of vibration of a water body can cause resonant oscillations. Such oscillations are called *seiches*. Lakes as well as seas, bays, and channels are subject to seiches. One of the commonest causes of seiches is earthquakes. Large earthquakes often generate long trains of waves with periods of a few seconds to a few tens of seconds. Lakes, ponds and harbors often have free periods in this range. As the earthquake tilts the ground forward and back, seiche amplitudes build up, sometimes to several meters in height. The 1755 Lisbon, Portugal earthquake caused seiches at many places between 980 and 2900 km from Lisbon. The most distant observations of water motion definitely correlatable with the earthquake were made at the Dal River in central Sweden.

Near the epicenter of the earthquake, seiches are rarely observed, possibly because the ground motions are too irregular to resonate at the natural frequencies of large bodies of water. Miniature seiches are sometimes reported in small bodies like swimming pools and goldfish bowls. Hanging objects like chandeliers are set into oscillation by a similar process of resonance.

8.7 ICE.

One of the peculiarities of pure water is that it expands on freezing. Pure ice at 0°C has a density of 0.9167 g/cc compared to 0.999 gm/cc for water. When salt water freezes, the density situation is complicated by the differentiation of the water into ice crystals, in which the salt occurs only as impurities in the crystal lattice, and into salt-enriched water in the interstices between the ice crystals. Bubbles of dissolved gases also separate out as the ice freezes, and further decrease the overall density. The density of sea ice is sometimes as low as 0.857 gm/cc.

Ocean water of 35 parts per thousand salinity begins to freeze at –1.9° C. The thickness of the ice layer is determined by the surface temperature, duration of the cold season, and the rate of air and water convection above and below the ice. Ice is a poor conductor of heat, so that generally ice on an open stretch of ocean never freezes to a thickness of over 5 meters (16 feet). The surface layer of ice on a large body of water is called *pack ice*. It protects the underlying water from the effects of wind, so that waves cannot form. Swell from distant storms may enter an ice-covered sea, and will tend to break up the ice pack until damped out by the restrictive effect of the ice layer. Wind, however, acts on an ice as well as a water surface, and pack ice is moved about by winds, often piling up on one side of a constricted sea. This produces ridges and hummocks which greatly thicken pack ice locally. Winds and currents also break pack ice into separate pieces called *floes*, and open up ice-free channels, called *leads*, between individual ice sheets.

The Arctic Ocean is largely ice-covered all year. Because the Arctic is a closed sea with only a narrow opening (Bering Strait) to the Pacific and limited access to the Atlantic, the pack ice tends to remain fixed in position, circulating from place to place as it is moved by wind and water currents. It never completely melts or breaks up (Figure 8.24).

The Antarctic ice pack, in contrast, breaks up in large part, and reforms the following year (Figure 8.25). In October it extends north almost everywhere to latitude 55° (Figure 8.26). In March it is confined to narrow shelves at places along the edge of the continent.

The largest ice blocks found in either Arctic or Antarctic waters are icebergs formed from *shelf ice*. These are large flat blocks of ice up to several hundred feet thick. They were formed over a period of many years along the shore line, where they were protected from the annual breakup to which the ice pack in the open ocean is subject. Those in the Antarctic are fed by glaciers on the land. Antarctica is covered by a layer of ice which at places reaches 4200 meters (14,000 feet) in thickness.

The largest icebergs in the Arctic are up to 30 km (20 miles) long and up to 60 meters (200 feet) thick. They originate on the north coast of Ellesmereland. Arctic ice shelves of this thickness may no longer be forming, the present thick shelf being left over from the ice ages which ended about 10,000 years ago.

Some of these icebergs have a life of many years, and have been occupied by scientific expeditions, which are carried across the Arctic Ocean as

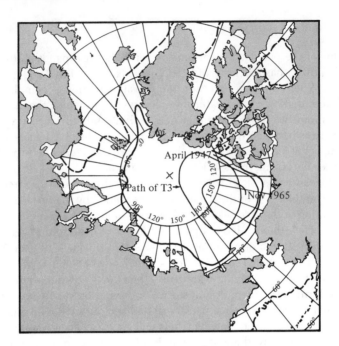

Fig. 8.24 *Typical extent of Arctic sea ice in February (dashed line), extent of permanently frozen ice pack and approximate path of Arctic Ice Island T3 from 1947 to 1965. (Data from U. S. Navy Hydrographic Office* Ice Atlas of the Northern Hemisphere *and L. Thomas, Jr.,* Natl. Geographic, *v. 128, p. 679.)*

if they were on a ship frozen in the ice pack. Some such ice islands have been given names (Figure 8.24).

Small icebergs usually come from the breaking up of glaciers (Figure 8.27). When more snow falls than melts annually, a glacier accumulates. As the snow layers thicken, the pressure compacts them. Gradually, thru recrystallization, the original snow-pack is transformed into ice. There are normally many entrapped air spaces, so that the average density is low. Large masses of ice of this sort are plastic (see Chapter 11) and flow slowly outward from the areas of greatest thickness. In mountainous areas, rivers of ice flow down valleys which they excavate for themselves, either melting as they reach lower, warmer elevations, or entering the ocean. In the North Atlantic Ocean, almost all icebergs are fragments of glaciers. In the spring and early summer they drift south, and are a serious hazard to ships. The main source of North Atlantic icebergs is Greenland.

The icecaps of Antarctica, Greenland and Ellesmereland could reasonably be considered as the uppermost layers of the rocks of Earth. From this point of view they are a precipitate from the air. Their ability to flow is greater but not fundamentally different from the ability of other solids such as rocks to flow, as will be discussed later. The pack-ice layer over the Arctic Ocean has no obvious counterpart in our experience on land except

Fig. 8.25 *Antarctic icebergs are often tabular pieces of shelf ice. (U. S. Coast Guard Official Photo.)*

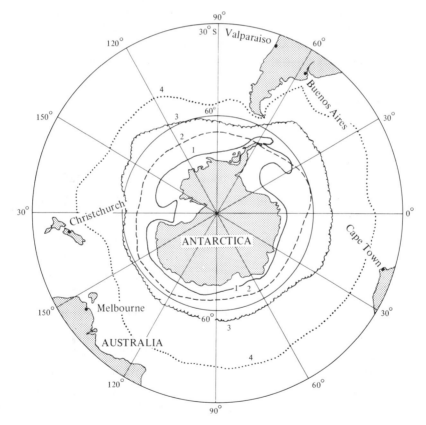

Fig. 8.26 *Extent of Antarctic pack ice. (After* Am. Geophysical Union Monograph 1.*)*

Fig. 8.27 *North Atlantic iceberg. (U. S. Coast Guard Official Photo.)*

in the case of lava flow, where a similar crust forms covering a liquid interior. But the presence of volcanic lava must mean that the interior of Earth is at least partly liquid, just as the ice-covered Arctic Ocean is liquid. To determine if there are other similarities between ocean structures and rock structures, it is necessary to examine in detail the features of solid Earth.

BIBLIOGRAPHY

Bowditch, N., 1966 *American Practical Navigator*. U. S. Govt. Printing Office. 1524 pp.

Defant A., 1958, *Ebb and Flow*. University of Michigan Press. 121 pp.

Defant, A., 1961, *Physical Oceanography*. Pergamon Press. v. I, 727 pp. v. II, 598 pp.

Gross, M. G., 1971, *Oceanography*. (2nd Edition) Charles E. Merrill Publishing Co., Inc. 160 pp.

Heezen, B. C.; M. Tharp; M. Ewing, 1959, *The Floors of the Oceans I. The North Atlantic. Geol. Soc. Am. Sp. Paper 65*, 122 pp.

King, C. A. M., 1963, *An Introduction to Oceanography*. McGraw-Hill Book Co., Inc. 337 pp.

Leet, L. D. 1948, *Causes of Catastrophe*. McGraw-Hill Book Co., Inc. 232 pp.

Mason, B., 1966, *Principles of Geochemistry*. John Wiley and Sons, Inc. 399 pp.

Miller, D. J., 1960, The Alaskan Earthquake of July 10, 1958: Giant Wave in Lituya Bay. *Bull. Seis. Soc. Am.*, v. 50, pp. 253–266.

Rankama, K.; T. G. Sahama, 1959, *Geochemistry*. University of Chicago Press. 912 pp.

Shepard, F. P., 1963, *Submarine Geology*. Harper and Row Publ. 557 pp.

Spaeth, M. G.; S. C. Berkman, 1967, The Tsunami of March 28, 1964, as Recorded at Tide Stations. *U. S. Coast and Geodetic Survey Tech. Rpt.* 33, 86 pp.

Strahler, A. N. 1963, *The Earth Sciences.* Harper and Row Publ. 681 pp.

Sverdrup, H. U.; M. W. Johnson; R. H. Fleming, 1942, *The Oceans.* Prentice-Hall, Inc. 1087 pp.

Thomas, J. L., 1965, "Scientists Ride Ice Islands on Arctic Odyssey." *Natl. Geographic,* v. 128, p. 671–691.

Turekian, K. K., 1968, *Oceans.* Prentice-Hall, Inc. 120 pp.

U. S. Coast and Geodetic Survey, 1965, "Tsunami, The Story of the Seismic Sea Wave Warning System." U. S. Govt. Printing Office. 46 pp.

9 SURFACE FEATURES OF EARTH

9.1 PRINCIPAL PHYSIOGRAPHIC PROVINCES.

If you were to stand on Moon and attempt to map the principal features of Earth's surface, the first division you would probably make would be into land and water. Water covers 70.8% of Earth's surface. Land is the exception. The continents are special areas, observable from Moon only because they rise above the sea surface. For this reason, most of your attention in mapping Earth would be given to the oceans.

If your observatory had a device which could see thru water, you could make a good classification of Earth's surface based primarily on elevation. Earth would be divided into two principal regions and three additional special divisions. The largest area would be the *ocean basins* occupying over half of the area of Earth. These are broad surfaces with only gentle undulations. They lie almost always at least 2500 meters but no more than 6000 meters beneath sea level. The average elevation is about minus 4500 meters (measured down from the sea surface).

The second principal division is a system of high plateaus, also quite flat, but broken locally by stream valleys with associated steep slopes and sometimes vertical cliffs. Most of the surface, however, is nearly level plain or gentle hills. The lowest elevations of this division are almost everywhere about 200 meters beneath sea level. The highest portions sometimes rise to several thousand meters above sea level, but generally lie below 1000 meters elevation. This division, the *continental platform*, occupies a little over 25% of Earth's surface. It embraces most of the area of the continents. A little over one fifth of it is water-covered (Figure 9.1). The ocean basins and the continental platforms constitute the two principal divisions of Earth's surface.

Between these two principal regions is a *transition zone* extending from 200 to 2500 meters depth. In contrast to the broad plateaus and the ocean basins, this region occurs either as narrow belts separating the basins and plateaus or as mid-ocean ridges lying between two basins. Occasional isolated mountains on the ocean floor make a small contribution. This transition zone occupies a little under ten percent of the total area of Earth.

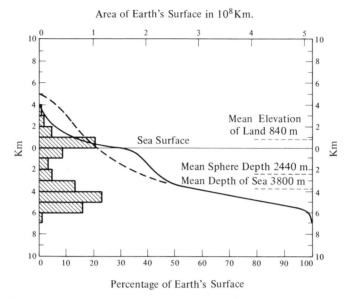

Fig. 9.1   *Hypsographic curve of solid-surface elevation (After Kossina,* Verof. Inst. Meer. *v. 9:1–70.)*

About nine percent of Earth's surface is covered by irregular systems of ridges and valleys, usually in long belts, and rising above the platform areas often by thousands of meters. These *mountain regions* tend to have steep slopes. They lie often (but not always) along the very edge of the continental platform, parallel to the transition to the ocean basins.

Less than one percent of Earth's surface consists of long narrow *troughs* deeper than 6000 meters beneath sea level. These are found most often along the edges of the ocean basins, especially where their borders are steepest and narrowest, and where mountain ridges lie along the edges of the adjoining continents. Less often these trenches, which are also called *deeps,* lie along one side of a mid-oceanic ridge.

All five of these categories can be further subdivided (Table 9.1). The subdivisions are made partly on the basis of elevation, slope and degree of irregularity of the surface and partly on the nature of the underlying rocks.

9.2 OCEANIC PROVINCES.

The main subdivisions of the ocean basins are abyssal plains, continental rises and oceanic rises (Figure 9.2 and 9.3). These subdivisions are not sharp, usually grading gradually into one another. This is due in part to the nature of the processes which form the rocks underlying the oceans and in part to the difficulty of mapping the ocean floor in detail. Large areas of the ocean floor have not yet been mapped except for widely-spaced profiles. Depths are normally determined by measuring the time it takes an echo to return to a ship at the surface. In deep water the echo is returned

Table 9.1. *Classification of surface areas.*

Area	Typical Elevation (meters)	Slopes	Commonest Rock Types
I. Ocean basin			
A. Abyssal floor	-5000 to -6000	almost flat	turbidites
B. Continental rise	-2500 to -5000	.005 - 1.4°	turbidites
C. Oceanic rise	-2500 to -5000	gentle	organic oozes
II. Ocean trough	-6000 to -10000	steep sides, flat floors	clastic sediments
III. Transition zone			
A. Continental slope	-200 to -2500	1.4-6°	sedimentary rocks
B. Mid-ocean ridges	+3000 to -2500	variable	igneous rocks
C. Seamounts	100 to -4500	flat tops, steep sides	limestone + volcanic rocks
IV. Continental platform			
A. Continental shelf	0 to -200	almost flat	clastic sediments
B. Covered shield	+1000 to 0	generally gentle	flat-lying sediments
C. Exposed shield	+1000 to 0	gentle	igneous rocks
V. Mountain belts			
A. Continental ranges	8500 to 1000	irregular to steep	folded sediments, volcanics, igneous intrusives
B. Island arcs	4000 to -200	irregular to steep	volcanic rocks

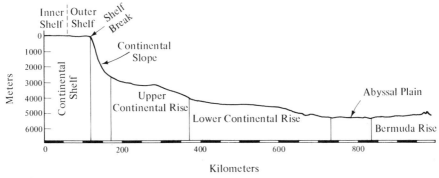

Fig. 9.2 *The continental margin province. (After Heezen et al.,* Geol. Soc. Am. Sp. Paper 65, *p. 26).*

by a large area, and details of surface irregularities are lost. Soft bottom muds do not provide a sharp reflecting surface, and sometimes echos are received from layer boundaries beneath the surface as well as from the true bottom (Figure 9.4). This can be an advantage, as it allows the structure of the underlying rocks to be mapped.

The compositions of the oceanic rocks are determined from samples. These are usually obtained by driving a tube into the bottom, closing off the end of the tube, and then pulling it back to the surface (Figure 9.5). Where the bottom is soft mud, cores up to several tens of meters long are sometimes obtained. Stiffer sands yield only short cores. Solidified rocks are not penetrated by such devices at all, and can be sampled only by knocking off a piece from an exposed corner and catching it in a net of some kind or by lowering a drill to the ocean floor. Drilling is so costly that oceanic drill-core samples are extremely scarce.

The *abyssal plains* have large areas where the surfaces slope less than 1 part in 1000. Here the surface material is largely fine silts and muds brought from the continental platforms. In the shallow water near the edge of the continents the largest storms create such large waves that they stir the sediment on the bottom. Muddy water is denser than clean water, and tends to flow downhill as a density current. Such currents commonly extend all the way to the lowest points in the oceans. Each layer of silt deposited here presumably represents the material brought by one storm current. Layers of mud of this sort are found interbedded with layers composed entirely of the skeletal remains of marine plants and animals which have died and sunk to the bottom (Figure 9.6). Most of these organisms are so small that their nature can be recognized only using a microscope. Such organic muds must accumulate very slowly, only a small fraction of a millimeter per year.

Along the continental edges of the abyssal plains are sediment-covered slopes rising gently from the ocean floors to the foot of the steeper transition zone. These belts are called the *continental rises.* The sediments here are formed in the same way as on the abyssal plains. Most are brought by density currents from the continents, but a few beds of organic remains also occur.

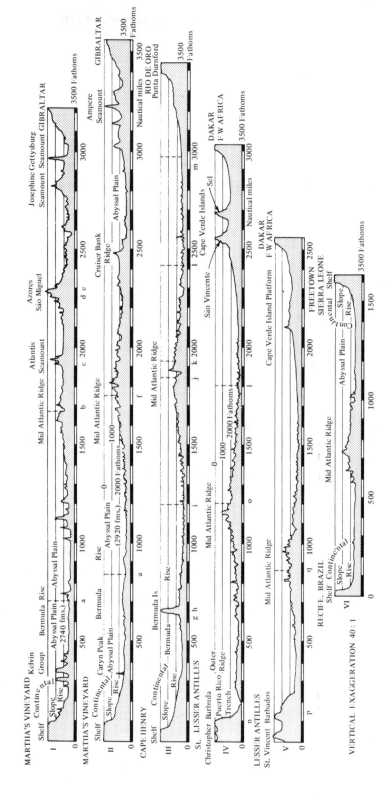

Fig. 9.3 Six trans-Atlantic topographic profiles. (Heezen et al., Geol. Soc. Am. Sp. Paper 65, plate 22).

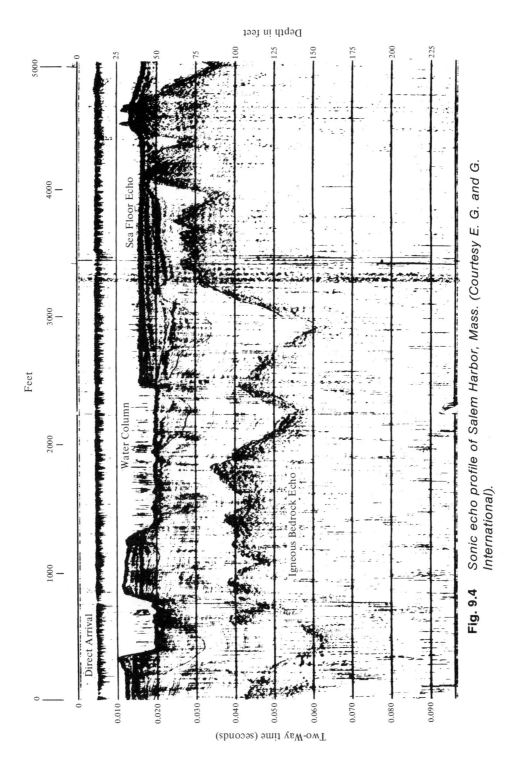

Fig. 9.4 Sonic echo profile of Salem Harbor, Mass. (Courtesy E. G. and G. International).

229

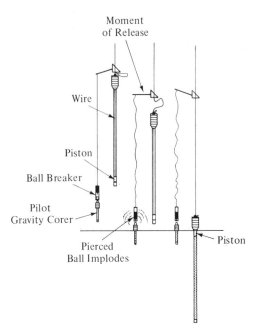

Fig. 9.5   *Principle of sampling tube for obtaining cores from the ocean floor. (After Shepard,* Submarine Geology, *p. 20. Courtesy Harper and Row Publ.)*

On the sides of the abyssal floors away from the continents there are again gently rising slopes. These are often much less smooth than the rest of the ocean basins. Hills from a few meters to a few hundred meters high are common. These *oceanic rises* are coated with organic muds. Many of them eventually grade into mid-ocean ridges, which belong to the transition-zone category, but others are isolated mounds entirely surrounded by abyssal plains. The composition of these muds depends in large part on the depth of the sea bottom. Most marine organisms have skeletons or shells composed of calcium and magnesium carbonate, but some utilize silica. In the deepest parts of the ocean, beneath 5000 meters, the carbonates are more soluble in water than at lesser depths, and only the siliceous materials accumulate. These are mixed with iron and manganese oxides deposited from solution in seawater. Fine dust of volcanic origin or wind-blown from the continents also forms an important component of this material.

In all the oceans there are isolated mountains which rise high above the floors of the ocean basins. These are called *seamounts* or *guyots* (Figure 9.7). Some are isolated peaks and others occur in groups. Although their tops usually do not at present reach sea level, they are flattened as though wave erosion had once truncated them. Often these flat summits are at depths of several thousand meters. This means either that sea level once lay that much below its present level, or the part of the ocean basin where the seamount now is once stood higher than at present. These peaks be-

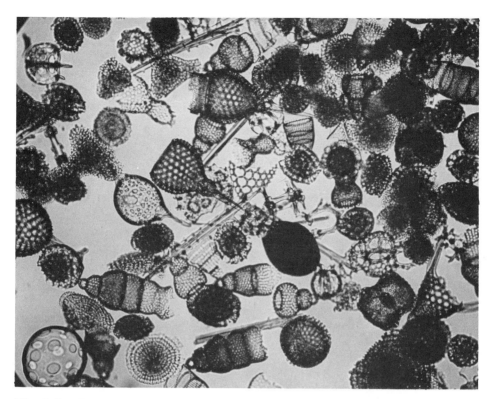

Fig. 9.6   *Deep-sea sediments are often composed almost entirely of the shells of animals. (Courtesy W. R. Riedel.)*

long, because of their elevations, to the transition zones rather than to the ocean province, in spite of their isolated locations and failure to be a transition to anything else.

Many of these seamounts are now topped by thick caps of limestone. Sometimes the limestone is exposed as a ringshaped atoll. Corals grow only in shallow water where sunlight penetrates. It is believed that the coral began to grow as a fringing reef which surrounded a volcanic island. As sea level rose or the volcano sank, the coral grew, keeping pace with the changing position of sea level (Figure 9.8). Eventually the top of the coral stood higher than the volcano. Islands in all stages of this process are found in the South Pacific Ocean. Under Eniwetok Atoll, where the United States had conducted nuclear-explosion experiments, volcanic rock has been found by drilling at a depth of 1267 meters.

The boundaries of the ocean floors are usually not sharp. As their edges are approached, the slopes steepen and the surfaces become more irregular, grading into the transition zones. The continental rises are built up by turbidity currents, which deposit their suspended muds largely over areas where the slope is less than 1 in 40 (1.4°). Over steeper slopes, the currents probably pick up as much material as they drop, depending on the frequency and speeds of the largest currents which flow. The point where the

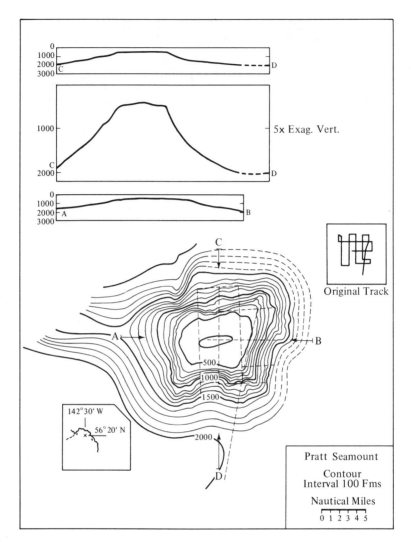

Fig. 9.7   *Profiles and contour map of Pratt Seamount, a guyot in the Gulf of Alaska. (After Menard,* Bull. Geol. Soc. Am. *v. 66, p. 1157.)*

slope reaches 1 in 40 provides a more meaningful means of defining the border of the ocean floor than defining it in terms of depth.

Along the mid-ocean ridges, the bottom topography is much more irregular and the boundary correspondingly difficult to define. The ridges themselves are very variable in height, sometimes rising to tall volcanoes which may emerge as islands, and elsewhere cresting at thousands of meters depth. The distinguishing feature of the ridge is not so much its height as the irregularity of the slopes of the mountains which constitute it. The gentle, sediment-coated slopes of the oceanic rise are replaced by steep mountainsides and cliffs. The few rock samples which have been obtained

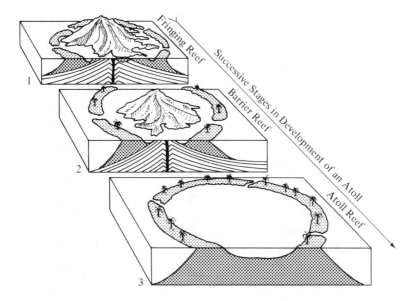

Fig. 9.8 *Three stages in the origin of a coral atoll.*

from these mountains are almost all rocks which crystallized from a hot melt, or are limestones such as are found on atolls. Organic sediment often but not always covers flat surfaces. Photographs of the bottom show that bare rock slopes are common.

The transition from ocean to continent is not everywhere gradual. Where oceanic troughs occur along the borders, the boundary is sharply defined. There is no sediment-covered rise as the continent is approached. The ocean basin along the deep may retain the hilly nature that it has along the mid-ocean ridges right up to the point where the floor dips down into the trench. The Puerto Rico trench at the west end of profile IV of Figure 9.3 is an example of this. The trenches are floored with layers of sediment carried from the adjoining land areas.

9.3 THE CONTINENTAL PLATFORMS.

The continental platforms are easier to subdivide because they can be examined directly. The principal divisions are based on the nature of the exposed rocks. In some regions, crystalline material is found immediately beneath a thin veneer of soil and partially weathered rocks. Individual crystals of the separate minerals of which the rock is composed are recognizable (Figure 9.9). These crystals have the appearance of having formed by solidification from the molten state. Laboratory studies show that the temperature at which this would have occurred was of the order of 1000°C or more. It is clear that these rocks were formed beneath the sur-

Fig. 9.9 *An igneous rock is typically an assemblage of crystals of several different minerals. (Courtesy C. E. Nehru.)*

face either by recrystallization of other rocks or directly from the liquid state. Areas where such rocks are found immediately beneath the soil veneer are called *shields*. In North America, the Canadian shield includes most of the provinces of Ontario and Quebec and parts of the states of Michigan, Wisconsin and Minnesota. Figure 9.10 shows the principal shield areas of the world.

Large parts of the shield areas consist of partially recrystallized rocks which obviously once were composed of beds of sediment. These have been deeply buried and deformed so that they have partially recrystallized. Such rocks are said to be metamorphic. At rare places there are patches of sedimentary rocks which have been relatively little disturbed except for compaction such as might have been caused by deep burial. Elsewhere, the process of recrystallization has proceeded so far that the original nature of the material is no longer recognizable. The shield rocks are what one would expect to find if erosion progressed for a long period of time, exposing rocks which once were deeply buried. Their relatively flat surfaces can be explained by the tendency of erosion to reduce any land mass to an even, low elevation.

Where crystalline rocks are not exposed at the surface, they can usually be found by drilling. Depth to the crystalline basement is variable, rarely exceeding five thousand meters (16,000 feet). The surface rocks in these

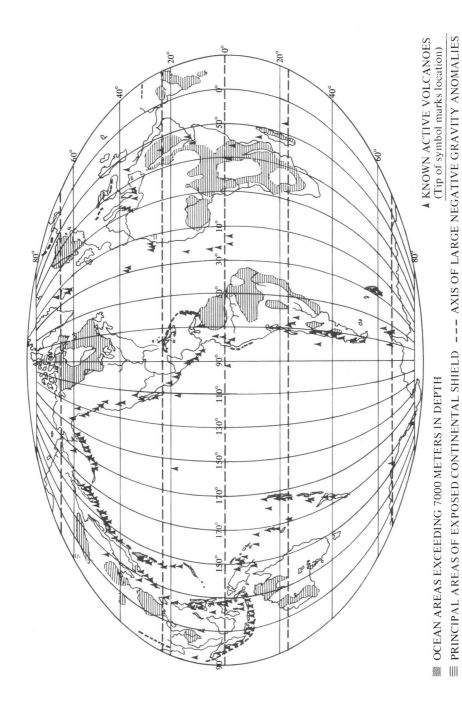

■ OCEAN AREAS EXCEEDING 7000 METERS IN DEPTH

▤ PRINCIPAL AREAS OF EXPOSED CONTINENTAL SHIELD — — — AXIS OF LARGE NEGATIVE GRAVITY ANOMALIES

▲ KNOWN ACTIVE VOLCANOES
(Tip of symbol marks location)

Fig. 9.10 *World map of principal tectonic features.*

areas are composed of three main classes. The first class consists of fragments of crystalline rocks of the types found in the shield areas. The second is made up of fine clays composed of minerals into which crystalline rocks decompose. The third type of material has crystallized from aqueous solution with or without the aid of living organisms. Limestone is the commonest example. The predominant characteristic of all these materials is that they have been carried by wind or water currents to their present locations and deposited there. Such rocks are called sedimentary and are found usually in layers, one on top of the other. The platform areas of Earth are underlain largely by sequences of such layers of varying thickness, usually in the nearly horizontal attitude in which they were deposited (Figure 9.11). Studies of the distribution of present rock types indicate that sedimentary formations once extended over much of the areas where shield rocks are now exposed. The distinction between exposed and covered shield is a result primarily of the degree to which the processes of disintegration of the rock and its removal by erosion have succeeded in removing the sedimentary veneer.

Most of these sediments were deposited by water in shallow seas. Deposits of this type can be seen forming today along the shores of the continents wherever rivers enter the ocean (Figure 9.12). The parts of the continental platforms beneath the ocean are called the continental shelves. They are floored by sediments, often thousands of meters in thickness, carried here from the exposed continents and dropped by river currents when their flow stops on reaching sea level, or carried to their present locations by ocean currents from the main depositional regions opposite river mouths (Figure 9.13).

The presence on the exposed continents of sediments similar to those forming today on the continental shelves shows that large parts of the continents were once covered by the sea. The presence of these marine deposits on what is now dry land means that one or more of the following three alternatives must have occurred: (1) the volume of ocean water has somehow been reduced; (2) the oceans have receded because their floors have sunk, leaving the continents bare; or (3) the continents have been raised.

The continental shelf and covered-shield areas are continuous with one another, and are considered separate units in this classification mainly because the presence of the oceans provides such an easily marked boundary between them. They also differ in that beneath sea level accumulation of deposits generally is occurring at present, whereas above sea level wind and streams are generally removing material. The boundary between areas of accumulation and of erosion often does not exactly follow the seashore. In the case of the Mississippi delta the shifting course of the river builds the pile of sediment above the sea level first in one location then in another (Figure 9.12).

9.4 MOUNTAINS.

Not all sedimentary rocks are found as flat-lying beds in or near the attitudes in which they were originally deposited. Often they are tilted

Fig. 9.11 Flat-lying sedimentary rocks cover shield-type crystalline rocks at many places, in this case the Grand Canyon of the Colorado River, U.S.A. (Courtesy U. S. Geological Survey. Photo by L. F. Noble.)

Fig. 9.12 *Growth of the exposed delta of the Mississippi River. Submerged deposits coat the surrounding parts of the shelf. (Shepard,* Bull. Am. As. Petr. Geol. *v. 40, p. 2541.)*

steeply or bent into folds (Figure 9.14). This occurs most commonly in regions of irregular topography. The mountain belts of the continents appear to be regions where the original rocks have been greatly disturbed. Often such rocks have also been intruded by molten material from Earth's interior. Volcanoes form where this molten material escapes to the surface. The characteristic features of these mountain belts are their higher elevations than surrounding areas, rapid erosion occurring as a result of this elevation, the prevalence of present or former volcanism and the formation beneath the surface of new crystalline rocks.

The mountain belts can be further subdivided on the basis of the stage of development they are in. Mountains such as those along the Pacific coast of North America are still being raised. Here the height is great because the mountain creation process has a good lead on the destructive processes of erosion (Figure 9.15).

Mountains such as the Appalachians in Eastern North America represent a later stage. Here erosion is proceeding at least as fast as uplift. Great contrasts in elevation are rare. The patterns of the exposed sedimentary rocks show that large quantities of material which once covered the areas have been removed. In many places crystalline rocks are exposed. These rocks do not differ greatly from the rocks of the shield areas. There is no sharp distinction between the old mountain areas and the shields. It appears that if erosion has time to operate, all exposed continental areas will be converted to shields in time.

One other type of area deserves a special place in the classification. Occasionally mountain ranges are found along the edges of the continental

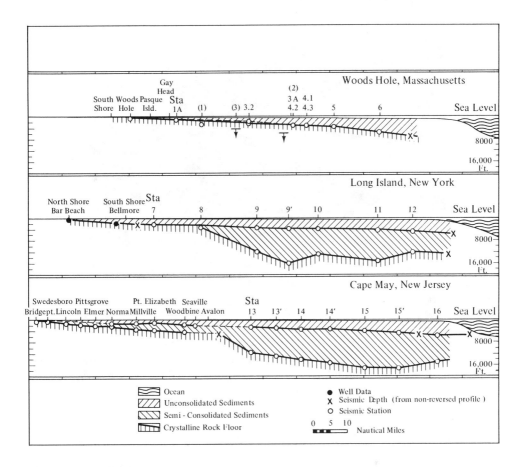

Fig. 9.13 *Three cross-sections of the Atlantic shelf and coastal plain showing the great accumulation of sediments. (Ewing et al.,* Bull. Geol. Soc. Am. *v. 61, p. 884.)*

platforms but separated from the main part of the continents by shallow seas. These are the *island-arc mountain ranges.* Japan is an example. Such mountains differ from the mountains of the continents proper in many details. Often they are much poorer in sedimentary rocks and richer in volcanoes and volcanic rocks. They are generally in a very active stage of formation. Oceanic troughs are most commonly found along the edges of continents which are bordered by such mountains. Some ranges of this sort are separated from the nearest continent by an ocean basin. New Zealand is such a mountain range. Others are mere strings of volcanic islands such as the Mariana Islands of the Western Pacific.

Such mountains are not easy to distinguish from mid-ocean ridges on the basis of their elevation, slopes or composition. They are classed as continental because of the nature of the deformational processes observed to be occurring in them. They are found usually in arcuate groups parallel to the

Fig. 9.14 *Exposed ends of folded sedimentary layers near Hancock, Mary-land, U.S.A., the remainder of which have been eroded away. (Photo by C. D. Walcott. Courtesy U. S. Geological Survey.)*

edge of a continent. In a few cases, such as the Aleutian Islands of Alaska, the island arc is an extension of a continental mountain range. These arcs may represent an early stage in the formation of continents. Their charac-teristics will be discussed in more detail later when the origin and evolu-tion of continents is taken up.

9.5 EROSION OF THE CONTINENTS.

The classification of the continents presented above divides Earth's surface into areas partly according to elevation and partly to whether they are pri-marily areas of accumulation of sediments or of denudation. Continental mountains and exposed shields are characterized by denudation. Continen-tal shelves and rises are areas of sedimentation. Ocean floors, oceanic rises and continental slopes are areas where neither process is proceeding rap-idly. Covered shields have alternated between deposition and erosion, and belong logically to the stable, slowly changing group.

The relationships of these areas can be shown by a graph. If the fraction of Earth's surface above a given elevation is plotted against elevation, the solid line of Figure 9.1 is obtained. This is called a *hypsographic profile*. The predominance of two elevations represented by the ocean floors and the continental plateaus can be seen by plotting the percentages of the sur-face within successive elevation ranges as a bar chart, shown at the left in Figure 9.1.

Remembering that the rocks of the covered shield and the continental shelf, slope and rise are largely sedimentary, and noting that the thickness is greatest under the continental shelf (Figure 9.13), the continental plateau can be seen to be due to the action of erosion and sedimentation.

Fig. 9.15 *Wave-cut terraces of the Palos Verdes Hills, southern California were formed at sea level. They show successive uplifts of the land at a faster rate than erosion can destroy it. (R. C. Frampton and J. S. Shelton photo.)*

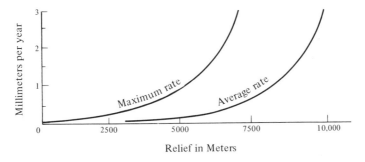

Relief in Meters

Fig. 9.16 *Dependence of rate of denudation on relief. (After Schumm, U. S. Geol. Surv. Prof. Paper 454H.)*

The great thicknesses of sediment on the shelves are found in this location because most of the material removed from the continents is carried downhill only to the edge of the oceans. Transport of material tends to be more rapid in streams on land than in the currents in the oceans. Rivers are pulled downhill with negligible resistance from the air they must displace as they fill the ocean basins. The density currents which carry sediment from the shelves to the ocean basins are driven by the contrast in density between muddy water and clear water. This is much smaller than the contrast between water and air, and the currents are slower for the same slope. Slower currents are less able to carry suspended solids, and the larger rock fragments are dropped as sedimentary deposits sooner. The shoulder at sea level in the hypsographic profile can be considered as a special effect of the presence of the oceans. Without erosion and sedimentation, the profile would presumably be smoother, rising more like the dashed curve in Figure 9.1.

On land, the rate of erosion depends on the steepness of the land surface. On the shield areas, both exposed and covered, where there are largely gentle slopes and a narrow range between maximum and minimum elevations, the erosion rate is slow. In the mountainous areas, where slopes are steep, it is high. *Relief*, the difference in elevation between the tops of the hills and the bottoms of the valleys, is a measure of the degree of irregularity of a surface. Rates of erosion depend on relief and on many other factors, including type of rock, rainfall, extent and nature of the vegetative cover, and the amount of snow which accumulates in the winter. Relief, however, is the major factor in how rapidly material can be carried away (Figure 9.16).

The overall rate of erosion in any area is measured by estimating the amount of material carried out of it by each stream flowing across its border, including the wind as one stream. Samples must be taken in all seasons, particularly during periods of peak flow (floods), because the ability of a stream to transport sediment increases at between the third and fourth power of the velocity of flow. The amount of material carried in solution as well as the fragmental matter carried in suspension must be included. All substances will dissolve a little in water. Material in solution constitutes about three tenths of the total matter removed from the continents.

About 8×10^{12} kilograms of material is carried off the continents every year. This sounds like a lot, but assuming that the average density of the rock from which it was taken was 2.67 gm/cc, this is only 3 cubic kilometers. The total area of the continents is 1.5×10^8 square kilometers, so about 2×10^8 kilometers or .02 millimeters (.0008 inches) is removed on the average per year. Small though this figure is, if it persists for very long, the continents will soon be worn completely away. Their average elevation above sea level is 840 meters. At .02 mm/year, it will take a little over 40,000,000 years to reduce all the land to sea level.

The oldest rocks are a little under 4×10^9 years old. There has been time in Earth's history to wear the continents completely away around 100 times. Some process must renew them as fast as they are destroyed. The most obvious places where new land is being created are the mountain ranges. Study of the exposed rocks in older, less prominent mountains (e.g. the Appalachians) shows that the total amount of material removed here is generally much greater than in the mountains which are currently tall. Old shield areas look remarkably like the most deeply eroded sections of the older mountain ranges. It is likely that every part of the continents has suffered extensive erosion at some time. Their history, from this point of view, is one of constant renewal by the formation of new mountain ranges, balanced by continual destruction by erosion.

9.6 COMPOSITION OF EARTH'S CRUST.

The exposed rocks of Earth's crust can be divided into two main groups. The first group is made up of rocks which have formed by crystallization under conditions of high temperature and often high pressure (Figure 9.9). These are call *igneous* rocks. The second group is composed of a variety of rocks formed under near-surface conditions of low temperature and pressure. Many such rocks are simply aggregates of fragments of igneous rocks (Figure 9.17). Others are composed of igneous mineral grains which have been altered by reaction with air and water in the surface environment to form new minerals (Figure 9.18). Finally, there are chemical precipitates formed sometimes with and sometimes without the aid of living organisms (Figure 9.19). All three are called *sedimentary* rocks.

Fig. 9.17 *Enlargement (29 times) of a sandstone taken thru a microscope using polarized light showing quartz grains cemented by chalcedony (a variety of silica). (Courtesy U. S. Geological Survey. Photo by W. R. Hansen.)*

The igneous rocks can be divided into subgroups on the basis of their chemical and mineralogical composition and on the size of the individual crystals of which they are composed. Laboratory experiments show that the crystal size tends to be small for rocks which formed quickly and large for rocks which formed slowly. Lavas which have solidified at Earth's surface are fine grained. The edges of small intrusive bodies which were chilled by the rocks they penetrated are also fine grained (Figure 9.20). The centers of large bodies of intrusive rocks are coarse grained. Grain size is also affected by the water content of the rock, abundant water encouraging the formation of large crystals.

Much can be learned of the detailed history of Earth's crust by studying small differences in composition and crystal size of the rocks. For the purposes of the overview presented here, it will be sufficient to recognize three principal subdivisions of the igneous rocks based on composition alone, granite, basalt and ultrabasic (Table 9.2), and to ignore the finer subdivisions. The average chemical composition of Earth's crust is also shown in Table 9.2. The continental composition is obtained by averaging over one thousand individual samples. By far the commonest elements are oxygen and silicon. Not only is oxygen the commonest element by mass, but it

Table 9.2. *Average composition of common rocks in parts per thousand. Unlisted elements constitute less than one part per thousand all columns (after Mason,* Principles of Geochemistry *and Bullard p. 57–137 in* The Earth as a Planet*).*

Element	A typical granite	Average continental rock	A typical basalt (diabase)	A typical ultrabasic (dunite)
Oxygen	485	466	449	439
Silicon	340	277	246	196
Aluminum	74	81	79	5
Iron	14	50	78	64
Calcium	10	36	78	5
Sodium	25	28	15	1
Potassium	45	26	5	----
Magnesium	2	21	40	287
Titanium	1.5	4	6	----
Hydrogen	0.4	1.4	0.6	----
Phosphorus	0.4	1	0.65	----
Manganese	0.2	1	1	----
Barium	1	0.4	0.2	----
Sulfur	0.2	0.3	0.1	----
Carbon	0.2	0.2	0.1	----
Lead	.049	0.013	.008	----
Thorium	.05	.0072	.0024	----
Uranium	.0037	.0018	.00052	----

also is the largest atom in physical size (Table 9.3). The crust has been likened to a great slab of oxygen in which the other elements are packed in the cracks between the oxygen atoms.

Oxygen is such a dominant element in igneous rocks that the composition of different rock types can be described by stating the percentages of different metal-oxygen compounds they contain. Table 9.4 lists typical compositions of the three principal contrasting types of igneous rocks. The *granite* family of rocks is characterized by a very high silica (SiO_2) content, and relatively large amounts of alumina (Al_2O_3), soda (Na_2O), and potash (K_2O). It is found exclusively on the continents. The *basalt* family, in contrast, is poorer in silica and richer in iron and magnesium oxides (FeO, Fe_2O_3, MgO). Basalts are generally richer in lime (CaO) and titania (TiO_2), poorer in soda and much poorer in potash than granites. They are found both in the continents and in the ocean basins. The basaltic rocks are generally heavier and darker in color due to the presence of more iron and magnesium-rich minerals.

Rocks of the third category are much rarer than granites and basalts. Occasionally exposures of igneous rocks which are much richer in magnesium and poorer in silica and alumina than basalts are found. The occurrence of these rocks is such that they appear to have formed only at exceptionally great depths within the crust. They are rarely exposed by erosion.

Table 9.3 *Relative size and abundance by volume of eight common elements in Earth's crust (after Mason,* Principles of Geochemistry*).*

Element	Weight per cent	Atomic radius $\times 10^8$cm	Volume per cent
Oxygen	46.6	1.40	93.8
Silicon	27.7	0.42	0.86
Aluminum	8.1	0.51	0.47
Iron	5.0	0.74	0.43
Magnesium	2.1	0.66	0.29
Calcium	3.6	0.99	1.03
Sodium	2.8	0.97	1.32
Potassium	2.6	1.33	1.83

Table 9.4 *Typical per cent compositions of principal components of igneous rock types (after Clark,* Geol. Soc. Am. Memoir *97 and Mason,* Meteorites*).*

Component	Granite	Basalt	Ultrabasic (dunite)	Stony meteorite
Silica	70.2	48.8	40.5	38.0
Alumina	14.5	14.0	0.9	2.5
Iron Oxide	3.4	13.4	8.4	25.0°
Magnesia	0.9	6.7	46.3	23.8
Lime	2.0	9.4	0.7	2.0
Soda	3.5	2.6	0.1	1.0
Potash	4.1	0.7	0.04	0.2
Titania	0.4	2.2	0.02	0.1

°half as iron sulfide and free iron

They are called *ultrabasic* rocks. The particular variety whose composition is listed in Tables 9.2 and 9.4 is called a dunite, and is composed almost entirely of the single mineral variety, olivine, an iron-magnesium silicate. The ultrabasic rocks are also poorer in soda, potash, and titania than either the granitic or basaltic rocks. They are more similar in composition to the stony meteorites than to the common rocks of the crust.

When any of these rocks is exposed to air and water at Earth's surface, changes occur in most of the minerals. The particular minerals in which any assemblage of chemical elements will occur depends on the temperature and pressure of the environment. If the environment is changed, the elements will tend to regroup themselves into that suite of minerals which is least subject to change under the new conditions. This obvious trend is called the *environmental law*. When a rock is exposed at Earth's surface, both temperature and pressure are much less than within Earth. In general, it takes more energy to pack elements into compact crystal lattices than into open lattices. Hence, under surface conditions, new, less dense minerals tend to form.

The presence of plentiful water has a profound affect on the mineral species found at Earth's surface. Under surface conditions, many of the metallic oxides regroup to include water in their crystal lattices. Aluminum, iron and potassium oxides have this tendency. The minerals which result are called *clays*. Sedimentary rocks composed largely of clay are called *shales*. Shales are the commonest of all the sedimentary rocks (Figure 9.18).

The lime, magnesia, and soda in rocks are relatively easily dissolved, and most of these elements are carried to the oceans in solution. Sodium is the principal metallic element dissolved in seawater. Lime and magnesia tend to be removed by organisms living in the sea or by direct precipitation from seawater in combination with carbon dioxide. The resulting rocks are composed of calcium and magnesium carbonate and are called *limestone*[°] (Figure 9.19).

Silica is the least affected by weathering of the igneous rock components. Silica is stable under both surface conditions and deep burial. It occurs by itself in igneous rocks in the form of the mineral quartz. When the rest of an igneous rock decomposes, the quartz is often left unchanged. When rocks are eroded, fragments of quartz are commonly one of the principal components carried away. Because the fragments are larger and commonly denser than the clay minerals, the quartz is carried less far by streams and is dropped first as sand, which, when cemented together, becomes *sandstone* (Figures 9.17, 9.21). Silica is also a component of most clay minerals, and a little silica is carried in solution by water, from which it can be deposited as a cement between the grains of sedimentary rocks or absorbed by plants and animals to form parts of skeletons and shells.

The sediments form a surface veneer on top of the main mass of igneous rocks. The latter have been found by drilling and by seismic surveys (see next chapter) everywhere they have been sought on Earth's surface. Although sedimentary thicknesses reach several tens of kilometers at a few places, usually they are limited to a few kilometers out of the total 6370 km radius of Earth. Since sedimentary rocks are derived ultimately from the destruction of igneous rocks, their overall average chemical composition must be the same as that of the exposed continental igneous rocks.

The mixture of crystalline rocks which constitute the bulk of the continental platform is often called *sial*, short for silica and alumina, its principal components. The basaltic and ultrabasic rocks which underlie the thin sedimentary veneer on the ocean floors and mid-ocean ridges are called *sima*, because of their richness in silica and magnesia.

In addition to the common elements, traces of almost all the other elements are found in the rocks. These occur as impurities in all rocks. Usu-

---

[°]The term limestone as used here refers to all calcium and magnesium carbonates. Some authors restrict the term to rocks composed of the mineral calcite, $CaCO_3$. Rocks composed of the mineral dolomite, $(CaMg)CO_3$, are called dolomites.

Fig. 9.18 *The more easily decomposed igneous minerals weather into clays which form shales, the commonest of the sedimentary rocks. (Courtesy U. S. Geological Survey. Photo by F. C. Calkins.)*

ally they will be found associated with the common elements whose chemical nature is most similar to their own. Thus manganese, nickel and cobalt tend to occur with iron. Strontium accompanies calcium; and thorium behaves somewhat like the metal zirconium. Occasionally, due to their special chemical properties, the rarer elements will become concentrated in deposits due to some peculiar environmental condition. Such ore deposits are of great economic importance.

9.7 ISOSTASY.

The most distinctive feature of the continental platforms is that they stand higher than the ocean basins. There is a greater volume of rock between the surface and the center of Earth under each square kilometer of the continents than under the same area of the oceans. It would be expected, therefore, that the gravitational attraction would be greater on the continents than over the ocean basins. The higher the surface elevation, the more rock there should be directly beneath the observer, and the greater the resulting pull of gravity would be expected to be. At least this is the prediction of Newton's universal law of gravitation, which indicates that the force per unit mass should increase with the attracting mass. Actually, however, gravity is not greater on the continents. As a result of this, we are led to an important discovery about the distribution of rocks.

Fig. 9.19 *This limestone reef is composed of carbonates extracted from seawater by corals. The corals grow from a bed of coral and shell fragments. (Courtesy R. F. Schmalz.)*

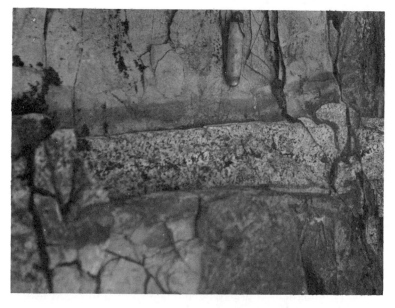

Fig. 9.20 *Crystal size in an igneous rock is smaller near the contact with the cold rocks which the lava intruded. (Courtesy U. S. Geological Survey. Photo by W. T. Holser.)*

The force of gravity on Earth is the resultant of two components: gravitational attraction and centrifugal effect. The gravitational component is due largely to Earth itself, modified slightly by periodic terms due to the tidal attractions of Moon and Sun. The effects of all other bodies are too small to detect. Neglecting the small contributions to gravity from bodies other than Earth, gravity can be expressed mathematically as (Figure 9.22):

$$g = \sum_n \frac{GM_n c_n}{R_n^2} - W_c R_c c_c \qquad (9.1)$$

where G is the universal constant of gravitation; M_n is the mass of any small part of Earth at point N a distance R_n from the point of measurement, P; c_n is the component fraction of the gravitational pull of the mass at N which is in the direction of the resultant total pull of gravity; W_c is the angular velocity of M_n around Earth's axis of rotation; R_c is the distance of P from Earth's axis of rotation, and c_c is the component of the centrifugal effect in the direction of the total pull.° $\sum$ means to sum the gravitational pulls of all the masses of which Earth is composed. The first term of Equation 9.1 is the purely gravitational term (compare Equation 5.2). The second term provides for the effect of centrifugal acceleration.

If Earth were a perfect ellipsoid composed of layers, each of the same density throughout, with no surface variations in elevation, then gravity would vary only with latitude, and could be calculated from Equation 9.1 and a knowledge of the density variation with depth. To a close approximation at sea level, gravity in newtons would be

$$g_p = 9.78049 \,(1 + .0052884s_1^2 - .0000059s_2^2) \qquad (9.2)$$

where s_1 and s_2 are factors which depend on the latitude.

Earth is not the perfect ellipsoid assumed in obtaining Equation 9.2. It has mountains, valleys, plateaus and ocean basins on its surface. Also the rocks vary in density from place to place. In this more complex case, Equation 9.1 can still be used to calculate gravity provided one knows the locations of the mountains and valleys and the densities of the rocks everywhere. The calculation is tedious; but with the aid of big digital computers, it is not difficult. The labor is greatly eased if it is broken into four parts

$$g = g_p + g_m - g_v + g_d \qquad (9.3)$$

° In trigonometric terminology, c_n is cos A_n and c_c is the cos A_c, so that (9.1) becomes

$$g = \sum_n \frac{GM_n \cos A_n}{R_n^2} - W_c R_c \cos A_c \qquad (9.1A)$$

° Similarly s_1 is sin A_c and s_2 is sin $2A_c$, so that (9.2) can be written

$$g_p = 9.70849 \,(1 + .0052884 \sin^2 A_c - .0000059 \sin^2 2A_c) \qquad (9.2A)$$

Note that A_c, the geographic latitude, is not exactly the direction from a point on the surface of Earth to the center, but is the angle a vertical line at the surface makes with the plane of the equator (Figure 9.22).

where g_p (Equation 9.2) is the theoretical gravity that would exist at the point P if Earth were the ideal ellipsoid mentioned above, g_m is the extra pull of any mountains or plateaus which rise above the theoretical ellipsoid surface; g_v is the lack of pull due to the rocks missing in any valleys beneath this surface; and g_d is the effect of any density variations within Earth other than the general variation with depth for which allowance is made in Equation 9.2. Gravity, g, can be measured; and g_p, g_m and g_v calculated using Equation 9.1. Therefore,

$$g_d = g - g_p - g_m + g_v \qquad\qquad (9.4)$$

In practice, g must be measured at the actual Earth surface, which is sometimes at sea level and sometimes above it. At sea, gravity can be measured, with some difficulty, on a moving ship at the surface, standing still on the ocean bottom, or in a submarine at any depth. For convenience in studying g_d, it is useful to eliminate the effect of changing elevation from one place to another. This is done by adjusting g to compensate for the decrease in gravity with elevation. The value of g used in Equation 9.4 is, therefore, the measured value plus the calculated decrease in gravity due to the elevation of the point of observation. This is not necessarily the value that would be measured at sea level, because, if the measurement were made underground, the overlying rocks would pull upwards instead of downwards. With g defined in this way, g_d is a measure of the departures of the density of the rocks from a uniform condition. It is called the *gravity anomaly*—the departure of gravity from the expected value.

Figure 9.23 is a contour map showing the variation of gravity anomaly in the United States calculated using Equation 9.4. It is obvious that there is a close relation between gravity and surface elevation. Gravity is low over the high, mountainous parts of the western states and along the Appalachian mountains. Similar results are found in most parts of the world. Under the oceans the gravity anomaly is high by almost exactly the amount of g_v, the term added to allow for the lack of rock filling the ocean basins.

What this pattern suggests is that the average density of the rocks underlying high mountains is less than that of rocks under the shield and covered-shield areas, and the density of the rocks under the ocean basins is greater than the density of the rocks under the continents. This comes as no surprise. It was noted previously that, aside from the relatively thin veneer of sediments, the continents are made up of a mixture of granitic and basaltic rocks, but the oceans are underlain only by basalts. Basalt is denser than granite.

The question then arises, can we find a density distribution in Earth which will explain all gravity anomalies? The problem is very complex because, as Equation 9.1 indicates, the gravity at each point on the surface is

Fig. 9.21 *In Monument Valley, Arizona, U.S.A., a layer of sandstone caps each butte. Because of its resistance to erosion, sandstones often stand with steeper cliffs than underlying shales can maintain.*

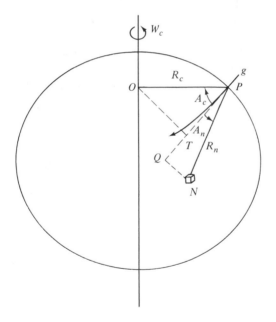

Fig. 9.22

affected by the density (amount of mass) at every point within Earth. With only a finite number of measurements of gravity at and above Earth's surface, there is no unique solution to this problem. The range of possible density variations can be narrowed by remembering that most of the anomaly at any observing point must be due to the rocks near that point,

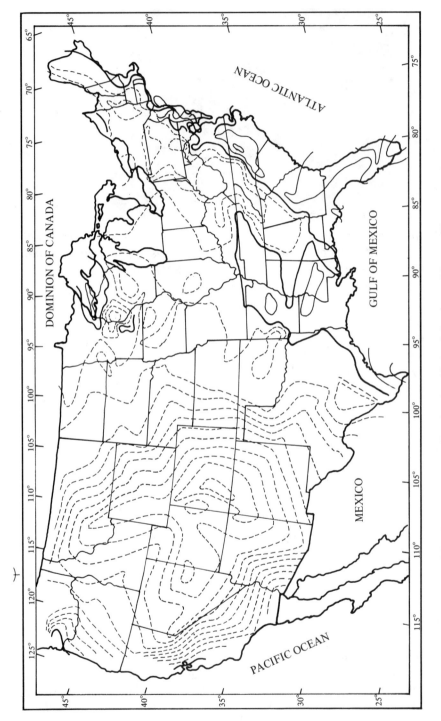

Fig. 9.23 *Gravity anomalies in the United States with no allowance for variations of the density of the buried rocks. Units are dynes / gm. (After Bowie, U.S.C.G.S., Sp. Pub. 40.)*

because their effect dies off with the square of distance. Thus the low values of gravity at high elevations are due to low densities of the immediately underlying rocks.

Many attempts have been made to find distributions of density which will produce as nearly as possible the observed gravity anomalies all over the world. One of the simplest approximations was proposed by J. H. Pratt in 1859, and was tested by the United States Coast and Geodetic Survey thru the work of J. F. Hayford and W. Bowie. The Pratt-Hayford theory postulates that anomalies can be explained by a deficiency in density of an amount, d_n, extending to a depth, H, beneath the surface (Figure 9.24) in such a manner that

$$d_n H = dE_n \tag{9.5}$$

where d is the normal average density of the surface rocks for a column extending to sea level and E_n is the elevation above sea level of the point beneath which there is the density deficiency, d_n. Hayford and Bowie found that a large portion of the anomalies observed in the United States (Figure 9.23) could be explained in this fashion. If the anomaly, g_i, expected using $d = 2.67$ gm/cc and $H = 113.7$ km is subtracted from the observed g_d and the remaining anomaly contoured for the United States, Figure 9.25 is obtained. The amount, g_i, subtracted from g_a is called the *isostatic correction*. The anomaly, g_{di}, shown in Figure 9.25 is called the *isostatic anomaly*.

$$g_{di} = g_d - g_i = g - g_p - g_m + g_v - g_i \tag{9.6}$$

There is no clear relation between isostatic anomalies and surface elevation. The average isostatic anomaly found in this way is about one-third of the anomaly that would be found if no allowance were made for the distribution of rocks of differing densities.

Some notably large anomalies remain in Figure 9.25, as in the northwest corner around Seattle, Washington and in the central portion along the Minnesota-Wisconsin border. These anomalies are so sharp that their causes must be within a few tens of kilometers of the surface. Removal of the isostatic correction from the gravity anomaly emphasizes local anomalies of this sort. Local anomalies are very useful in locating buried rock bodies which may be associated with ore deposits or petroleum reservoirs.

The Pratt-Hayford method of calculating the isostatic anomaly is based on the concept that beneath the depth H below sea level, pressure in Earth is everywhere the same at any one depth. In an anomalous area, the deficiency of mass to the depth H is exactly equal to the amount of material in the column of length E_n beneath this depth (Figure 9.24). The bottom of a column of length $E_n + H$ is at a constant depth, H, below sea level, and is at a constant pressure, dH, (neglecting any variation in gravity with depth in the Earth). H is called the *depth of compensation*.

Pratt's density arrangement is not the only one which will explain the observed gravity anomalies. It is not even the best one. G. B. Airy in 1855

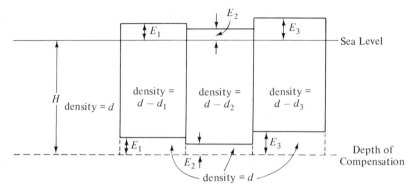

Fig. 9.24 *The Pratt-Hayford density postulate.*

suggested that gravity anomalies could be explained by a light surface layer of varying thickness, such that the length of the block beneath a standard depth was proportional to the height of the surface above sea level (Figure 9.26)

$$d_1 E_n = (d_2 - d_1) L_n \tag{9.7}$$

where d_1 and d_2 are the densities in the upper and lower layers, E_n is surface elevation and L_n is the length the root of the upper layer extending into the lower. For oceanic areas,

$$d_w E_w = (d_2 - d_1) L_w \tag{9.8}$$

where d_w = seawater density, E_w is ocean depth, and L_w is the height of the lower end of the block above the standard depth.

The Airy hypothesis is equivalent to assuming that there is a light surface layer of material floating in a denser substratum. It results in the same equal-pressure condition postulated by Pratt and Hayford, but the depth to the mass deficiency compensating for the excess surface rocks is different. The isostatic anomalies which remain after subtracting the effect of an Airy density distribution are slightly less than those obtained by Hayford and Bowie using the Pratt assumptions. From this, it appears that the reason for the high elevations of the continents is that they are composed in large part of lightweight granitic rocks which are missing under the oceans.

Even smaller isostatic anomalies can be found by using more complicated density distributions. An example is the variation of density with depth and surface elevation shown in Figure 9.27. The next step beyond this is to postulate variations which are not systematic with surface elevation and to take into account data on variations from place to place in the layering of Earth obtained by means other than gravity surveys. Such density distributions are being actively sought by geophysicists all over the world. There has to be a density distribution which will explain every variation in the measured values of g_d, because the anomalies are reflections of the real mass arrangement in Earth.

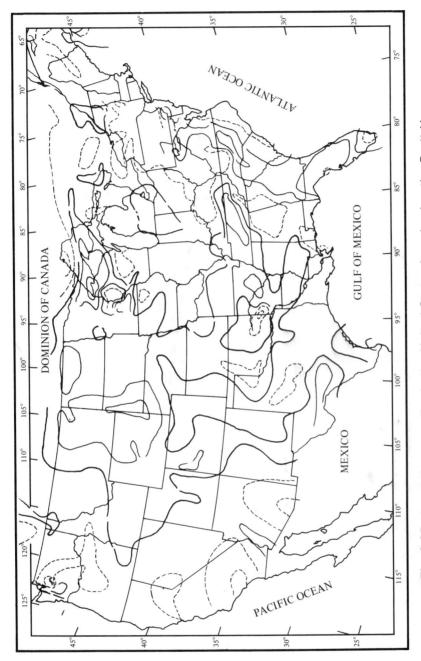

Fig. 9.25 *Isostatic anomalies in the United States found using the Pratt-Hayford method for H = 113.7 km. Units are dynes/gm. (After Bowie, U.S.C.G.S. Sp. Pub. 40.)*

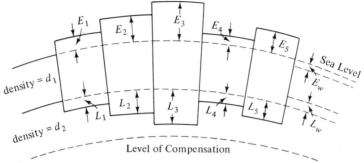

Fig. 9.26 *The Airy density concept.*

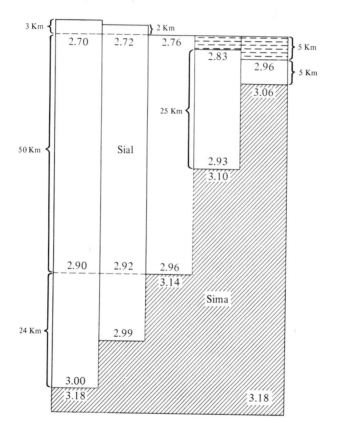

Fig. 9.27 *Modern density concepts provide for an increase of density with depth which varies with crustal thickness. (After Heiskanen, Ann. Ac. Sci. Fenn. A36(3).)*

Formulas 9.5 and 9.7 both are based on density arrangements which will describe closely the observed values of gravity everywhere on Earth. But, earlier in this chapter, it was pointed out that large amounts of material were being eroded off the continents and deposited along the edges of the ocean basins. One would expect that in the depositional areas there must be an excess mass and in the source areas a mass deficiency reflected in the gravity pattern. This is not the case. This cannot reasonably occur by chance. It is known that in some regions many thousands of meters of sediments are being deposited, and in other regions equal amounts of material are being removed. The only way this can occur without there being large isostatic anomalies associated with these regions is for there to be a compensating back-flow of material beneath Earth's surface. When matter is transferred from land to sea at the surface, an equal amount of mass must be shifted in the opposite direction beneath the surface (Figure 9.28).

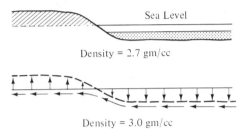

Fig. 9.28 *Erosion of 1000 meters of surface rock causes the crust to rise 900 meters. Filling the oceans causes the underlying crust to sink. Subcrustal material flows to compensate.*

The surface layers of Earth behave as though they were floating on the underlying layers like ice on a pond. When you skate on such an ice layer, it bends down beneath you, displacing water equal to your weight. As you skate away from any point on the ice, it will rise behind you as the displaced water flows back. The same process is implied in Earth. Beneath the surface layers, there must be currents of matter which flow in under the eroded area, raising them as the surface is removed.

This process is called *isostatic compensation*. The law which controls it is called the *Law of Isostasy*. This law states: all large land masses on Earth's surface tend to sink or rise so that, given time for adjustment to occur, their masses are hydrostatically supported from below, except where local stresses are acting to upset equilibrium. The term *isostasy* itself refers to this ideal state of hydrostatic balance. Isostasy need not be perfectly achieved everywhere. Local anomalies like the one near Seattle, Washington may be the result of density irregularities in the crust. Individual mountains and valleys at the surface are not compensated by corresponding variations in crustal thickness, but are supported by the crust locally.

Also it takes time for the inflow of material beneath the crust to occur. During the Ice Age which ended about 10,000 years ago, all of Scandanavia was covered by a layer of ice several thousands of meters thick. This layer of ice depressed the surface rocks, causing material to flow out from under the region. The ice melted off faster than the material could flow back. The region is still rising at a rate of about one centimeter per year (Figure 9.29). This was first recognized by noting the slowly changing sea levels recorded by tide gages. It is estimated that this area still has several

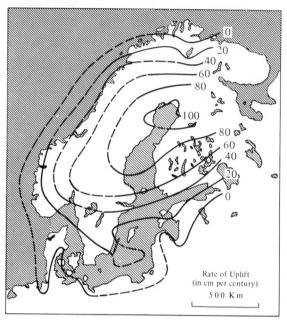

Fig. 9.29 *Present rate of uplift of Scandinavia. (After Gutenberg,* Arch. Met. Geoph. Biokl. *v. 7, p. 247.)*

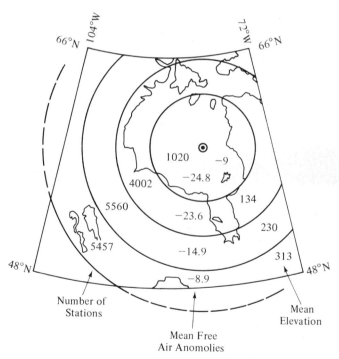

Fig. 9.30 *Mean gravity anomalies in dynes/gm in the Hudson Bay area. (After Innes, et. al.,* Science, History and Hudson Bay *v. 2, p. 724, Dominion Observatory of Canada.)*

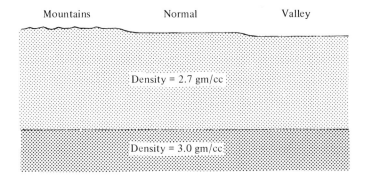

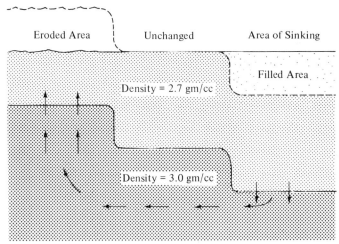

Fig. 9.31 *In eroding an area to a level plane, it takes 10 km of material to fill a valley 1 km deep because the valley floor sinks under the weight of the fill.*

hundred meters to rise before it reaches isostatic equilibrium. The gravity deficiency in the center of the region is .0005 newtons per kilogram (.05 dynes/gram). A similar situation exists around Hudson Bay in Canada, which was the center of the ice sheet which once covered North America (Figure 9.30).

   If the material flowing in compensation at depth had the same density as the matter transported at the surface, mountains could continue to be raised indefinitely as fast as they were eroded away. The deep-seated rocks, however, are generally denser than surface rocks, both because of the greater pressure at these depths and because of differences in chemical composition. (The evidence for this will be developed in Chapter 10.) Therefore, a region being raised by isostatic compensation does not come up as far as its surface has been lowered. If the rocks at the surface are ten per cent lighter than those moving in below, then the surface will rise only 90 meters for every 100 meters removed from the top. Similarly, a region being filled sinks 90 meters for every 100 meters of fill (Figure 9.31).

This will greatly extend the length of time it takes for a continent to be eroded away, but it will not prevent its ultimate destruction. The process of isotatic adjustment explains why mountains continue to rise long after their original formation. As the crests are removed by erosion, the region rises, until eventually even its roots are exposed.

The mechanisms by which the mountains were caused to form in the first place will be discussed later. To understand what may have happened it is necessary to know something of what the interior of Earth is like.

BIBLIOGRAPHY

Bullard, E., 1954, "The Interior of the Earth." Chapter 3 of *The Earth as a Planet.* University of Chicago Press.

Foster, 1969, *General Geology.* Charles E. Merrill Publ. Co., 631 pp.

Heezen, B. C.; M. Tharp; M. Ewing, 1959, *The Floors of the Oceans.* Geol. Soc. Am. Sp. Paper 65. 122 pp.

Holmes, A., 1965, *Principles of Physical Geology.* Ronald Press Co., 1288 pp.

Howell, B. F., Jr., 1959, *Introduction to Geophysics.* McGraw-Hill Book Co., Inc. 399 pp.

Mason, B., 1962, *Meteorites.* John Wiley and Sons, Inc., 274 pp.

Mason, B., 1966, *Principles of Geochemistry.* John Wiley and Sons, Inc., 329 pp.

Pettersson, H., 1954, *The Ocean Floor.* Yale University Press. 181 pp.

Shepard, F. P., 1963, *Submarine Geology.* Harper and Row Publ., 557 pp.

Stacey, F. D., 1969, *Physics of the Earth.* John Wiley and Sons, Inc., 324 pp.

Sverdrup, H. V.; M. W. Johnson; R. H. Fleming, 1942, *The Oceans.* Prentice-Hall Inc., 1087 pp.

10 INTERIOR OF EARTH

10.1 EXPECTED CHANGES.

In the previous chapter it was shown that the surface rocks tend to be lighter and different in composition from those formed at great depths. The deeper the place of formation of the rock the denser and more iron and magnesium rich it tends to be. This is consistent with the requirement established in Chapter 5 that density increase with depth to provide for an average density of 5.52 gm/cc for the whole Earth and a moment of inertia of 0.33078 $M_e R_e^2$.

Even if there were no change in composition with depth, we would expect density to increase due to the compaction caused by the weight of overlying rock. At a few thousand meters depth, all openings in the rock such as caves and even the tiniest pores are pressed shut, except when they are filled with a liquid such as petroleum or water. Eventually, at great depths, even liquid-filled pores become very scarce.

In addition, laboratory studies of the behavior of the types of minerals found in rocks indicate that phase changes are to be expected. It has been impossible so far to duplicate in the laboratory pressures as great as those which must exist in the center of Earth. Because of this, it is not known exactly what would happen to a granite or a basalt at the pressure of Earth's center. The best estimates, however, suggest that pressure alone would never increase the density of surface rocks enough to account for the whole mass of Earth. It must be expected, therefore, that there will be some change in composition with depth toward heavier elements than oxygen, silicon and aluminum, the most common ones in Earth's crust.

10.2 VARIATION IN SEISMIC VELOCITIES WITH DEPTH.

The next line of evidence comes from the study of earthquakes and large blasts. When an earthquake occurs, a series of vibrations is generated. For a large earthquake, these vibrations travel to all parts of Earth. Some of the vibrations are *sound waves,* and travel through the interior as compressions and expansions of the rock (Figure 10.1). Some are *shear waves,* and travel through Earth as distortions in the shape of the rock. Some are *surface waves*, and travel only along the surface.

The speed with which these waves travel is determined by three properties of the rock: the density, the bulk modulus, and the rigidity. *Density* is simply the amount of mass per unit volume. *Bulk modulus*, b, tells how much the volume, V, of a rock changes as a change in pressure, ΔP, is applied to it.

$$b = -\frac{\Delta P}{\Delta V/V} \qquad (10.1)$$

In this formula "Δ" stands for "change in." The formula says that bulk modulus is the ratio of change in pressure to change in volume per unit of volume. The minus sign means that volume decreases as pressure increases.

Since density, d, increases at the same rate that volume decreases, bulk modulus can also be defined as

$$b = \frac{\Delta P}{\Delta d/d} \qquad (10.2)$$

Rigidity is the property which relates how much a rock bends to the force bending it. If a square frame is visualized in the center of a large body of rock, and a set of forces F_s are exerted across the rock so that its shape is changed from square to diamond-shaped as shown in Figure 10.2, then the sides of the square will be rotated an amount A. F_s and A are related by the formula

$$F_s = 2rA \qquad (10.3)$$

where r is called rigidity. Rigidity is a measure of how much a rock is bent by forces exerted on it, whereas bulk modulus is a measure of how much its volume is changed.

Experiments made on rock samples show that sound waves travel at a velocity given by the formula

$$v_p = \sqrt{\frac{b + \frac{4}{3}r}{d}} \qquad (10.4)$$

whereas shear waves travel at a velocity

$$v_s = \sqrt{\frac{r}{d}} \qquad (10.5)$$

where b is bulk modulus, r is rigidity, and d is density. Comparison of Equations 10.4 and 10.5 shows that sound waves travel faster than shear waves. The two types arrive in sequence, first the sound waves, then the shear waves.

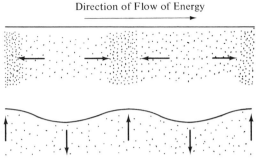

Fig. 10.1 *Particle motions in the ground during the passage of sound waves (top) and shear waves (bottom).*

The velocity of sound waves in rocks found at Earth's surface varies from less than one kilometer per second for soils to four to six kilometers per second for igneous rocks such as granite and basalt.

When an earthquake occurs, the sound waves reach the opposite side of Earth in about twenty minutes and nine seconds. The average velocity thru Earth, therefore, is

$$v_{av} = \frac{\text{Diameter}}{\text{Time}} = \frac{12{,}740 \text{ km}}{1209 \text{ sec}} = 10.5 \text{ km/sec} \qquad \textbf{(10.6)}$$

Since this velocity is greater than that of surface rocks, the velocity of sound waves, like the density, must be greater in Earth's interior than at the surface.

Observations in many parts of the world have shown that almost everywhere on land beneath the veneer of soil and sedimentary rocks there are rocks with a sound velocity of around six kilometers per second. This can be proven in the following way (Figure 10.3). Detonate a blast at point *0* on the surface and observe the ground vibrations at successive distances *A*, *B*, *C*, *D*, *E*, etc. A pulse of sound will be observed to pass *A*, *B*, *C*, etc. at times predicted by the formula

$$T_1 = \frac{\text{Distance}}{v_1} \qquad \textbf{(10.7)}$$

where v_1 is the sound velocity in the surface rocks. Beginning at a short distance from *0*, a second pulse will be observed to arrive. It will be found that a second straight line can be drawn thru a plot of these arrivals. Pulse 1 will define a straight line thru the origin given by Equation 10.7. Pulse 2 will define a second straight line which can be described by the equation

$$T_2 = T_{D2} + \frac{\text{Distance}}{v_2} \qquad \textbf{(10.8)}$$

It is exactly as if pulse 2 had travelled with a velocity, v_1, diagonally down to a second layer along the top of layer 2 with velocity, v_2, then diagonally

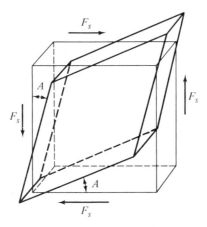

Fig. 10.2

back up to the surface again at velocity v_1. Velocity v_2 can be found from the difference in arrival times at any two stations. If station D is 60 kilometers from Station B, and pulse 2 reaches station D ten seconds after it reaches B, then velocity v_2 is $\dfrac{60 \text{ km}}{\text{'10 sec.}} = 6.0$ kilometers per second. The depth to layer 2 can be determined from the time it takes the pulse to go down to layer 2 and back to the surface.

Additional arrivals are commonly observed. Some but not all of these are shear waves. In particular, almost everywhere there is observed a third sound arrival with a velocity of around eight kilometers per second. Sometimes, but not always, there are other layers with velocities between six and eight km/sec.

From this evidence, it is seen that Earth can be divided into three parts: (I) A thin veneer of soil and sedimentary rock of varying thickness up to about 10 km thick, but usually not over 3 km thick and sometimes entirely absent. (II) A layer or series of layers with velocities from 5.5–7.6 km/sec. These velocities are what is observed for crystalline rocks such as granite and basalt. Under the floor of the ocean this layer is about 5 km thick. Under the continents it is usually about 30 km thick, but it is often thicker under high mountain ranges. (III) A layer with a velocity at the top of 7.7–8.5 km/sec. This third layer is called the *mantle*, and the layers above it are called the *crust*. The boundary between the crust and the mantle is called the *Mohorovicic discontinuity* or just the "Moho," after Andreija Mohorovicic, a Yugoslavian scientist, who first recognized its existence from studying earthquake pulses.

If the recorder is moved farther out, the curvature of Earth's surface complicates the problem. The arrivals define curved lines instead of straight lines (Figure 10.4). The same principles of interpretation, however, apply. Each layer will give a distinctive return. There are no simple, sharp boundaries in the mantle which lead to separately recognizable pulses of the type found for the crust. However, the curvature of the travel-time plot (Figure 10.4) is greater than can be explained by the curvature of

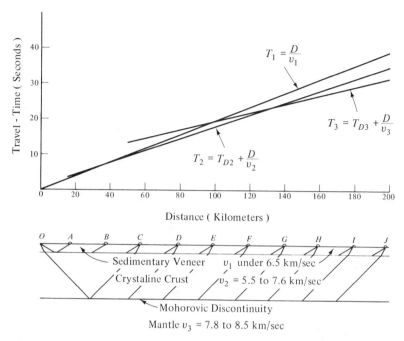

Fig. 10.3 *Seismic-wave paths and travel-time graph for a flat, layered earth.*

Earth's surface alone. Its shape requires in addition that there be an increase in velocity with depth. Thru laboratory studies it is known that waves travel from one point to another by the path which takes the least time. This rule is called *Fermat's principle*. The curvature of the travel-time plot in Figure 10.4 results from two factors. First, distance in Figure 10.4 is measured along Earth's surface, but the waves can take a shortcut by going thru the interior. Second, by penetrating more deeply than the straight-line path, the waves can take advantage of the higher velocity of transmission at greater depths. The actual path is always the minimum-time path.

The travel-time curve can be approximated by a series of straight lines. This is equivalent to assuming that Earth is made up of a series of layers. If the velocity increases with depth continuously, then the velocity as a function of depth can be calculated from the travel-time plot.

If at any depth the velocity decreases, then additional information is required. This can be obtained from surface waves. Surface waves travel at velocities which depend on the wavelength of the wave. If the wavelength is small, the waves travel thru a thin section of near-surface material. If the wavelength is large, the wave travels in a thick section. But the thicker the section, the greater will be its average velocity because it includes more of the deep, high-velocity material. Thus long-wavelength waves usually (although not always) arrive before short-wavelength waves. The times of arrival of surface waves of different wavelengths are very sensitive to the velocities at different depths, particularly if there is one depth in which the velocity is lower than in the overlying material. From the times of arrival

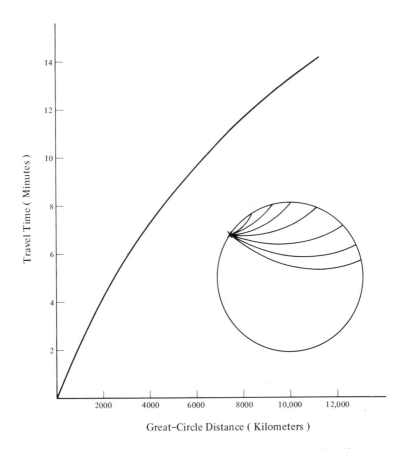

Fig. 10.4 *Times of arrival of sound waves. In circle: paths the waves take thru the earth.*

of the surface waves, combined with data from sound and shear-wave arrivals, the velocity as a function of depth can be calculated. Figure 10.5 is an average plot of velocity of sound and shear waves as a function of depth. Velocity varies horizontally as well as vertically, though by a much smaller amount. The averaging process smoothes much of the details of variations with depth which can be mapped by thorough analysis of the travel-time curves. Note the minimum velocity between 150 and 200 km depth.

A problem is encountered in calculating the velocity-depth relationship from the travel-time curves. Beyond 11,400 kilometers from the source, the sound (*P*) and shear (*S*) pulses, which were easily recorded at shorter distances, are not observed (Figure 10.6). Beyond this distance, there is a shadow zone where these pulses are missing. It is just as though there were a core of some kind in Earth's center which blocked them out.

This core can be detected in another way. At short distances from the source, there is a pulse which arrives eight minutes and 34.3 seconds after the earthquake (*PcP* in Figure 10.6). If this pulse is followed to successively

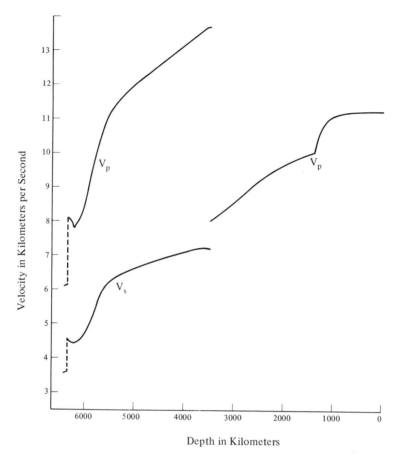

Fig. 10.5 *Velocity of sound (P) and shear (S) waves as a function of depth.*

greater distances, it gradually gets closer and closer to the principal pulse travelling directly from the source to the receiver. These two pulses completely merge at a little beyond 11,000 km. This new pulse is an echo from the core. The depth to this reflecting core can be calculated from the velocity of sound waves in the mantle and the travel-time of the echo. It can also be found from the depth of penetration of the waves which emerge at 11,400 km. The depth is found to be 2900 kilometers. This is the bottom of the mantle.

Determining velocity in the core is a great deal more difficult than for the mantle. The principal reason for this is that the velocity of sound waves is lower in the core than in the mantle. When velocity increases with depth, the rays along which sound waves travel are bent back toward the surface, because of Fermat's principle of least-time path (Figure 10.3). The energy which penetrates to a high-speed layer is returned to the surface, and at large distances arrives before energy travelling entirely in the low-velocity surface layers. If a deep layer like the core has a lower veloc-

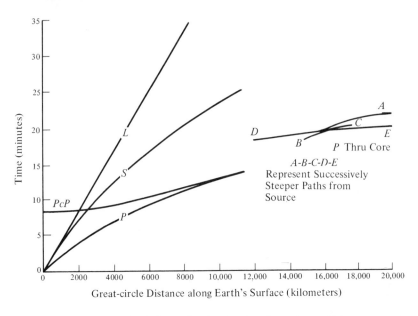

Fig. 10.6 *Times of arrival of the principal seismic pulses. P is a sound wave; S is a shear wave; R is a surface wave; PcP is an echo from the core boundary.*

ity than overlying layers, then it bends the rays down toward Earth's center. Such down-bent rays reach the surface only because of the curvature of Earth's surface. No matter how sharply they are bent downward as they cross the core boundary, they must ultimately return to the surface of a round Earth. The core acts like a lens and focuses the sound waves in the shadow zone. The pattern of this focusing is determined by the velocity structure in the core.

In the shadow zone, rays may reach a single point on the surface by as many as four different paths thru the core (Figures 10.6 and 10.7). The observed pattern of emerging rays is best explained by dividing the core into two parts: an outer part in which the velocity of sound waves gradually increases from about 8 km/sec to 10 km/sec and an inner core with a velocity of about 11 km/sec. The two parts are separated by a transition layer perhaps as much as 200 km thick at a depth of about 5000 km.

One peculiarity which the outer core has is that no waves have ever been observed which travelled thru it as shear waves. Solids transmit both sound and shear waves, but fluids transmit only sound waves. This means that not only do we know the velocity of seismic waves in the crust and mantle, but we know also that these two layers are solid. The outer core, on the other hand, appears to be liquid.

No shear waves have been identified which passed thru the inner core either. This would seem to suggest that the inner core is also a liquid. However, if one calculates the strength of the waves which would reach the surface after passing thru the inner core as shear waves, they turn out

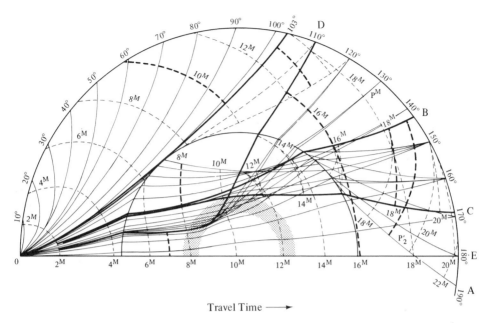

Travel Time ⟶

Fig. 10.7 *Paths of seismic pulses thru Earth's interior showing complexity of paths which pass thru the core. (After Gutenberg,* Geoph. Jour. Roy. Soc. *v. 3, p. 252).*

to be so weak that it would be almost impossible to record them except for a very large earthquake. The seismograms of the largest earthquakes each year are being studied, and it is possible that shear waves thru the inner core will soon be identified.

The jump in velocity going from the outer to the inner core is reasonable for a phase change from a liquid to a solid of the same composition. This is one reason why the search for shear waves in the inner core is being pressed. The shear-wave velocity would provide a check on the likelihood of the inner core being a solid of the same chemical composition as the outer core.

On the other hand, the drop in velocity from the mantle to the core is too large to explain by a phase change from solid to liquid. Therefore, it is believed that the core must be of a different chemical composition from the mantle. The compositions of all layers will be discussed in more detail later.

10.3 DETERMINATION OF DENSITY AS A FUNCTION OF DEPTH.

The profiles of velocity of sound and shear waves as a function of depth are the most fruitful of all sources of data on Earth's interior. They divide the solid Earth into four principal layers: crust, mantle, outer core and inner core. They tell which layers are solid and which liquid. They also can be used to give information on the other physical properties on which velocity depends (Equations 10.4 and 10.5). Laboratory experiments tell us that density, rigidity, and bulk modulus all increase with pressure. They also tell us about how rapidly these three quantities increase with pressure. Generally, rigidity and bulk modulus increase with pressure more rapidly than density. Therefore, it can be expected that velocity will increase with depth, as is observed at most depths. If we knew what the value of density was at every depth, we could tell whether the observed increase of velocity can be reasonably explained solely by the increase in pressure, or whether we should expect also that there are changes in composition or crystal phase.

An exact answer for density cannot be obtained, but we now have enough data to determine a maximum and a minimum value of density at all depths in the mantle. To do this, start by solving Equation 10.2 for change in density, Δd:

$$\Delta d = \frac{d\Delta P}{b} = \frac{\text{density} \times \text{change in pressure}}{\text{bulk modulus}} \tag{10.9}$$

Now square both sides of Equation 10.4 and solve for bulk modulus, b

$$v_p^2 = \frac{b + \frac{4}{3}r}{d} \tag{10.10}$$

$$b = v_p^2 d - \frac{4}{3}r \tag{10.11}$$

Square (10.3) and solve for rigidity, r

$$v_s^2 = \frac{r}{d} \tag{10.12}$$

$$r = v_s^2 d \tag{10.13}$$

and substitute this into (10.11) obtaining

$$b = v_p^2 d - \frac{4}{3} v_s^2 d = d\left(v_p^2 - \frac{4}{3}v_s^2\right) \tag{10.14}$$

Substitute this into (10.9) and cancel d in the numerator and denominator

$$\Delta d = \frac{d\Delta P}{d\left(v_p^2 - \frac{4}{3}v_s^2\right)} = \frac{\Delta P}{v_p^2 - \frac{4}{3}v_s^2} \tag{10.15}$$

Equation 10.15 says that the change in density with depth is equal to the change in pressure divided by a function of the sound and shear-wave velocities.

On the average, throughout Earth, the change in pressure is due to the weight of the overlying rocks. There may be local variations where some rocks are supported from the side, but at any depth the pressure over a large area is the sum of the weights of all the rocks above that depth. Weight, however, is gravity times mass. Thus, we can say that the increase in pressure at any depth is given by

$$\frac{\text{change in pressure}}{\text{change in depth}} = \frac{\Delta P}{\Delta R} = -g_R d \qquad (10.16)$$

where g_R is gravity at radius R from Earth's center and d is density. The minus sign is needed because pressure increases as the radius decreases.

Now Equation 5.2 (p. 119) gives us the value of g at Earth's surface. It can also be shown that the gravitational attraction of a spherical shell at points inside that shell is zero. As one goes down in Earth, therefore, only the rocks beneath the observer contribute to the observed gravity. Within Earth at radius R, gravity has the value (compare Equation 5.2):

$$g_R = GM_R / R^2 \qquad (10.17)$$

where M_R is the mass inside of the radius R and G is Newton's universal constant of gravitation.

Dividing both sides of Equation 10.15 by $\triangle R$

$$\frac{\Delta d}{\Delta R} = \frac{\Delta P / \Delta R}{v_p^2 - \dfrac{4}{3} v_s^2} \qquad (10.18)$$

and substituting (10.16) in (10.18)

$$\frac{\Delta d}{\Delta R} = \frac{-g_R d}{v_p^2 - \dfrac{4}{3} v_s^2} \qquad (10.19)$$

Substituting (10.17) into (10.19)

$$\frac{\Delta d}{\Delta R} = \frac{-GM_R d}{R^2 \left(v_p^2 - \dfrac{4}{3} v_s^2\right)} \qquad (10.20)$$

Equation 10.20 tells how fast density, d, increases with depth at any given depth.

Since G is a constant and we know v_p and v_s as functions of depth, we can calculate $\dfrac{\Delta d}{\Delta R}$ at any place where we know density d and M_R, the mass beneath that depth. For instance, we know both M_R and density at Earth's surface, so we can calculate $\dfrac{\Delta d}{\Delta R}$ at the surface. Using 2.800 grams per cu-

bic centimeter as the value of density at the surface and 5.977×10^{27}

grams for M_R, 6.000 km/sec for v_p and 3.500 km/sec for v_s, $\dfrac{\Delta d}{\Delta R}$ is .0014

grams per cubic centimeter per kilometer of depth. Therefore, at ten kilometers depth the density will be 2.800 + .0014 × 10 = 2.814 gm/cc. Using an average value of 2.807 gm/cc for the density of the top ten kilometers, the mass of a spherical shell of this thickness is found to be $.014 \times 10^{27}$ grams. The mass beneath ten kilometers is therefore $5.977 \times 10^{27} - .014 \times 10^{27} = 5.963 \times 10^{27}$ grams. With these new values of mass and density, the rate of increase in density for another ten-kilometer-thick layer can be calculated, from which the mass and density for a third calculation can be obtained, and so on repeatedly to the center of Earth.

There is one difficulty with this method. It assumes that density increases continuously without any jumps. At any place where there is a change in composition or phase, a jump in density can be expected. Equation 10.20 is good only within each layer. It does not allow for a jump in density such as would be expected at the Moho, the core boundary, or the top of the inner core.

A reasonable estimate can be made of the density immediately beneath the Moho. For a wide range of materials, it has been found experimentally that rocks with a sound-wave velocity of around 8 km/sec have densities close to 3.3 grams per cubic centimeter. If this value is assumed for the layer immediately beneath the Moho, then Equation 10.20 can be used to find density in the mantle.

Additional information is needed to determine the density jump at the boundary of the core. One such piece of data is that the moment of inertia of Earth about its axis of rotation is $0.33078\ M_e R_e^2$, where M_e is the mass of Earth and R_e is its radius (Section 5.6).

The correct density variation must satisfy three conditions: (1) the density in each spherical shell must be such that the sum of the masses of all the shells adds up to the whole mass of Earth; (2) the density in each shell must be such that the sum of the moments of inertia of all the shells adds up to the moment of inertia of Earth; (3) within each shell of uniform composition, the density must increase as predicted by Equation 10.20.

One other assumption is normally made: that density never decreases at any level down into Earth. Lighter material always overlies heavier. If this were not so, then the heavier material would tend to sink and the light to rise. The assumption is reasonable because it is believed that, except in the crust, Earth is plastic and can flow given enough time. The reasons for believing that the mantle is plastic will be discussed later. There may be small local exceptions to the rule of continually increasing density; but if there are large regions where this assumption is violated, then the conclusions reached below are not necessarily valid.

With these limitations, we can now try to calculate a possible density variation starting with a density of 3.30 gm/cc beneath the Moho. When this is done, an unexpected difficulty is encountered. No matter what size jump in density is assumed at the outer boundary of the core, either the moment of inertia which results from using Equation 10.20 in the core is too small, or the mass is too great.

The only way around this is for density to be greater than has been assumed up to now somewhere in the crust or mantle. The necessary increase could be introduced starting at the Moho, but this would require a greater density at the top of the mantle (at least 3.67 gm/cc) than seems likely considering the seismic-wave velocities observed. One might ask, however, if there is any chance that we have overlooked a discontinuity somewhere within the mantle. We have assumed up to now that the mantle is all one layer because there is nowhere within it a jump in seismic-wave velocities. Examination of Figure 10.5 shows that between 150 and 900 km the velocity increases much faster than at depths greater than 900 km. Below 900 km the increase is roughly what one might expect as the result of the increase in pressure in a material of the density predicted by Equation 10.20. From 150 to 900 km, the increase is greater than would be expected from the increase in pressure alone.

Equation 10.20 is derived starting with Equation 10.9, which requires that the change in density over any interval is due entirely to the effect of pressure in compacting the rock. If there is either a gradual change or a number of small steps in composition, then density can increase more rapidly than predicted by Equation 10.20. The rapid increase in seismic-wave velocity from 150 to 900 km suggests that such changes do occur in this region.

With an extra increment in density allowed in the central mantle, it is possible to meet all four conditions: mass, moment, Equation 10.20 in each layer, and density always increasing with depth. Figure 10.8 shows one possible solution. With the assumptions made, the true situation cannot differ much from this in the mantle. The most the density could depart from the values given is only about 0.2 gm/cc.

In the core the density is much less certain. By studying the ratio of the reflected energy to the energy passed thru the boundary of the core, it can be concluded that there must be some increase in density at this point, and that the most likely ratio of the density inside to outside the core boundary is close to 1.8.

The range of possible values of density is also limited by data obtained from studying the free oscillations of Earth. Very long-period earthquake vibrations (200 seconds or longer) have resonances which are stimulated by very large earthquakes. The frequencies of these resonances depend critically on the variations of seismic-wave velocity and density with depth.

Neither total mass nor moment of inertia are highly sensitive to what happens at the inner core boundary, but the small step in seismic-wave velocity suggests that this step is not large. It is difficult to find a reasonable solution in which the density at the center of Earth is less than 12 gm/cc. This means that the densities shown in Figure 10.8 are probably correct to

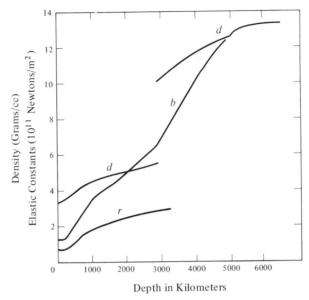

Fig. 10.8 *Density, d, bulk modulus, b, and rigidity, r, within Earth.*

within ten percent for the outer core. Density could be much greater than is shown for the inner core. Solutions can be found in which the density at the center rises to twice the value shown. Considering how little the seismic-wave velocity changes in the inner core, it seems more likely that the density also varies very little and is only slightly higher than in the outer core.

10.4 PRESSURE, GRAVITY AND ELASTIC CONSTANTS.

Knowing density and the seismic-wave velocities, bulk modulus and rigidity can be calculated as a function of depth using Equations 10.13 and 10.14. For the mantle, their values are shown in Figure 10.8. In the outer core, the rigidity is zero because the material is liquid. The elastic constants cannot be calculated for the inner core because the velocity of shear waves is unknown.

Gravity inside Earth is calculated using Equation 10.17 (Figure 10.9). It is nearly constant throughout the mantle, and decreases to zero at the center.

Pressure at any depth is the sum of the weight of the overlying layers. In the laboratory, pressures and temperatures equivalent to those at a few tens of kilometers can be maintained. Pressures corresponding to depths of 100 kilometers or more can be applied to cold materials in laboratory tests or can be produced momentarily in explosions, but they cannot be maintained on hot samples because the chambers holding the hot material melt and flow too easily.

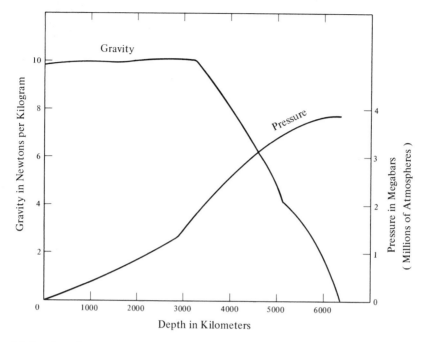

Fig. 10.9 *Gravity and pressure in Earth.*

These facts—density, pressure, elastic constants (including seismic velocities), moment of inertia and gravity—are almost all that is known with certainty about Earth's interior. These are known with assurance within narrow limits. Almost all else that can be said about the interior is to some degree speculative. The extensive discussion of compositions, conditions and processes which fills most of the rest of this book rests on one or more uncertain assumptions. In evaluating the theories presented, the reader is warned to *remember the assumptions*. The most beautiful edifice of conclusions may tumble if it is built on the sands of faulty assumptions. Having issued this warning, the author feels free to present his own speculations as reasonably probable theories as long as these are consistent with the density and elastic constants, etc. presented above.

10.5 THE EFFECT OF TEMPERATURE.

One other important factor influences the density variation in Earth. This is temperature. The problem of the effect of temperature has been deliberately avoided while discussing density because of the large uncertainties as to what the temperatures are. Density is less sensitive to temperature than to pressure except for the important role of temperature in controlling phase changes. Typical basalts and granites expand linearly five to eight parts in a million per °C change in temperature compared to one to two parts in a million per atmosphere change in pressure. Temperature varies over a few thousand °C in Earth (see Chapter 12) but pressure increases by a factor of over three million. For any reasonable range of temperatures, density would not differ greatly from Figure 10.8.

Temperature does have an important effect on the rate of increase in the seismic velocities with depth. Rigidity is strongly temperature dependent, and even bulk modulus is affected. Increased temperature tends to decrease the seismic-wave velocities. Where temperature is increasing rapidly, it may cause velocity to decrease with depth in spite of the effect of increasing pressure. This is the most likely explanation of the velocity minimum at 150 km (Figure 10.5). The physical properties are particularly sensitive to temperature near the melting point. If one percent of the rock at 150 km were molten, as has been proposed by some investigators, this could explain the minimum of velocity observed, while still permitting the transmission of waves.

10.6 COMPOSITION.

Where they crop out, we observe that the crystalline rocks of the crust range in composition from granite to basalt and dunite (an ultrabasic rock composed almost entirely of the mineral olivine), which are all rocks rich in oxygen and silicon with varying amounts of the metals aluminum, iron, magnesium, calcium, sodium, potassium and titanium plus minor amounts of other elements. It is reasonable to suppose that mantle material is similar, but not identical. In Hawaii, volcanic eruptions begin with a series of small earthquakes at a depth of 60 km. Over a period of weeks, the depth of focus of the earthquakes gradually decreases. Just as the depth reaches zero, lava starts extruding at the surface. This suggests that at least part of the extruded material came from 60 km depth in the mantle.

The lava and other fluids escaping from volcanoes are presumably a mixture of mantle materials and constituents of the crust. The part escaping from great depths may be different in composition from the part left behind. Because the degree to which crustal and mantle materials are mixed in this fashion is unknown and because the degree to which mantle materials differentiate into rising and remaining parts is uncertain, it is impossible to state precisely the composition of the upper mantle. Basalt is by far the most common volcanic rock. Most rocks believed to have been formed from the solidification of volcanic materials are composed largely of the silicate and oxide minerals also found in basalt, though the proportions vary. Any reasonable upper-mantle composition must be one which can produce a magma which rises in the volcanic process, possibly dissolves some crustal material, and yields finally a range of volcanic rocks of which basalt is a typical example.

Because the predominant minerals in basalt are silicates, it is hard to escape the conclusion that the upper mantle is also composed in large part of silicates. Because iron-magnesium silicates are denser than aluminum silicates and density increases with depth, it is reasonable to suppose that the mantle is richer than the crust in iron and magnesium and poorer in aluminum. Because these elements are by far the commonest elements in the crust, it is hard to go beyond this in predicting composition.

There are chemical rules which can be used to predict the behavior of each element in a melt of complicated composition. If any element other than oxygen, silicon, aluminum, magnesium or iron made up more than a small percentage of the composition of the mantle, it would be expected to be found in greater abundance than it is in igneous rocks. It can, therefore, be concluded that the upper part of the mantle is composed largely of iron and magnesium silicates with lesser amounts of other elements. A rock of this composition would be expected to have the density and to transmit seismic waves with the velocities observed.

This composition is remarkably close to the composition of the stony meteorites. It is entirely reasonable that the composition of the upper mantle is close to the composition of stony meteorites (Table 9.4, p. 245). It is also entirely reasonable that there should be a considerable difference between the two compositions. All that can be said with assurance is that the similarity of the composition of stony meteorites to what is expected for the upper mantle is consistent with the theory that the meteorites are fragments of a planet which was like Earth in composition, and which broke up at some time in its history.

The reverse argument, that the meteorites are a sample of the composition which might be expected in Earth, is much flimsier. It requires that one assume the meteorites came from a planet which was formed from the same raw material as Earth, went thru a similar process of formation from this raw material, and evolved similarly up to the time of break-up, so that the source planet of the meteorites was composed of layers similar to those of which Earth is composed. If at any stage of development, the two planets differed in their evolutionary history, these differences might result in differences in the composition of the meteorites and the corresponding Earth layers. They differ today in that one planet has already broken up and the other has not. Their distances from Sun differ, implying different surface histories. There could easily have been other differences. Nevertheless, the predicted upper-mantle compositon is so similar to stony-meteorite composition that the average composition of meteorites provides one of the best lines of evidence available for estimating the average composition of Earth.

The next question is how far down into the mantle a composition of iron-magnesium silicates extends. The seismic and density evidence indicates that between 150 and 900 km there is a change in mineral composition. Could there be a change in chemical composition as well? This question is unanswerable. There is no way of being sure that there are not considerable changes in chemical composition. Experimental knowledge of the physical properties of minerals of different compositions at the temperatures and pressures expectable at these depths is scarce, and we have to depend largely on theoretical predictions and extrapolations of data gathered at lower temperatures and pressures. Many materials could have the known physical properties of the lower mantle.

This is where the analogy with meteorites becomes important. If the parent body of meteorites was similar to the present Earth, and if the chemical composition of Earth's lower mantle differed from that of the upper mantle then there should be meteorites with the composition of Earth's lower mantle. Tektites are far too light in density and siderites too heavy. It is possible that the siderolites are similar to the lower mantle. This would mean that the lower mantle differs from the upper in being much richer in iron.

It is not necessary, however, that the lower mantle differ at all in chemical composition from the upper mantle. The increase in density and seismic velocity between 150 and 900 km could result from mineral phase changes to dense, high-pressure forms. Phase changes of this sort are expectable, and the amount of the density and velocity increases is well within the range of what is reasonable. From 900 km to the core boundary the increases in density and velocity are close to what would be expected as a result of the increase in pressure with depth. It is, therefore, not necessary that there be any change in chemical composition from the top to the bottom of the mantle.

It is probable that the core has a different chemical composition from the mantle. At its boundary density increases greatly. Normally, when a material melts, the component atoms become less tightly packed and density decreases. If the core and mantle were of the same composition, a decrease in density at the core boundary would be expected.

The only condition which would explain an increase in density without a change in composition would be that the core boundary is a transition from a solid to a plasma. When pressure and temperature are great enough to break electrons out of the outer orbits, the individual atoms can be more tightly packed than in the solid state. Temperatures and pressures as high as those at the core boundary cannot be reproduced in the laboratory, so it is not known when silicate materials would be converted to plasmas. Theoretical considerations predict that this change would be expected only at much more extreme conditions than what is likely in Earth even considering the uncertainties in the temperature.

Accepting the probability of a change at the core boundary, what is the most likely composition? If the core is a plasma, the composition could be any of a wide variety of substances including a hydrogen-rich mixture of the primordial material from which the planets formed. If the core is an ordinary liquid, it must be composed largely of heavier elements than the crust and mantle. From the density prediction (Figure 10.8), the most likely candidates are pure metals or mixtures of metals with lesser amounts of other elements. Here again the analogy with meteorites suggests the possibility of an iron core. This is an entirely reasonable possibility. The density of liquid iron at Earth's surface is about 7.8 gm/cc. Silicate materials are increased in density from around 3.3 to around 5.5 gm/cc in going

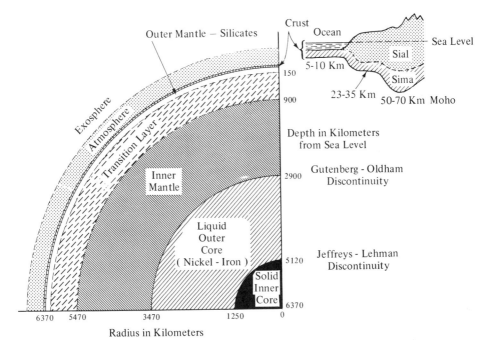

Fig. 10.10 *Principal layers of Earth.*

from the top to the bottom of the mantle. A similar increase for iron would produce a density of 13 gm/cc. This is larger than the calculated density for the core, so the composition is probably not pure iron. It is more likely an alloy or compound of iron and some lighter element such as silicon or sulfur or even hydrogen. Almost certainly there is much less oxygen than in the mantle. Oxygen atoms are big compared to metal atoms. The core density is too great for oxygen to be a major ingredient.

All of this is speculation. There is no direct evidence for the composition of any material beneath the upper mantle. The model described above is reasonable in the light of the evidence obtained from the density and elastic constants (i.e. seismic velocities). It is common to use this model in discussing processes such as convection in the interior. It must be remembered, however, that this model is at best an educated guess as to what is really there.

In particular, there is good reason to believe that the temperature, physical properties and composition may be different under different parts of the crust. Such differences obviously exist at the surface between the continents and ocean basins and from place to place on the continents. Nothing but our ignorance leads us to think of the interior as being any less complicated in detail.

BIBLIOGRAPHY

Clark, Jr., S. P.; A. E. Ringwood, 1964, "Density Distribution and Constitution of the Mantle." *Rev. Geophysics* 2:35–88.

Gaskell, T. F., 1967, *The Earth's Mantle.* Academic Press. 509 pp.

Gutenberg, B., 1959, *Physics of the Earth's Interior.* Academic Press. 240 pp.

Howell, B. F., 1959, *Introduction to Geophysics.* McGraw-Hill Book Co., Inc. 399 pp.

Stacey, F. D., 1969, *Physics of the Earth.* John Wiley and Sons, Inc. 324 pp.

11 DEFORMATION OF ROCKS

11.1 DIFFERENT TYPES OF STRENGTH.*

What do we mean when we say a rock is strong? The nearly vertical cliff of El Capitan facing Yosemite Valley in California stands 890 meters (2918 feet) above the valley floor (Figure 11.1). Mt. Everest stands 8805 meters (29,000 feet) above sea level. These mountains stand high above their surroundings because the rocks of which they are composed are strong. Yet in Chapter 9 it was shown that Scandinavia is rising isostatically even though it is only about 200 meters below equilibrium (Fig. 9.29, p. 258). In deep mines and bore holes, the walls sometimes gradually swell inward, closing the shaft (Figure 11.2). When will rocks stand and when do they yield?

To understand what can happen, consider the different types of stress to which a rock can be subjected. Consider first compressive stress. If a rock is squeezed from all directions simultaneously, every atom of which it is composed is pressed closer to every other atom. The rock is reduced in volume by an amount determined by its bulk modulus. As pressure goes up, the atoms may rearrange themselves into more compact crystal lattices, until finally, under very extreme pressures, the atomic structure may collapse into the plasma state. None of these changes, except possibly the last, is likely to involve a violent change. As the atoms become more and more tightly packed, their resistance to further change increases. Rocks exhibit great resistance to compressive stress. Large forces are needed to cause small compactions.

---

*There is no generally accepted nomenclature for many of the terms used to describe the deformation of rocks. The situation is further confounded by different customary usages of words in different professions, as by physicists, engineers and geologists. The meanings used here are a compromise between the need to outline a complex subject in simple but rigorous terms, which are understandable to the beginning student, without introducing a lot of unfamiliar words, and the desire not to complicate a student's more advanced study by fixing in his mind oversimplified definitions which must later be unlearned. Wherever possible, words will be used as nearly as possible in their ordinary, non-technical meaning. For example, "yield" will mean deform whether elastically, by flow (to which some writers would restrict the term) or by fracture.

The reader is warned that he may encounter in the literature terms such as "strength," "plastic," and "coefficient of internal friction" with meanings which differ slightly or greatly from the usage employed here.

Fig. 11.1 *The south face of El Capitan, Yosemite Park, California. (Courtesy U. S. Geological Survey. Photo by F. E. Matthes.)*

If compressive stress is applied to a rock in only one direction, the rock tends to shrink along the direction of the force and to expand at right angles to this. It bulges like the sides of a basketball if you sit on it, except that the amount of bulge of the rock is many orders of magnitude less for the same pressure. If a cylinder of rock is squeezed in the laboratory, its sides will at first swell as its length decreases. If the pressure is increased, a stress will eventually be reached where the rock will crush into fragments (Figure 11.3). The final collapse is sudden. We say the rock breaks. The pressure required to cause such failure is called the *crushing strength*.

Examination of Figure 11.3 will show that although the cylinders of rock were squeezed parallel to their axes, failure (except for zero confining pressure) occurred at an angle to this direction, and usually in the form of a shear displacement wherein adjoining parts of the rock shifted parallel to the surfaces along which the rock fractured. The rock has not really failed due to compression. Rocks are immeasurably strong in pure compression. What has happened is that the rock has failed by shear. Deep within Earth, where pressure has closed all the openings in the rock, there are no spaces into which a rock can conveniently bulge one way as it is squeezed in other directions. It can, however, still shear if stress is not equal in every direction.

Fig. 11.2 *Timbers buckled by pressure in a mine shaft. (From Obert and Du-vall*—Rock *Mechanics and the Design of Structures in Rock, p. 510, John Wiley and Sons, Inc. Photo courtesy U. S. Bureau of Mines.)*

The situation is very different with respect to tension. If a bar of rock is grasped by the ends and pulled, it will at first stretch. Each atom will feel force drawing it apart from adjoining atoms. The amount of pull needed to part a rock along the boundaries of grains is much less than the stress needed to crush or shear the same rock. Rocks are weak under tension. The force per unit area needed to open a gap is called the *tensile breaking strength*.

There are thus several breaking strengths, each defined as the force per unit area needed to produce a particular type of sudden failure: by crushing, by tensional parting or by shearing (Table 11.1). In all these cases the rock separates along one or more distinct surfaces.

Solids can deform in other ways than by breaking. If you walk on a layer of mud or soft earth, you will often leave a footprint. Sometimes the footprint has distinct sides where the ground is broken down along a minute cliff, but often the only mark made is a broad depression of no distinct shape. In the latter case the ground has clearly suffered permanent deformation without any recognizable parting. The ability of the ground to bear your weight has been exceeded even though it has not broken on any distinct surface. This is a case of a different type of strength. Such ground may feel soft and springy to walk on. A man with big broad shoes may leave only faint tracks or no tracks, while a girl with high heels makes distinct marks. Only when sufficient stress is applied to the ground does it de-

Table 11.1 *Typical values of breaking strength. (From Howell,* Introduction to Geophysics.*)*

Type of rock	Crushing strength kg/m^2	Shearing strength kg/m^2	Modulus of (tensile) rupture kg/m^2
Granite	$3.7 - 37.9 \times 10^6$	—	—
Basalt (diorite)	$9.6 - 26 \times 10^6$	$2.77 - 3.26 \times 10^6$	$1.01 - 3.86 \times 10^6$
Sandstone	$1.1 - 25.2 \times 10^6$	$3.5 - 14 \times 10^6$	$0.5 - 1.6 \times 10^6$
Limestone	$0.6 - 36 \times 10^6$	$1.8 - 20 \times 10^6$	$0.35 - 1.4 \times 10^6$
Slate	$6 - 31.3 \times 10^6$	—	$4.2 - 10.6 \times 10^6$

form permanently. It recovers from lesser stresses without damage. The pressure required to cause permanent deformation by flow without parting is called the *yield strength* or, commonly, simply the *strength*.

Yield strength is an entirely different property from breaking strength. A rock which deforms by flow at one pressure will break at some higher pressure. A rock which breaks before it will begin to flow is called *brittle*. A rock which flows at a lower pressure than is required to break it is called *ductile*.

Many rocks have a very narrow ductile range between their yield strength and their breaking strength. Since flow tends to relieve the stress causing it, even a narrow ductile range is likely to allow a rock to flow without breaking. Under conditions in Earth, stresses often change only slowly, and there is ample time for yielding by flow. Consequently, rocks which appear to be brittle in laboratory experiments, where stresses are quickly applied, may behave ductilely in nature. One theory holds that all rocks are ductile, but that the breaking strength is so close to the yield strength for brittle rocks that measurable flow is never observed for them.

All processes of non-recoverable yielding without fracture are here lumped together and called *plastic flow*. Plastic flow can take place in a number of ways. One variety is independent of time but increases in amount with the size of the stress which causes it. Another variety, viscous plastic flow, increases indefinitely with time, the rate of flow depending on the stress. Other types are intermediate between these two.

A stress which is too small to cause a rock to either break or flow will still make it deform. If a force is applied to a rock deforming it, then is removed, and the rock returns to the condition it was in before the force was applied, the yielding is said to have been *elastic*. Elastic deformation can take place instantaneously or spread over a period of time.

The simplest type of elastic behavior is *proportional elasticity*. In this case the deformation occurs just as fast as the pressure can be applied. The strain is always proportional to the pressure causing it. A graph of strain against stress is a straight line (Figure 11.4).

Many common commercial plastics exhibit a time-dependent elasticity. Take a thin, straight 30-centimeter (12-inch) plastic ruler and bend it into a U-shape (Figure 11.5). Hold it thus for several minutes, Then set it on edge

Fig. 11.3 *Failure of a rock sample undergoing compression. At left, undeformed sample. Continuing pressure was increased from left to right. (Photo courtesy M. S. Paterson, compare* Bull. Geol. Soc. Am. *v. 69, p. 468, plate 1.)*

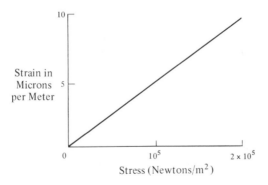

Fig. 11.4 *In proportional elasticity, strain is always proportional to stress.*

on a smooth, flat surface. At first, if you set it down carefully, it should be possible to set it upright on its edge, and to see a distinct curvature in it. Within a short time this curvature will noticeably decrease or even disappear. This is delayed elastic response, also called *elastic after-working*. Rocks sometimes have this property.

The property of the rock which controls the rate of recovery is called the *coefficient of internal friction*. The larger the coefficient of internal

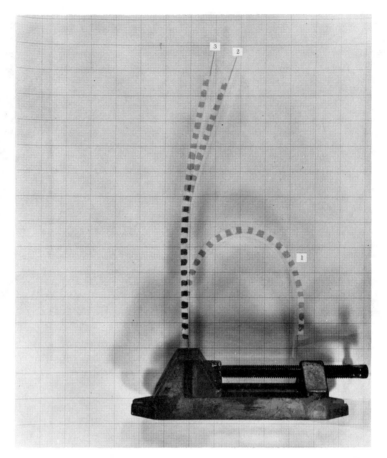

Fig. 11.5 *Black and white tabs were attached to this plastic ruler to make it easily visible. This triple photograph shows the ruler in outline: (1) bent in a U; (2) released recovering proportional elastic strain; and (3) several minutes later as elastic afterworking causes further return toward a straight condition.*

friction, the more slowly the deformation or recovery from a state of strain takes place. Where internal friction is very high, it may be hard to distinguish between delayed elastic response and plastic flow.

11.2 PHYSICAL PROPERTIES WHICH CONTROL DEFORMATION.

The proportional elastic effects of tension and compression are described by two constants. *Young's modulus* is the ratio of the pressure, P, to the proportional change in length it produces (Figure 11.6).

$$y = \frac{P}{\Delta L/L}$$

(11.1)

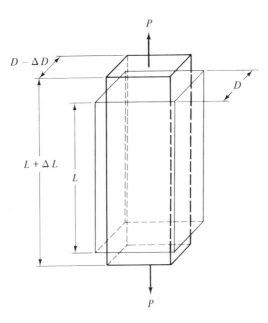

Fig. 11.6

where ΔL is the change in the length, L, produced by the pressure, P. *Poisson's ratio*, z, tells how much a body which stretches an amount $\Delta L/L$ will shrink at right angles to this (or how much it bulges laterally when it is squeezed lengthwise)

$$z = \frac{\Delta D/D}{\Delta L/L} \qquad \textbf{(11.2)}$$

Poisson's ratio is normally greater than zero but no greater than $1/2$. A material which has a Poisson's ratio of $1/2$ is said to be *incompressible* because its volume does not change when it is squeezed along one direction. This does not mean that its volume cannot be changed by compressive forces exerted in all directions at once. It means only that unidirectional forces will not change its volume. Liquids like water are incompressible in this sense. Water has a measurable bulk modulus (compare Equations 10.1 and 10.2, p.262), which tells how much it compresses under hydrostatic stresses.

The amount a body deforms by proportional elasticity under shear stress is given by the *rigidity* (compare Equation 10.3, p.262)

$$r = \frac{F_s}{2A} \qquad \textbf{(11.3)}$$

where F_s is the shearing stress producing the angular deformation, A (Figure 10.2, p.262).

Typical values of these elastic properties are given in Table 11.2. A bulk modulus of 10^6 atmospheres means that the volume of a rock is decreased

Table 11.2 *Typical proportional elastic constants of some common rocks. (Data from* Geol. Soc. Am. Memoir 97.)

Rock type	Pressure atmos.	Bulk modulus (Atmos.)	Young's modulus (Atmos.)	Rigidity (Atmos.)	Poisson's ratio
Basalt (Diabase)	4000	8.6×10^5	11×10^5	4.4×10^5	.27
Granite	4000	5.4×10^5	8.6×10^5	3.4×10^5	.27
Dunite	4000	12×10^5	17×10^5	6.8×10^5	.27
Limestone	500	6.4×10^5	6.3×10^5	2.4×10^5	.34
Sandstone	500	2.9×10^5	4.6×10^5	1.9×10^5	.23
Shale	1	1.2×10^5	2.7×10^5	1.2×10^5	.12

by one part in one million for each increase in pressure equal to normal atmospheric pressure applied to it. A column of such rock of density 3.0 grams/cc (3000 kg/cubic meter) and 10 km long will exhibit a fractional decrease in volume at the bottom of the column of .0029 due to the weight of the overlying rock. This is calculated as follows. Solving Equation 10.1 (p. 262) for $\Delta V/V$, and remembering that ΔP is the weight of the overlying material:

$$\frac{\Delta V}{V} = -\frac{\Delta P}{b} = -\frac{gHd}{b}$$

$$= -\frac{(9.80 \text{ newtons/kg}) (10^4 \text{ meters}) (3000 \text{ kg/m}^3)}{(10^6 \text{ atmos}) (1.01325 \times 10^5 \text{ newtons/m}^2/\text{atmos})} \qquad \textbf{(11.4)}$$

$$= -.0029$$

where ΔP is the pressure increase with depth, H; d is the density of the rock; and g is the force of gravity.

Similarly a Young's modulus of 10^6 atmospheres would indicate a shortening of one part in 10^6 per atmosphere of pressure. If the same rock had a Poisson's ratio of 0.25, it would bulge 0.25 as much as it shortened if not prevented by surrounding rocks.

Young's modulus and bulk modulus are not numerically equal because they describe different types of deformation. The pressure beneath a 10 km column is not necessarily hydrostatic. The weight of the column of rock produces a vertical stress. The attempt of the rock to bulge horizontally under this pressure is resisted by the surrounding rocks. Within Earth there is no place for the rock to go horizontally, so no bulge may take place. Instead, the tendency to bulge produces horizontal stress. Unless the yield strength or breaking strength of the rock is exceeded, only a vertical compression can occur. At the bottom of a 10 km column, the rock may be in a state of stress which is different in the horizontal and vertical directions. Shear stresses as well as compressive stresses will then act on the rocks. Because of the different strengths of the rock to different types of stress, the patterns of yielding tend to be complicated. The pressure of overlying rocks precludes there ever being a state of tension except at very shallow depths. Because the resistance to compression is great, most yielding will take the form of some variety of shearing.

As temperature rises, yield strength generally decreases. Consequently, rocks which would be brittle at surface temperatures may be ductile at depths of several tens of kilometers. Confining pressure, on the other hand, generally causes strength to increase. In the first few tens of kilometers from Earth's surface, the effect of temperature is greater than that of pressure, and yield strength is believed to decrease greatly. Beneath 100 km, this may no longer be so, as temperature changes more slowly.

If material is moved from one place to another at the surface, this changes the stress conditions at all depths. The expectable result of a change in surface load will be compression and expansion of the underlying rocks plus some flow from beneath the loaded region toward the unloaded region. This flow occurs, presumably, largely at a depth of some tens of kilometers where the yield strength is less than near the surface. Isostatic adjustment, according to this view, is the result of the low yield strength of rocks below Earth's crust.

Viscous plastic flow is one of the commonest types of plastic flow. It is determined by the solid viscosity, which is measured in poises. A viscous material, either solid or fluid, is said to have a *viscosity* of *one poise* if a shear stress of one dyne/cm^2 (0.1 newtons/m^2) between two surfaces one centimeter apart causes the surfaces to slip past one another at a rate of one centimeter/second

$$\text{viscosity in poises} = \frac{\text{stress gradient in dynes/cm}^2\text{/cm}}{\text{shearing velocity in cm/sec/cm}} \quad \textbf{(11.5)}$$

Table 11.3 compares viscosities in some solids and liquids. Clearly rocks will flow only slowly compared to water or air.

Solid viscous flow differs from liquid flow in several respects besides the generally much larger values of viscosity. First, a solid will deform elastically as well as plastically. Thus, if a stress is applied to a solid, part of the deformation is elastic. This part of the deformation is recovered if the stress is removed. This is not so for a liquid undergoing shear. Liquids have zero rigidity, and resist shear stress only by their viscosity and inertia. Second, a solid will begin to flow only when the stress applied exceeds the yield strength, and the rate of flow is proportional to the difference between the applied stress and the yield strength. Liquids have zero yield strength, and deformation occurs for all levels of stress.

It is possible that below the level of isostatic compensation yield strength becomes negligibly small. Even if the yield strength does go to zero here, any deformation which occurs will be partly elastic and partly plastic.

Viscosity and yield strength may depend on the length of time that the stress is applied. Rocks are sometimes found which have obviously suffered flow deformation, but which do not deform plastically in laboratory experiments. This may in some cases be because the natural deformation

Table 11.3 *Typical values of viscosity.*

Material	Temperature, °C	Viscosity, poises
Air	18°	.00018
Water	20°	.0100
Mt. Vesuvius lava	1100°	28,300
	1200°	2,760
	1400°	256
Pitch	15°	5.1×10^9
Soda glass	575°	1.1×10^{13}
Glacier ice	—	$10^{13} - 10^{14}$
Rock salt	25 – 400°	$10^{14} - 10^{18}$
Limestone	25 – 500°	$10^{13} - 10^{22}$
Crustal rocks	—	$10^{22} - 10^{23}$
Upper Mantle	—	$10^{17} - 10^{22}$

took place under temperature and pressure conditions different from those reproduced in the laboratory, or it may be because the flow occurred so slowly that the amount of deformation produced in the relatively short laboratory experiment was too small to measure, or it may be that after stress is applied for a long period of time the properties of the rock change, allowing flow to begin. One theory is that all rocks really are ductile, but that brittle rocks have such high viscosities that measurable flow occurs only in periods measured in thousands of years.

11.3 YIELDING IN ROCKS.

The result of all these different ways in which a rock can deform is that it is not easy to predict exactly how a rock will behave. There are too many ways in which it can yield. Furthermore, the processes of deformation tend to bring about changes in the internal structures of the minerals of which the rock is composed, changing their physical properties with time (Figure 11.7). Both elastic afterworking and plastic flow absorb energy from the forces which cause them. This energy is transformed thru internal friction into heat, which will raise the rock's temperature unless it is quickly carried away. Heat moves slowly thru most rocks, because they are poor conductors. The physical properties change with temperature.

The following much oversimplified picture may help to indicate the way rocks behave. Consider a rock to which a gradually increasing shearing stress is being applied in small steps (Figure 11.8). When the first step of stress is applied, there is an instantaneous elastic deformation by an amount which is proportional to the applied force. As time passes, without increase in stress, the amount of yield increases, but at a decreasing rate as delayed elastic yielding occurs. Each successive step in stress creates a new step in proportional response plus a temporarily accelerated delayed response.

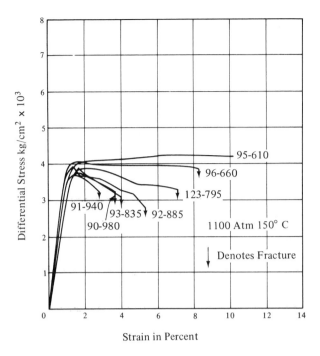

Fig. 11.7 *Typical experimental stress-strain curves for limestone. Note that after plastic flow begins, yielding is often so rapid that the applied stress falls, and that breaking (shown by vertical arrow) occurs at a lower stress than the rock withstood earlier. (After Heard,* Geol. Soc. Am. Sp. Pap. 79, *p. 217.)*

When the stress has been sufficiently increased, the yield strength of the rock is exceeded. At this point, there is a sharp increase in the rate of deformation. There continues to be elastic deformation, both instantaneous and delayed. Also, there is an additional yielding which steadily increases with time, and does not slow down as did the delayed elastic response. From this point on, each successive step in stress increases the rate of plastic flow.

Finally a stress jump will be applied which causes the rock to break along one or more surfaces. Shortly before breaking occurs there may be an acceleration of the plastic flow without any increase in stress. This represents reorganization of the molecular structure of the minerals of which the rock is composed as individual linkages between atoms are torn apart by the flow. Destruction of these linkages weakens the rock allowing both the increased flow and the final parting. Breaking can be expected soon after this accelerated flow begins, and may occur without there being any increase in stress.

After this fracturing, it may be impossible to maintain a high stress level on the rock because the fault planes will be weak and will permit displacement at much lower stresses than previously. Only a long period of

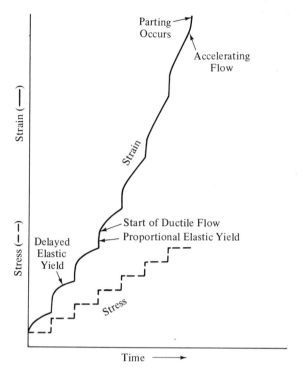

Fig. 11.8 *Idealized shear strain under successive increments of stress.*

quiescence which permits chemical processes in the rock to reseal the several blocks across the broken surfaces will again permit a high build-up of stress.

The role of flow is central in this picture. In nature, flow tends to relieve the stress which causes it, preventing further build-up of stress, or even causing it to decrease after a time. Thus, rocks which flow will seldom break. If this is the case, then there should be more breaks in the rocks of the crust than in those of the mantle, where flow is most likely to occur. This is exactly what is observed.

11.4 EVIDENCE FROM EARTHQUAKES.

When one examines rocks, displacements are commonly found showing that stresses have at some time been sufficiently strong to break them. Some of these breaks are very extensive, being traceable for many kilometers. Breaks of this sort are called *faults*. In large earthquakes, displacements on such faults sometimes reach Earth's surface (Figure 11.9). One theory of their origin postulates that nearly all earthquakes are caused by the breaking of rocks on such faults. The earthquake is the vibration of the ground which results when a parting occurs between two stressed blocks of rock. A similar vibration can be felt in a wooden baseball bat if

Fig. 11.9 *Cliff formed by fault displacement, Dixie Valley-Fairview Peak, Nevada, Earthquake of December 16, 1954. (E.S.S.A. Photo.)*

you hit the ball so hard that the wood splits. As the wood (or ground) splits, the sudden displacement spreads thru the bat (or Earth) as a distinctive elastic-wave pulse which feels quite different from the normal impact of bat on ball.

Assuming that most earthquakes are produced by faults in this fashion, the relative frequency of earthquakes with depth can give evidence as to how rocks deform. Figure 11.10 shows the relative frequency of large earthquakes as a function of depth at which the earthquake occurred. The point where the energy is released is called the *focus* of the earthquake. By far the majority of earthquakes occur within 60 km (36 miles) of Earth's surface. The frequency of their occurrence drops off with depth to about 300 km, then flattens with possible minima at 300 and 450 km. Earthquakes have not been observed to occur below 720 km. Earthquakes occurring within 60 km of the surface are called *shallow*. Those between 60 and 300 km are called *intermediate depth*, and those deeper than 300 km are called *deep focus*.

In the previous chapter it was shown that rigidity went thru a minimum between 50 and 300 kilometers depth. This is the same range in which strength and viscosity are thought to be least and the frequency of occurrence of earthquakes is decreasing. It was also shown that between 300 and 900 km the composition changes either mineralogically or chemically or

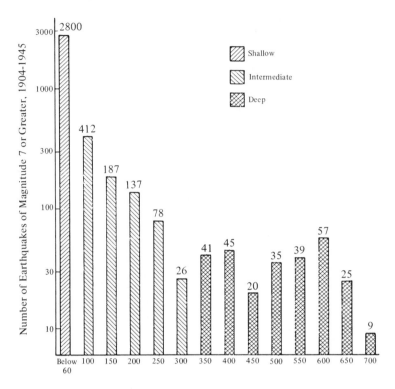

Approximate Depth of Focus in Kilometers (± 25 Km)

Fig. 11.10 *Plot of earthquake frequency of occurrence versus depth of focus. (Note that vertical scale is logarithmic.)*

both, but that beyond 900 km, the seismic velocities were consistent with little or no change in composition. Earthquakes do not occur in this lowest part of the mantle.

It is probable that the frequency of earthquakes is related to these changes of composition and physical properties, but the nature of this relationship is uncertain. There are three possibilities. One is that the strength of rocks steadily decreases with depth so that stresses are relieved more often by flow than by faulting as depth increases from 60 to 300 kilometers. Beneath 700 km, strength could be so small that faulting never occurs.

A second possibility is that breaking strength increases with depth. Although true within the crust, this is unlikely for the mantle. If it were so, there would be a tendency for the size of earthquakes to increase with depth. There may be some difference in the average size of the smallest earthquakes. These are not thoroughly studied because it is hard to record small earthquakes, especially deep ones. Earthquakes are always distant from the nearest observatory by an amount equal to their depth of focus. Thus a small shallow earthquake has a better chance of being near enough an observatory to be recorded than does a small deep earthquake. If there is any difference in the average size of large earthquakes, shallow earthquakes probably tend to be larger than deep ones. Here again the available data are limited, this time by the small number of large deep earthquakes. The existing evidence favors the largest deep earthquakes tending to be smaller than the largest shallow ones, which would be consistent with the postulate that strain is relieved more by flow than by breaking except near Earth's surface.

The third possibility is that stresses may be less often or more slowly built up at great depths than near the surface. It is obvious that erosion and sedimentation create stresses at the surface. It is also known from the patterns of gravity anomalies that relief of isostatic stress is accomplished at depths of around 100 km or less. If surface-generated stresses were all that were involved, there would be no need for deep earthquakes at all. The pace of isostatic adjustment is fast enough to relieve the surface stresses at shallow depths. The fact that earthquakes persist to 700 km indicates that differential stresses persist to that depth also. They may well cease below that depth because conditions are more stable. The rapid decrease in the number of earthquakes at intermediate depths could be due at least in part to a decrease in rate of stress generation with depth. To say more than this we need to know more about what causes the stresses.

11.5 FLOW AT SHALLOW DEPTHS.

The preceding discussion has suggested that flow is the typical mode of deformation for ductile rocks except where stresses accumulate rapidly, and that fracture is the mechanism for brittle rocks. Studies of rock properties have shown that igneous rocks, which are generally crystalline, tend to be brittle under near-surface conditions. Sedimentary rocks, however, are often ductile. Shales flow especially easily. In sections of deformed rocks, the shales often exhibit much more complex deformation than adjacent layers (Figure 11.11).

Water in a rock may affect the way it yields. Clay minerals commonly contain water as part of their composition, and are usually soft and often plate-shaped. They slip readily over one another, making it easier for the rock to deform by flow. The clay mineral glauconite is so soft when wet that persons and animals walking on outcroppings of this material occasionally become mired.

Water pressure may act to force mineral grains apart, allowing them to slip easily over one another. Even sands may become incapable of supporting heavy objects when they are sufficiently wet. A quicksand is a sand and water mixture, usually one where water is seeping upward thru the sand, into which a person will sink if he tries to walk on it. In such a material, the individual sand grains slide so easily past one another that the whole mass has almost no rigidity.

Fig. 11.11 *Deformed sediments in which clays have flowed and broken more than adjacent limestone layers. (Courtesy F. A. Donath, compare* Bull. Geol. Soc. Am. *v. 75, p. 51, plate 8.)*

Water can also help to redistribute stress in hard rocks, causing fractures. Near Denver, Colorado, a series of small earthquakes began when water was being pumped into an underground reservoir, where it was stored in the open pores of a sandstone. Examination of the frequency of these shocks showed that they correlated well with the rate of water injection into the ground (Figure 11.12). It was theorized that as the water filled the pores in the rocks it tended to exert pressure widening small cracks and concentrating stress already present in the rocks onto a few strong points. Finally, as the water filled more and more small cracks, the rocks began to slip along small faults, causing the earthquakes.

Rocks which are firm under normal circumstances may weaken under added stress and yield unexpectedly. Flow occurred in the ground on which the buildings in Figure 11.13 sat when subjected to earthquake vibrations.

11.6 VISCOSITY WITHIN THE EARTH.

Viscosity is the property which determines how fast a rock will flow under stress. Because viscosities of rocks are very high, allowing only very slow flow rates, they are hard to measure directly by laboratory experiments. Nature, however, provides a number of processes which can be used to calculate viscosity. The gradual post-glacial rise of Scandanavia (Figure 9.29, p. 258) is consistent with a viscosity of 10^{22} to 10^{23} poises in the upper-mantle rocks.

Another line of evidence is the damping of the motion of Earth's axis of rotation with respect to its axis of shape (Figure 4.8, p. 116). This oscillation has a period of about 420 days. The amplitude changes continually, there being a tendency for it to be reduced due to the energy absorbed in deformation of the solid earth. If this deformation is assumed to

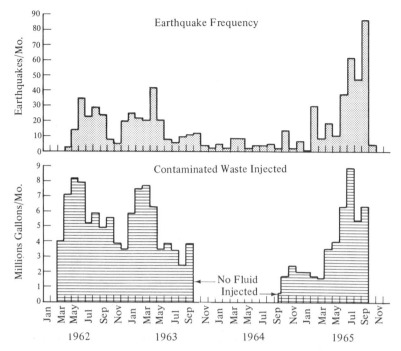

Fig. 11.12 *Local earthquake frequency correlates with fluid injection rate to an underground reservoir near Denver, Colorado. (D. M. Evans,* The Mountain Geologist, *v. 3, nr. 1, p. 27.)*

Fig. 11.13 *The ground under these rigid buildings flowed due to their weight during an earthquake. (E.S.S.A. Photo.)*

be viscous in nature and uniformly spread thru the mantle, then the necessary viscosity to explain the observations is roughly 3×10^{17} poises.

Observed rates of maximum gradual deformation of Earth's surface are of the order of magnitude of one to ten centimeters per year change in elevation and horizontal position. Such rates are expectable for rocks of viscosity 10^{17} to 10^{22} poises under reasonable stresses.

BIBLIOGRAPHY

Badgley, P. C., 1965, *Structural and Tectonic Principles*. Harper and Row, Publ. 521 pp.

Gutenberg, B., 1959, *Physics of the Earth's Interior*. Academic Press. 240 pp.

Gutenberg, B., 1951, *Internal Constitution of the Earth*. Dover Publ., Inc. 437 pp.

Howell, B. F., Jr., 1959, *Introduction to Geophysics*, McGraw-Hill Book Co., Inc. 399 pp.

Obert, L.; W. I. Duvall, 1967, *Rock Mechanics and the Design of Structures in Rock*. John Wiley and Sons, Inc. 650 pp.

Scheidegger, E. A., 1958, *Principles of Geodynamics*. Springer-Verlag. 280 pp.

12 VOLCANISM AND EARTH'S HEAT

The source of the energy which deforms rocks must be within Earth itself. Solar heat is radiated away too soon to provide such a source. However, temperature nearly everywhere increases with depth. Heat flows down this gradient, showing that the interior is a source of energy. Beneath a few meters depth, this steady efflux of heat plays a dominant role in determining the environment.

12.1 SURFACE TEMPERATURE.

The temperature of surface soil and rock is controlled primarily by five factors: (1) radiant heat received directly from Sun or back-reflected and reradiated down from the atmosphere; (2) the *albedo* (reflectivity) of the ground, which is the ratio of the reflected energy to the incident energy flux; (3) the heat rising from Earth's interior; (4) the *emmissivity*, which is the efficiency with which the exposed material radiates heat away; and (5) the convection loss thru the air. The first of these effects, the influx of radiant energy, is by far the most important in controlling surface temperature. The variation of surface temperature with latitude results from the variation in solar-energy influx per unit of area (Figure 6.7, p. 141). The dependence on elevation results from the variation in back radiation from the atmosphere, which decreases as the air gets thinner.

The flux of heat rising from Earth's interior is so much less than the average solar energy flux that it can be neglected as a factor in controlling surface temperature except in areas of volcanic activity. It does, however, almost everywhere cause a steady increase in temperature with depth into Earth. On this thermal gradient are superposed the fluctuations due to the changing surface temperatures. The thermal conductivity of rocks is so low that the daily variation of surface temperature, which may reach several tens of degrees Celsius, is rarely detectable at a depth of one meter (three feet). The annual variation is only about 20% as large at five meters depth as it is at the surface, and it is usually barely detectable at 20 meters. Long-term climatic changes are detectable to greater depths. The temperature at 28 meters depth at Paris, France has gone down 1°C over the past century.

In a few areas, the heat flux from below plays an important role in surface temperature. Around active volcanic vents, particularly on recent lava flows, the ground is noticeably warm to the touch and the radiant heat can be felt. Even if the heat flux is too small to have an obvious effect, it may be detectable by measuring the temperature gradient in the soil. High soil temperatures have been used to locate sources of natural steam which can be harnessed to generate electric power (Figures 12.1 and 12.2).

12.2 THE TEMPERATURE GRADIENT.

The temperature gradient (rate of change of temperature with depth) in Earth varies greatly from place to place. On fresh lava flows and in volcanic vents, temperature may rise hundreds of degrees in less than one meter (Figure 12.3). The average temperature gradient at the surface, however, is about 3°C/100 meters (1°F/60 ft). It usually lies between 0.6° and 6°C per 100 meters. If this rate continued to the Earth's center the temperature there would be 30°/km × 6370 km, or roughly 200,000°C.

The melting temperature would be reached somewhere around 35–60 km (1050–1800°C). This is consistent with evidence from volcanoes, whose lava appears to originate at about this depth (Section 10.6).

Since the mantle is solid except for local pockets of magma (molten rock), it can be concluded that the gradient sharply decreases somewhere in the crust of upper mantle. Evidence for where this decrease occurs is very scanty; but it seems reasonable to suppose that it occurs no deeper than 35–60 km since common silicate rocks would be molten at this depth if the surface gradient persisted unchanged this far. Such evidence as is available from measurements in deep wells indicates that, in the majority of cases, the temperature gradient tends to steepen with depth down to the greatest depth where measurements have been made. The evidence from the Hawaiian volcanoes suggests that the decrease does not occur much before 60 km on the assumption that the earthquake activity marks the beginning of the rise of molten magma in the volcano. The flattening of the temperature gradient may not occur everywhere at the same depth.

12.3 SURFACE VOLCANISM.

One of the results of a high temperature gradient is that the rocks involved will be lighter than normal. If a solid material is heated, its component atoms vibrate more actively. More active vibration means that they occupy a larger volume. For most silicates, an increase in temperature of 1000°C will cause a volume expansion of one to six per cent, or a linear expansion in each of three dimensions of one third of this. This means that a column of hot rock weighs less than a column of cold rock of the same length. The pressure, P, in a column of rock of length, L, is given by

$$P = gLd \qquad \textbf{(12.1)}$$

Fig. 12.1 *Natural steam escaping from a well being developed in Italy as a source of power for generating electricity.*

where d is density of the rock and g is gravity, which is close to 9.9 newtons/kg throughout the crust and mantle (Figure 10.9, p. 257). Assuming a typical coefficient of linear expansion (increase in length per unit of length), c_e, of 10^{-5} per °C, the difference in weight of a 60 km-(60,000 meter-) long column of rock of average density 3.0 gm/cc (3000 km/m³) with an average temperature difference, T, of 500°C would be

$$\Delta P = gL\Delta d = gLc_e Td$$
$$= (9.9 \text{ newtons/kg}) \, (60,000 \text{ m}) \, (10^{-5}/°\text{C}) \, (500°\text{C}) \, (3000 \text{ kg/m}^3)$$
$$= (9 \times 10^6 \text{ newtons/m}^2) \tag{12.2}$$

or about 900,000 times the force of gravity at Earth's surface.

If such a temperature difference does exist, there will be this much extra pressure at the bottom of the column. Isostasy requires that the crust of Earth bulge upwards in response to such a pressure. The height of the bulge can be found by solving Equation 12.1 for L

Fig. 12.2 *Two 28,000 kilowatt electrical generating plants run by natural steam in the Geysers area, California. (Courtesy Pacific Gas and Electric Co.)*

$$L = \frac{P}{gd} = \frac{9 \times 10^6}{9.9 \times 3000} = 300 \text{ meters} \qquad \textbf{(12.3)}$$

Here, then, is one reason why the surface of Earth is not even. Temperature differences from place to place will cause the surface to be higher or lower.

Rocks can respond in another way to this great pressure. If the rocks are molten at the bottom of our 60 km-long column, and if there is a crack extending to the surface, the magma will be forced up this crack. Imagine for simplicity that the crack extends from 60 km to the surface and is filled with molten magma throughout. At the surface the magma may be 1000°C or more hotter than the adjoining rocks and at the bottom of the column it may be at the same temperature as the adjoining rocks. Our requirement is only that the average temperature difference in the whole column be 500°C. Unless somehow we push down on the top of our column, the 9×10^6 newtons of force of Equation 12.2 are going to push the magma up out of the crack.

This is how a volcano works. Magma is going to keep flowing out of our crack until it piles up a big enough mass of rock to hold the rest of the magma in the ground.

Fig. 12.3 *Fresh lava flow on Mt. Etna, Italy. Red-incandescent lava lies within sight of the surface in the steaming vents.*

Because the magma is liquid, it will flow to the side away from the crack. If it flows easily (has a low viscosity) it will spread out horizontally over a large area before it solidifies. Such lava flows are common, e.g. in Oregon and Washington in the United States. If the lava is highly viscous, it will flow outwards more slowly, and may freeze before it has travelled far. Such lavas will form mounds. If lava continues to flow out of the top of the mound, a cone may be formed by successive layers of lava flowing down the sides and solidifying one above the other (Figure 12.4). More commonly, the lava breaks out of the sides of the mound. It may flow first from one vent, around which a mass of lava hardens. Later, eruption may shift to another vent, forming a new flow. Sometimes, the magma will flow outwards in cracks which never reach the surface, hardening into crystalline igneous bodies which will be seen only if uplift and erosion of the region eventually expose them.

This model of the process of formation of a volcano is good as far as it goes. However, if one tries to explain a volcano like Kilauea in Hawaii a difficulty arises. The top of Kilauea is 4170 meters above sea level and the whole island of Hawaii stands an additional 5000 meters above the level of the surrounding ocean floor. This is over thirty times the elevation produced by our example, which was based on an estimate of the depth to the source of the Hawaiian lavas. What have we underestimated?

It is conceivable that the temperature difference could be twice or even three times as great as postulated. More than this would require that the

Fig. 12.4 *a. Popocatepetl volcano near Mexico City erupted in 1523. b. Ix-taccihuatl volcano is either dormant or inactive.*

temperature at 60 km be well above the melting temperature of the rocks, which we know from their ability to transmit shear waves to be generally solid. The column would have to extend to 1850 km to provide the pressure difference solely due to the coefficient of expansion. Considering that isostasy indicates that horizontal flow occurs at depths around 100 km, it seems unlikely that a discrete column can extend to so great a depth.

However, the coefficient of expansion is not the only source of pressure available to drive lava out of a volcano. Up to now we have assumed that density differences resulted only from the effect of temperature and the coefficient of expansion. But magma is a liquid, not a solid. There is a change in density going from a solid to a liquid, and the additional possibility that the liquid magma at the bottom of the crack may be different in composition from the surrounding rocks. The density difference between liquid and solid silicates at the melting point is around ten per cent. This means that a liquid column 66 km long would weigh the same as a solid column 60 km long. Here are six of the nine kilometers we need to explain our volcano.

The remaining density is available from possible differences in composition between the two columns. Lava is not the only material escaping from the volcano. If one examines a pool of molten rock in a volcano it will be seen that vapors are bubbling out of the liquid. Samples of these vapors show that they are largely water, carbon dioxide, oxygen, nitrogen and smaller amounts of several sulfur compounds, hydrochloric acid, argon and other minor constituents (Table 12.1). The exact composition of the average volatile material escaping from within Earth is one of the major uncertainties of our knowledge of the planet. This is because much of the escaping material may be contaminants picked up by the magma on its way to the surface. The most common of these impurities is groundwater, although a wide variety of other constituents may be acquired as various crustal rocks are partially or wholly dissolved by the magma. These lightweight constituents help to decrease the density of the magma.

Table 12.1. *Typical compositions (volume percent) of gases escaping from Kilauea, Hawaii.*[*]

Gas	Composition of Hawaiian samples	Composition Pure dry air
Water	4.7 – 99.8	(0 – 4)
Hydrogen	0.0 – 0.58	.00005
Carbon dioxide	.00023 – 11.61	.033
Carbon monoxide	0.0 – 0.37	——
Methane	0.0 – .047	.00002
Hydrogen sulfide	0.0 – 0.12	——
Sulfur dioxide	0.0 – 6.24	——
Carbon disulfide	0.0 – .026	——
Oxygen	0.0 – 21	20.9
Nitrogen	.0001 – 78	78.1
Argon	0.0 – .064	.9

[*]Heald et al., Jour. Geoph. Res. v. 68, p. 546; and Holland, p. 93 in the *Origin and Evolution of the Atmospheres and Oceans*, John Wiley and Sons, Inc.

The density decrease takes two forms. At high pressures in the lower crust and in the mantle, the volatile materials are dissolved in the magma. Each atom occupies only its small share of the space. The volatiles help hold density down only because they are lightweight atoms. As the magma approaches the surface, however, a time will come when the pressure in the liquid becomes so low that bubbles of the gas begin to form. This will occur when the pressure equals that of 2400 meters (7900 feet) of water. At larger pressures water does not boil. Thus lava erupted on the deep ocean floor does not produce steam. When bubbles do form, they will be much lighter than the magma, and they reduce its density by expanding and occupying more space.

The bubbles will tend to rise thru the magma because of their lower density. As they rise, pressure decreases further, and the bubbles expand more and more. As the bubbles expand, the average density of the magma-bubble mix correspondingly decreases. Under these conditions, lava is blown out of the mouth of the volcano like water out of a garden hose. At Kilauea volcano at the height of the 1959 eruption, the lava column rose 580 meters (1900 feet) into the air in a column of red-hot molten rock (Figure 12.5). In other volcanoes, where it is largely gas which escapes from the vent, pieces of lava of a wide range of sizes are carried high into the air. Fine dust may be carried downwind across the countryside covering hundreds or thousands of square miles with layers of volcanic dust tens of centimeters thick. Larger chunks of volcanic rock, either already solid or in liquid blobs, are thrown into the air and pile up around the vent to form a cone by themselves or are interbedded with flows.

The fraction of the original magma which was potentially volatile is unknown. It could easily average two percent or more, since some igneous rocks which solidified in the crust contain this much water. Some volcanoes erupt little material except gases, and in their final stages may liter-

Fig. 12.5   *Lava fountains during the 1959 eruption of Kilauea volcano, Hawaii. (Courtesy U. S. Department of the Interior, National Park Service.)*

ally blow themselves up. Krakatoa in Indonesia is such a volcano. In 1883 it erupted so violently that two thirds of the volcano above sea level was destroyed. The largest explosion was heard in Australia, over 2000 km away, and the pressure wave was recorded in Berlin, Germany over 11,000 km away. Fragments of pumice, a porous rock full of gas pockets, floating in the sea around the remains of the island impeded navigation in the Sunda strait between Java and Sumatra until they sank or were washed ashore. Ashes fell over a 750,000 square-kilometer area, and fine dust was blown high into the atmosphere where some of it remained suspended for over a year, causing sunsets to be exceptionally red and sometimes green.

Not all of the volatile material escapes from the volcano thru the surface openings. Once the magma has risen to a depth where the pressure is too weak to hold all the water in solution, steam and aqueous solutions begin to separate from it. Because they have a lower viscosity, they can move more readily thru the rocks than can magma. They move thru fine cracks and small pores which magma would not enter. The result is that, fanning out from any intrusive body, veins are commonly found in which minerals are deposited out of solution from the escaping volatiles. The rocks immediately adjoining an intrusive body are often found to be altered by the introduction of elements carried outward from the magma and may be recrystallized to form new minerals with the aid of the high temperatures induced by the hot material. The process of recrystallizing the rock without melting it, either with or without change in chemical composition, is called *metamorphism.*

The loss of material from the magma changes its nature. Volatiles in the magma help to keep it liquid. As they leave, the cooling magma tends to solidify. In a silicate melt, the crystals which form first are generally not of the same composition as the liquid. Laboratory experiments and the examination of rocks in the field have shown that the first crystals to form are generally richer in iron, magnesium and calcium than the melt as a whole. They are also denser. As a result, they tend to sink to the bottom of the magma body. In any one region, there is sometimes a regular sequence of compositions of the intrusive bodies, the last-forming being the richest in the lighter, sialic compounds.

The water, the sulfur and the other volatile elements all tend to be enriched in the melt as it crystallizes. Their escape from an entrapped magma occurs most extensively at the very end of solidification.

The minor elements tend to separate out in families, some collecting in the iron-rich minerals, some occurring most commonly in the aluminum-rich fraction, and others staying in solution to the very end and being carried off by the volatiles. This process of one magma forming rocks of many compositions is called *magmatic differentiation.* The process of the iron-rich minerals solidifying out first is called *fractional crystallization.*

If a magma solidifies at one location without being disturbed, it will form a body in which composition varies slowly from top to bottom and which is surrounded by a halo of metamorphic rocks. This halo is the source of many of our metallic mineral resources. It is rare, however, for solidification to procede to completion without disturbance. The pressures which forced the magma into the rock in the first place often continue to act over long intervals of time. As new cracks form in the surrounding rock, the unsolidified portion of the magma may move on, leaving behind most of the solid material. This process is called *filter-pressing.* The magma which moves on is usually enriched in sodium, potassium, aluminum, silica and the volatile components. The volatiles may continue to be mobile to the very end and may escape to the surface in hot springs and fumaroles (gas vents). Hot springs are far more common than active volcanoes (Figure 12.6). They tend to persist long after the times when lava ceases to escape. They show that the process of differentiation and solidification of the intrusive material is still going on at depth.

Not all, probably not even most, water in hot springs is of magmatic origin. Some comes from groundwater which seeps down to the molten

Fig. 12.6 *Mammoth Hot Springs in Yellowstone National Park, U.S.A. (Courtesy U.S. Geological Survey. Photo by W. T. Lee.)*

magma or to still-hot solid rock. This water may be turned to steam and then force its way upward again. Thus, circulating groundwater may help to redistribute the soluble components of the magma and to concentrate them into ore deposits.

The overall result of these processes is believed to be a gradual increase in the iron and magnesium content of intrusive rocks with depth and a devolatilizing of all this material. Most of the water escapes to the surface, adding to the oceans, and the gases build up the atmosphere.

The mantle material from which the magma was originally obtained is obviously subject to differentiation also. No direct evidence is available as to its composition. It is logical to suppose, however, that the differentiation process has the same chemical trend as at shallower depths. The residual material which is left behind is probably richer in iron, magnesium and related elements and is lower in silica, alumina, volatiles and the suite of minor elements which is associated with these principal components. As a result, beneath a volcanic region there must be a root of material which is denser and of different composition from the mantle as a whole. Its other physical properties can also be expected to be different. The loss of the volatile constituents may have a considerable effect here. Volatiles tend to hold down the melting point of a rock and to decrease its viscosity. Their effect on properties like rigidity and bulk modulus are almost entirely unmeasured, but it is probable that there is considerable effect. They would be expected on theoretical grounds to decrease rigidity. Thermal and electrical conductivity are probably increased. The coefficient of internal friction and the solid viscosity, which control how easily a solid deforms under stress, are probably decreased by the presence of volatile constituents.

12.4 METHODS OF HEAT TRANSFER.

Volcanoes provide reasonable evidence that temperature increases to somewhere between 1000 and 2000°C at 60 km depth, probably varying appreciably from place to place. Temperature cannot be much above 2000°C or the rocks would be molten. The lower limit is much harder to establish, even though the high temperature gradient at the surface makes 1000° seem reasonable. The problem is that the temperature gradient must decrease somewhere because the mantle is solid throughout. How shallow can this decrease in gradient begin? To make even a reasonable guess, it will help to understand what factors control the temperature gradient.

The temperature gradient is determined by two things—the amount of excess heat available to be transferred and the ability of the rock to transfer it. The more heat there is available to be transferred the hotter the rock will be, and hence the higher the gradient will be. The more able the rock is to transfer heat, the faster the temperature will change with time, and hence the lower the gradient will be.

Heat is transferred in three fundamental ways. In the rocks of the crust the most important means is by conduction. If one atom in a crystal lattice is more agitated (hotter) than an adjoining one, the less agitated atom will in general receive momentum from the more agitated atom each time they strike one another in their movements. In this way heat flows from warmer to cooler areas. Thermal conductivity generally increases with pressure, due to the more intimate contact between adjoining atoms; but it decreases with increasing temperature under conditions encountered near the surface. The temperature effect is generally the stronger one, explaining in part why the temperature gradient at first increases with depth. The increase is also influenced by the changing composition of the rocks, particularly by any variation in water content with depth. This is of particular importance in sedimentary rocks, where the fine cracks and pores, which are normally filled with water, tend to be closed by the increase in pressure with depth.

As temperature rises, however, another conduction mechanism begins to play a role, and conductivity can be expected to increase. Part of the energy of a solid is in the form of defects in the crystal structure. These may take the form of a hole where an electron is missing or of an extra electron beyond what is needed to balance the charge on the atom. Either a hole or an extra electron can move about, carrying energy with it. This adds to the conductivity. Such defects are much more mobile at high than at low temperatures. A loose electron is particularly capable of transferring energy. Wherever it is, it greatly disturbs the arrangement of the atoms because there is no place the electron can conveniently fit into the crystal structure. The electrons must all play musical chairs, and one is always left with no place to sit. At a high temperature, an extra electron upsets the equilibrium of the atomic community like a rabid dog loose in a nudist colony. It produces a great deal of agitation. Under these conditions the ease of transfer of heat is improved. As the melting temperature is approached, therefore, conductivity can be expected to increase.

Rocks in general are very poor conductors of heat. This is why the effects of daily atmospheric temperature changes are so rarely observed below a meter's depth. The atmosphere changes temperature easily because heat can pass thru it as infrared radiation. The mineral assemblages found at Earth's surface are very efficient absorbers and reflectors of infrared, and poor transmitters of it. This is true also of most of the silicates found in buried igneous rocks. However, even if chemical composition does not change with depth, phase changes are expected in the upper mantle. It is believed that, at the temperatures and pressures found in the mantle, the transparency of the rocks may be greater than at the surface, and as much or more energy may be passed by radiation as by conduction. This effect will increase the total conductivity, though by how much is uncertain.

In the air, convection also plays an important role. Hot air rises and cold air flows in under it. This process cannot be seen in the crust, but presumably exists in the liquid outer core. The mantle is solid, so is not subject to rapid convection. However, under large enough stresses even a solid rock will yield, and the evidence of isostasy shows that rocks do flow. Although movements must be very slow, convection cells in the mantle may play a role in the transfer of heat. The ability of a solid to flow can be expected to increase as the melting temperature is approached. Thus, much or all of the crust may be too strong to flow. The mantle, on the other hand, can be expected to be weaker. Again, the amount of heat transferred by convection in the mantle is unknown; but below the point where the temperature approaches the melting temperature, it is reasonable to expect convection to play a significant role. Here again, lack of knowledge of the composition makes it hard to predict rates of heat transfer. The role of the volatile constituents of the mantle may also be critical in decreasing the yield strength and viscosity of the material.

Since all three means of heat transfer—conduction, radiation and convection—increase in effectiveness when the melting temperature is approached, our original estimate of 1000–2000°C at 60 km may be a good figure for all areas of Earth, both continental and oceanic. About all that can be said is that there is no reason to expect the temperature to be below this at 60 km under any region. Nor is there any reason to expect it to average more than this over any large region. Beneath 60 km, the gradient would be expected to be less than above this depth.

12.5 THE HEAT FLUX.

The heat flux, H_f, at Earth's surface is determined from the temperature gradient and the thermal conductivity, c_h, because only conduction is important as a transfer mechanism in surface rocks.

$$H_F = c_h \frac{\triangle T}{\triangle L} \tag{12.4}$$

where $\triangle T$ is the temperature difference over the depth range $\triangle L$. Conductivity is measured on a sample of the rock and $\triangle T / \triangle L$ is measured in a well or by thrusting a temperature measuring device into the mud on the bottom of the ocean or of a lake.

Except in volcanic regions, the heat flux in general exhibits much less variability than the temperature gradient. The average heat flux is 1.5 × 10^{-5} ± 10% kilogram-calories/m²/sec (470 kg-cals/m²/year). There is no observable difference between the average heat flow on land and under the oceans, though the latter has been more thoroughly sampled. A heat flow of over 1.9 × 10^{-5} kg-cals/m²/sec is considered anomalously high. Sixteen such regions have been found, some on land but more associated with mid-oceanic ridges, where heat flow is often above average. A heat flow of less than 0.08 kg-cals/m²/sec is considered anomalously low. Eleven such areas are known, largely in the ocean basins. The covered shield areas of the continents generally show lower than average but not anomalous heat flow.

The total heat flux over the whole surface of Earth can be estimated by multiplying the average heat flux by the area, and amounts to 7.7 × 10^9 kg-cals/sec (2.4 × 10^{17} kg-cals/year). This is about .00019 times the total solar heat flux of 4.1 × 10^{13} kg-cals/sec over the whole Earth. This large difference explains why the internal heat flux plays no significant role in controlling surface temperatures. On the other hand, because the temperature gradient is so high, the surface temperature plays only a minor role in the internal temperature.

The range of average temperatures from the equator to the poles is only about 70°C (126°F), which is equal to the temperature increase in a little over 2 km depth. (Compare Figure 6:7, p. 141.) This does mean, however, that the surfaces of equal temperature (isogeotherms) are deeper under the poles than at the equator. Earth's radius is 21.5 km less at the poles than at the equator, so the average temperature gradient from the surface to the center of Earth must be 23.5/6370 or 0.37 per cent higher at the poles than at the equator.

In addition to the heat lost everywhere by conduction, there is a small additional loss in volcanic regions thru the escape of hot gases into the atmosphere, thru hot springs, and by the cooling of lavas at the surface. The total heat lost thru volcanism has been estimated to be about 10^7 kg-cals/ sec or about one thousandth of the total heat lost by conduction.

12.6 SOURCES OF TERRESTRIAL HEAT.

The second factor which controls the size of the temperature gradient is the amount of heat to be transferred. Here again it is difficult to be precise, but rough estimates of the relative importance of each potential heat source can be made.

One source of Earth's present heat is the original temperature of the materials from which it is formed. The current rate of heat loss is so small that original heat could be the only source. To prove this, it is only necessary to show how fast the present heat flow would drop the temperature if all parts kept the same relative temperatures that they have now. It is not necessary to know what the temperatures are at any depth. It is necessary, however, to estimate the heat capacities of the rocks. Heat capacity, h_c, is the amount of heat needed to raise one gram of rock 1°C. The change in temperature, ΔT, occurring in a time interval, T_s, is related to the heat flux, H_f, by the equation

$$\Delta T = \frac{\text{heat lost thru surface in time } T_s}{\text{heat required to change whole earth by } 1°C}$$

$$= \frac{H_f T_s}{M_e h_c} \tag{12.5}$$

where M_e is the mass of Earth.

Typical values of heat capacity for rocks lie in the range 0.12 to 0.33 kg-calories/kg/°C. Using 0.20 cals/kg/°C for the heat capacity, 7.7×10^9 kg-cals per second for the heat loss, 10^6 years (3.156×10^{13} seconds) for T_s and 6.0×10^{24} kg for M_e

$$\Delta T = \frac{7.7 \times 10^9 \times 3.156 \times 10^{13}}{6.0 \times 10^{24} \times 0.20} \tag{12.6}$$

$$= 0.2° \text{ per million years}$$

For the approximately five billion years that Earth is believed to have existed, this amounts to a potential change in temperature of 1000°C.

If initial heat were the only source of Earth's heat, this would mean that Earth could have started at an average temperature somewhat above 1000°C hotter than at present and cooled since then. (The heat flow would have been larger when Earth was warmer.) Since the present temperature at 60 km is near the melting temperature, Earth would initially have been molten at that depth, and presumably at most or all greater depths. One would, in this case, have to deal with an initially liquid Earth and explain how it solidified throughout most of its volume.

This is far too simple a model. If one examines the rocks of which the crust is composed (Table 9.2), it will be found that they contain a small but measurable amount of radioactive elements. In disintegrating, an element gives off gamma radiation and the particles into which it breaks fly apart violently. Both the gamma radiation and particle motion are largely converted into heat. Table 12.2 shows values of the heat generated per year in this manner for typical granites and basalts.

Table 12.2. Typical values of heat generated per year by principal radioactive elements. (*Data after GSA Memoir 97; Mason, Principles of Geochemistry.*)

Element	kg-cals gen. per year per kg	Granite		Basalt (diabase)		Stony meteorites		Iron meteorites	
		Conc. ppm	kg-cals gen. per 10⁶ kg	Conc. ppm	kg-cals gen. per 10⁶ kg	Conc. ppm	kg-cals gen. per 10⁶ kg	Conc. ppm	kg-cals gen. per 10⁶ kg
U238	0.71	3.7	2.6	0.52	0.37	.014	.00)9	10^{-5}	7×10^{-6}
U235	4.3	.026	0.1	.0035	0.02	.000098	.0004	7×10^{-8}	3×10^{-7}
Th232	0.20	52	10.4	2.4	0.48	.040	.0080	10^{-5}	2×10^{-6}
K40	0.21	5.4	1.1	0.64	0.13	.011	.0023	—	—
Total			15.2		1.00		.0206		9×10^{-6}

If the total heat generated per year is calculated using these figures, it is found that 10 km of granite of average density 2.7 gm/cc (2700 kg/m³) will generate 410 kg-cals/year/m² (10⁴m × 2700 kg/m³ × 15.2 × 10⁻⁶ kg-cals/kg/year). Similarly 20 km of basalt of average density 3.0 gm/cc will generate 60 kg-cals/year/m² (2 × 10⁴ × 3000 kg/m³ × 1.0 ×10⁻⁶ kg-cals/kg/year). Thus 10 km of granite plus 20 km of basalt will produce 470 kg-cals/m²/year of heat, which is the average heat flux thru Earth's surface. The heat generated in the thin basaltic crust under the oceans is much less than the surface heat flux.

The heat generated in the mantle and core is impossible to calculate because the compositions are not known. If these layers had the composition of meteoritic material, they would add substantially to the heat flow under both the continents and the oceans. Because the heat generated in basaltic rocks is less than that generated in granitic rocks and because composition is believed to change further beneath the crust in the same direction that it changes going down in the crust, it is reasonable to suppose that the concentration of radioactive elements in the mantle and core are less than in basalt. This is, however, almost entirely guesswork. The best that can be argued with assurance is that a concentration no higher than in basalt and probably at least as low as in meteorites is reasonable.

The following calculation will demonstrate that the radioactive element concentration must be much less than in basalt. If the concentration throughout the whole Earth were equal to the concentration in basalt, then 10^{-6} calories would be generated per kg per year. Since Earth's mass is 6×10^{24} kg, this would amount to 6×10^{18} calories per year for the whole Earth, or 12,000 calories per m² of Earth's surface. This is 26 times the observed heat flux. If the excess heat of 5.76×10^{18} cals/year accumulated in Earth, it would gain heat at the rate

$$\Delta H = \frac{5.76 \times 10^{18} \text{ cals/year}}{6.0 \times 10^{24} \text{ kg}} = 0.96 \times 10^{-6} \text{ cals/kg/year} \quad \textbf{(12.7)}$$

Assuming a heat capacity in the rocks of 0.20 cals/kg/°C, Earth would be warming at the rate

$$\Delta T = \frac{0.96 \times 10^{-6} \text{ cals/kg/year}}{0.20 \text{ cals/kg/°C}} = 4.8 \times 10^{-6} \text{ °C/year} \quad \textbf{(12.8)}$$

or 4800°C per billion years. Earth would long ago have become molten, which it has not. Therefore, it seems likely that the concentration of radioactive elements in the mantle and core is much less than in basalt.

This conclusion is reinforced by another factor. The rate of heat generation by the decay of radioactive elements is steadily declining. The number of atoms of any radioactive element disintegrating per second is proportional to the number of atoms present. In each successive second there are fewer atoms by the number that disintegrated in the previous second. Consequently, the amount of heat generated steadily decreases. The decline is exponential and follows the rule

$$N_t = N_0 e^{-d_c T} \qquad \text{(12.9)}$$

where N_t is the number of atoms at time T, N_0 is the number at time $T = 0$, and e has the value 2.7183. The quantity d_c is called the disintegration constant. Since the amount of heat generated per time unit, H_t, is proportional to the number of atoms disintegrating, and hence to the number of atoms present, the amount of heat generated per unit of time decreases according to the similar formula

$$H_t = H_0 e^{-d_c T} \qquad \text{(12.10)}$$

The time T_h for which N_t/N_0 (or H_t/H_0) is 0.5 is called the *half-life* of an element. Half-lives for the principal natural radioactive elements are given in Table 12.3. The shorter the half-life, the more rapidly the heat generation will change with time. Figure 12.7 shows the ratio of heat generated per year compared to the present rate as a function of past time for five elements. Note that the vertical scale is logarithmic. Heat generated for each element increases exponentially with past time.

One of the implications of Figure 12.7 is that Earth can have existed in its present form for only a limited time in the past. If the U235 curve is projected to 12 billion years ago, the amount of U235 would then have been 150,000 times as great as at present. But Pb207 is formed from U235. If disintegration had been going on for 12 billion years, there should today be at least 150,000 times as much Pb207 in rocks as U235. There is not nearly this much. The present ratio is only about 157 atoms of daughter Pb207 to each atom of parent U235. Equation 12.9 can be rearranged and solved for the length of time, T, it would take to reach the ratio of N_t (present uranium) to N_0 (original uranium) assuming there was no Pb207 present initially.

$$T = \frac{\ln_e (N_0/N_t)}{d_c}$$

$$= \frac{\ln_e 158}{9.72 \times 10^{-10}} = 5.2 \times 10^9 \text{ years} \qquad \text{(12.11)}$$

(Note: $N_0 = 1 + 157$ because the number of atoms of U235 originally present must equal the number of atoms of Pb207 now present plus the number of atoms of U235 now present.) Unless the crust has been enriched in U235 or depleted in Pb207 since Earth's origin, this event occurred less than 5.2×10^9 years ago.

The age obtained using the U238 series is only slightly larger (Table 12.4). The ages obtained using other series are substantially larger, but are less meaningful. In the case of thorium and rubidium, the half-lives are so large that the effect of any Pb208 or Sr87 originally present is very great, and tends to increase greatly the calculated age. The age obtained from

Table 12.3 *The principal Natural Radioactive Series (Data from 46 ed.* Handbook of Physics and Chemistry, *1965, Chemical Rubber Publ. Co.)*

Element	Mass number	Atomic number	Half-life	Radiation	Disintegration constant
			Uranium Series		
U	238	92	4.51×10^9 years	α	1.54×10^{-10}/yr
Th	234	90	24.1 days	β	
Pa	234	91	6.7 hr, 1.18 min (2 isomers)	β	
U	234	92	2.5×10^5 years	α	
Th	230	90	8×10^4 years	α	
Ra	226	88	1,622 years	α	
Rn	222	86	3.823 days	α	
Po	218	84	3.05 min	99.8%α, 0.02%β	
Pb	214	82	26.8 min	β	
or At	218	85	1.35 sec	α	
Bi	214	83	19.7 min	99.96%β; 0.04%α	
Po	214	84	1.6×10^{-4} sec	α	
or Tl	210	81	1.3 min	β	
Pb	210	82	21 years	β	
Bi	210	83	5 days	β	
Po	210	84	138.4 days	α	
Pb	206	82			
				$8\alpha + 6\beta$	
			Actinium Series		
U	235	92	7.13×10^8 years	α	9.72×10^{-10}/yr
Th	231	90	25.6 hr	β	
Pa	231	91	3.34×10^4 years	α	
Ac	227	89	21.6 years	99%β, 17%α	
Th	227	90	19 days	α	
or Fr	223	87	22 min	β (small α alternate)	
Ra	223	88	11.7 days	α	
Rn	219	86	4.0 sec	α	
Po	215	84	1.8×10^{-3} sec	α (small β alternate)	
Pb	211	82	36.1 min	β	
Bi	211	83	2.15 min	90.7%α, 0.3%β	
Tl	207	81	4.78 min	β	
or Po	211	84	25 sec	α	
Pb	207	82			
				$7\alpha, 4\beta$	
			Thorium Series		
Th	232	90	1.39×10^{10} years	α	4.99×10^{-11} yr
Ra	228	88	6.7 years	β	
Ac	228	89	6.13 hr	β	
Th	228	90	1.91 years	α	

Table 12.3 (cont.)

Element	Mass number	Atomic number	Half-life	Radiation	Disintegration constant
Ra	224	88	3.64 days	α	
Rn	220	86	51.5 sec	α	
Po	216	84	0.16 sec	α	
Pb	212	82	10.64 hr	β	
Bi	212	83	60.6 min	$6.65\%\beta$, $35\%\alpha$	
Po	212	84	3×10^{-7} sec	α	
or Tl	208	81	3.1 min	β	
Pb	208	82			
				6α, 4β	

Potassium Series

K	40	19	1.3×10^9 years	$\beta(89\%)$ or K-orbit electron capture (11%)	5.3×10^{-10}/yr
Ca	40	20			
or Ar	40	18			

Rubidium Series

Rb	87	37	4.7×10^{10} years	β	1.5×10^{-11}/yr
Sr	87	38			

Carbon Series

C	14	6	5770 years	β	1.2×10^{-4}/yr
N	14	7			

the potassium series is increased in uncertainty by the very large ratio of Ca40 to K40. It is, therefore, unusually subject to an augmentation of the calculated age due to Ca40 originally present and not derived from K40. It seems unlikely, therefore, that Earth can have existed for much longer than 5.2×10^9 years. A lesser age is, of course, possible.

Conditions in Earth are not proper for the formation of the heavy elements. Even in Sun, it is unlikely that there are many elements heavier than iron being formed. The only places in the universe where it is likely that such elements are forming today are in novas and supernovas and possibly in the centers of the superdense dwarf stars, where conditions are so different from anything we know about that it is hard to predict exactly what goes on. If the whole universe were once packed into a very small volume, then this super-concentration of matter could have been at so high an energy level that nuclear reactions could produce new radioactive elements from older materials. The age of 5.2×10^9 years calculated for Earth does not apply to the origin of the elements themselves because of the possibility of differentiation of parent and daughter elements during the formation of Earth.

If the relative abundances of the elements in Earth are known, then the relative rate of heat generation in the past can be calculated (Figure 12.8).

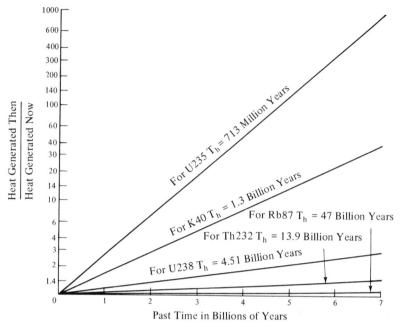

Fig. 12.7 *Increase in abundance of radioactive elements with past time.*

It is not necessary to know the actual abundances, only the relative abundances. Assuming that the relative abundances of the elements and isotopes in the whole Earth are the same as given in Table 12.4 for the crust, then 2.95 billion years ago the heat being generated was twice what it is today. Five billion years ago, it was five times what it is today. The role of K40 (potassium) is particularly important in determining this rate. Although at present K40 is less important than either U238 or Th232 as a heat source, its shorter half-life causes its rate of heat generation to double every 1.3 billion years of past time.

This increase in heat generation with past time means that 4 billion years ago there was 3.1 times as much heat being generated as at present. The present rate of heat generation in the continents is enough to provide the total heat flowing thru Earth's surface above them. If they existed four billion years ago, they would have been heating so rapidly that by now they would have melted. The continents must have evolved since this time. The fact that the amount of heat at present generated in the continents is very nearly equal to the heat flow at Earth's surface suggests the possibility that the thickness of the continents may be related to the rate at which they conduct heat to the surface.

By far the largest part of the mass of Earth is in the mantle. Hence its radioactive-element content is crucial in determining the amount of heat being generated today. If the figures presented in the last column of Table 12.5 are correct, then Earth is at present cooling. However, it would take only about a sixty percent larger radioactive heat generation in the mantle to make the average heat generation in Earth equal to the heat flux thru the surface. A little over two billion years ago, this much heat would have

Table 12.4. Time required to reach present isotopic abundance assuming there was no daughter material present initially

Parent	Daughter	Parts per million element abundance in crust°	Isotope abundance	Isotopic abundance relative to U238	Present daughter/parent atomic abundance	Half-life†	Age
U235	Pb207	2.7 / 12.5	.0072 / .215	.00725	157	0.713 × 10⁹ years	5.2 × 10⁹ years
U238	Pb206	2.7 / 12.5	.9928 / .252	1.00	1.35	4.51 × 10⁹ years	5.5 × 10 years
Th232	Pb208	9.6 / 12.5	1.00 / .52	3.59	0.755	13.9 × 10⁹ years	11 × 10⁹ years
Rb87	Sr87	90 / 375	.278 / .070	9.34	1.05	47 × 10⁹ years	48 × 10⁹ years
K40	Ca40	21,000 / 42,000	.000118 / .97	.925	19,400	1.3 × 10⁹ years	19 × 10⁹years

°Data from Ahrens, *Distribution of the Elements in our Planet*, McGraw-Hill Book Co., Inc.
†Data from Geol. Soc. Am. Memoir 97 and *Handbook of Chemistry and Physics*, 46th Ed., The Chemical Rubber Co.

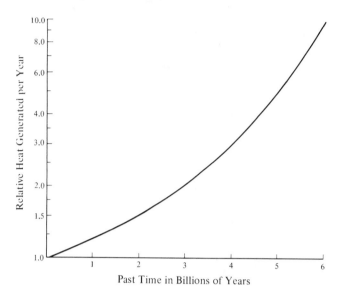

Fig. 12.8 *Relative heat generated assuming that at present U238: U235: Th 232: K40 = 1.00:0.00725: 3.59: 0.925.*

been generating if the mantle had a stony-meteorite composition. This means that one of the following must be true: (1) the interior of Earth was getting hotter until two billion years ago; or (2) the mantle today contains less radioactive material than stony meteorites; or (3) the heat flux was formerly higher than at present. These alternatives are not mutually exclusive. If the radioactive-element content of the mantle is more than sixty percent greater than that of stony meteorites, the mantle is still heating.

The possibility that the mantle may still be heating raises the question of how low its original temperature may have been. Either original heat or radiogenic heat can account for the total heat flow at the surface today. Plenty of heat is available to account for the present observations. What is uncertain is whether the history of Earth's interior is primarily one of heating, or cooling, or a combination of the two. In deciding this, it is important to know what is the minimum possible temperature at the initial conditions. To investigate this, assume first that, as one extreme, Earth originated by the gathering together of matter which possessed no heat whatever before it was added to Earth. As each piece of material fell to Earth it would gain energy dropping thru Earth's gravitational field. This energy would in large part be converted into thermal energy. If the incoming pieces were small enough, most of this energy would be imparted to Earth's atmosphere, from which it would be radiated into space. Larger pieces would strike the solid surface. When this occurred, seismic vibrations would be transmitted into the interior, where they would eventually be absorbed as heat. The surface materials would be compacted and often melted by the impact. An uncertain part of the energy of the incoming meteorite would be transferred to Earth. If even a small fraction of this

Table 12.5. Estimates of heat generated in Earth.

Assumed type of rock	Heat gen. cals/yr/kg (Table 12.2)	Average density in gm/cc allowing for pres.	Layer depth continental section (km)	kg/cals gen. cont. earth per year	Layer depth oceanic section (km)	kg-cals gen. oceanic earth per year	Kg-cals/yr .395 cont .605 ocean
Granite	15.2×10^{-6}	2.70	0 – 10	209×10^{15}		—	83×10^{15}
Basalt	1.00×10^{-6}	3.00	10 – 35	38×10^{15}	5 – 10	8×10^{15}	20×10^{15}
Stony Met.	$.0206 \times 10^{-6}$	4.50	35 – 2900	83×10^{15}	10 – 2900	83×10^{15}	83×10^{15}
Iron Met.	9×10^{-12}	13.0	2900 – 6370	$.02 \times 10^{15}$	2900 – 6370	$.02 \times 10^{15}$	$.02 \times 10^{15}$
Total				330×10^{15}		91×10^{15}	186×10^{15}
				650 cals/m^2		180 cals/m^2	370 cals/m^2

Average over area of Earth (5.1×10^{14} m^2)
Measured heat flow is 470 kg-cals/year/m^2 or 2.4×10^{17} kg-cals/year

energy of impact were retained by Earth, it is hard to escape the implication that the average temperature would be raised well toward the melting temperature. The minimum initial temperature must, therefore, have been the present surface temperature or above this. The present surface temperature is what would be achieved if nearly all the energy were radiated away as fast as it was supplied (assuming an atmosphere similar to the present atmosphere).

As the mass of Earth increased, the weight of the growing surface layers would compact the interior. As the center was compacted, all layers would fall thru the gravitational field of Earth. This field would be smaller initially because Earth was then smaller, but it would have increased as incoming mass was added.

To get some idea of how much energy is involved, consider how much energy, E, is gained by a body of mass, M, falling a distance, H, under Earth's present surface gravitational field, g.

$$E = MgH \qquad \qquad \textbf{(12.12)}$$

$$\frac{E}{M} = gH \qquad \qquad \textbf{(12.13)}$$

For a fall of one meter thru a field of 9.8 newtons/kg, 9.8 newton-meters or .00234 kilogram-calories per kg is gained. For a heat capacity of 0.2 calories/kg/°C this would produce a temperature increase of .0117°C. A one-kilometer fall would increase the temperature by 11.7°C. A ten-kilometer fall would increase temperature by 117°C. Assuming that the incoming material had the density of 3.4 gm/cc, the total fractional volume change from the start to the finish of the compaction process, at which time the average density is that of the whole earth, would be roughly $\dfrac{5.51}{3.4}$ or 1.6. A decrease in volume by a factor of 1.6 corresponds to a decrease in radius by $\sqrt[3]{1.6}$ or roughly a factor of 1.17. Seventeen percent in Earth's radius is 1100 km. A body dropping 1100 km and gaining 11.7°C/km would increase in temperature by 13,000°C. The average fall would be about half this much, so a temperature gain of 6500°C is a more reasonable estimate. Even this is probably more than the increase in temperature due to the compaction of Earth's interior, because gravity is less throughout the innermost layers, and because gravity was less during the early part of the compaction. These factors are in part compensated by the fact that differentiation has occurred in the interior, with the heaviest materials sinking toward the center to form the core. Although an exact calculation of the minimum temperature is not possible, it is not likely to be below 2000°C on the average throughout Earth. The initial temperature could be enough to melt the whole Earth.

The overall result of these arguments, especially those involving compaction and radiogenic heat, is that, although the temperature of Earth's interior may quite possibly have started low, it must have risen to above or to very near the melting point at some time in the past. Only if the radio-active-element content of the mantle is well below that of stony meteorites is this escapable.

The interior today consists of a mantle and a core. Unless Earth formed by the infall first of dense core material, then later of lighter mantle material, the layers must have separated from an original mix of one composition. It is much easier to see how this could have occurred if the interior were at one time liquid. The alternative is that differentiation occurred by partial melting and flow of the liquid portion or by solid diffusion. The speed with which an element can diffuse thru a solid is difficult to measure but it is certainly very slow except near the melting temperature. The interior must, therefore, have been at least near melting for some portion of Earth's history.

There are a number of other processes of heat generation within Earth. All are either known to be minor in comparison to those already discussed, or are minor today and not known to have ever been important. These sources include such things as chemical reactions and energy gained or lost thru crystallization. Chemical reactions at great depths are more a result than a cause of temperature changes. Everywhere in Earth, the suite of elements present tends to arrange itself into minerals which have maximum stability at the existing temperature and pressure. In general the high-pressure and high-temperature mineral phases require energy to form from low-pressure and low-temperature forms. The need for this energy uses up a small part of the energy creating the high pressures and temperatures.

The transfer of energy from Earth's rotation to Moon's revolution about Earth is accompanied by tidal friction. Almost all of this is released along the interface between the oceans and their floors, especially in shallow water, but a small fraction may be released thru solid friction in Earth's interior.

When an earthquake occurs, the disturbance spreads thru Earth as seismic waves. These are ultimately absorbed, the energy presumably being converted into heat. The amount of energy involved is about 8×10^6 kilogram-calories/sec. This is about .001 of the heat flow thru Earth's surface.

In addition to electromagnetic radiation, small, massless, uncharged particles called neutrinos are released by Sun and other stars. These particles are so small and nonreactive with matter that most of them striking Earth will pass right thru it. A few, however, will be absorbed and their energy released as heat. The total energy involved in the neutrino flux is quite large, of the order of magnitude of .04 kg-cals/m^2/sec, or about 3000 times the heat flow from Earth's interior (but only one eighth of the solar constant). However, the portion expected to be absorbed is thought to be negligibly small, several orders of magnitude too small to contribute significantly to terrestrial heat. Only uncertainty about the rate of neutrino absorption leads to its being considered at all.

Earth's magnetic field is another factor in the energy balance of Earth's interior. The best explanation for the magnetic field seems to be that it is caused by electrical currents in the core. These currents are opposed by the resistivity of the rocks thru which they flow. They must lose energy to the rocks continually. This energy appears as heat. Electromagnetic fields are a basic means of transferring energy from one place to another. A current in one place generates a magnetic field which in turn can generate a current elsewhere. In Earth, however, the magnetic field may resist energy transfer more than aid it. A temperature gradient in the outer core is the assumed driving mechanism of the convection cells which are the generators of the field. The electrical currents flow in the same region where the field is being generated, and hence return heat to the very place where a heat flow is needed to produce the field. The process cannot be indefinitely self-perpetuating, and it will not add to the internal heat except insofar as changes in the magnetic field are externally generated.

A part of the field is related to atmospheric electrical currents. It was shown in Chapter 8 that the daily magnetic field variations are generated by the influx of solar energy which ionized the atmosphere and caused it to expand and convect. The resulting magnetic-field fluctuations produced electrical currents in the solid part of Earth. These are opposed by the resistance of the rocks. The resulting absorption probably takes place largely in the upper mantle and crust, and makes a small contribution to the heat produced by other means.

12.7 ADIABATIC GRADIENT.

It would be much easier to estimate the temperature in the mantle if it were liquid instead of solid. If the mantle were liquid, then the temperature would almost certainly be near or below the adiabatic gradient (Section 6.2, p. 137). This is the rate of increase of temperature with depth needed to cause convection. Convection will not occur for very low gradients because of the effect of the increase of pressure with depth. To see why this is so, consider Figure 12.9. This represents a column of material all of one composition. Going from top to bottom the pressure increases from P_1 to P_h. Consider first the case where temperature, T, is the same at the top and bottom of the column. In this case, density will be greater at the bottom than at the top because of the effect of pressure. If a body of material is moved from region H to region L it will expand because of the removal of pressure. But to expand requires energy, which it will get from its internal energy, causing it to become colder. But the loss of temperature will result in contraction, increasing the density. Moved from region H to region L, it will be denser and heavier than the material in region L. Material moved from L to H will be similarly raised in temperature and hence will be lighter than the rest of the region-H material. Gravity will oppose any such change in position. Every part of the material in such a column will be stable, tending to remain just where it is and resisting any change in this arrangement.

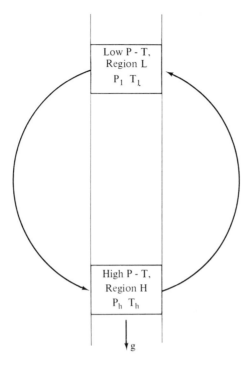

Fig. 12.9

Now consider the case when there is a temperature gradient in the column, with temperature increasing from L to H. As long as the temperature increase from L to H is less than the amount which a body warms on being moved from L to H, the column remains stable as in the previous case. If the temperature gradient is larger than this, however, a body moved from H to L remains hotter when it gets to L than the surrounding material, and hence is lighter. A body going from L to H is colder and denser than its surroundings when it reaches H. Such a column is unstable. Gravity will tend to make it overturn, moving the warmer material to the top and the cooler material to the bottom. The temperature gradient at which the parts of a column can be moved up or down without ever differing in temperature from the surrounding material is called the *adiabatic gradient*. In a fluid, this is the gradient where convection begins.

The rate of convective overturn depends on how much the actual temperature gradient exceeds the adiabatic gradient. The greater the excess temperature, the faster convection occurs. Flow is resisted by the viscosity of the material, which usually becomes smaller with increasing temperature. If the mantle of Earth were ever liquid, convection would tend to reduce the temperature gradient toward the adiabatic gradient. Convection alone would never reduce the gradient to this value, because the rate of convection would slow as the gradient decreased. In the liquid outer core of Earth, the gradient is presumably slightly above the adiabatic gradient.

A liquid mantle hotter than the adiabatic gradient would drop in temperature until the gradient approached the adiabatic rate. To cool below this point, conduction and radiation would have to remove part of the heat. But in rocks, conductive and radiative heat transfer is very slow. Calculations based on the best available estimates of thermal conductivity and transparency predict that the whole history of Earth is too short for the core to have lost enough heat to cool it by more than a few degrees without the aid of convection in the mantle.

This would seem to support the hypothesis that the interior must be warming. The mantle is solid, so it cannot cool rapidly. Internally generated heat will be unable to escape, and the interior will gradually warm. Eventually it will melt. Only when it melts, however, can it begin to differentiate rapidly. Yet Earth's interior seems to be differentiated.

The answer may lie in what happens in a solid just below the melting point. As this point is approached, solid diffusion speeds up allowing differentiation to begin. One or more components of the rock may melt, allowing liquid diffusion to start or even the flow of the liquid component thru the porous mass. Possibly most important, near melting even a solid rock will flow. The adiabatic gradient in a solid has the same meaning as in a liquid, except that it is the rate of increase in temperature where the rock begins to feel a convective stress. Flow will start only if this stress exceeds the yield strength of the rock. The evidence of isostasy shows that yield strength must be small in the mantle. Near melting it may be zero.

The rate of flow in a solid will be less than in a liquid for the same temperature gradient, because solid viscosities are generally much greater than liquid viscosities and because part of the stress is used in overcoming the yield strength before flow starts. Any convection, however, will assist in the transfer of heat and will permit the temperature gradient to continue to fall. The important effect of flow in a solid is that it provides a mechanism which will enable temperature to fall below the melting temperature. It can be concluded that if the mantle was formerly liquid, its temperature now has fallen below the melting point, but is still at all depths above a value found by projecting temperature downward at the adiabatic gradient from the temperature at 60 km depth. From large extrapolations of laboratory measurements of the effects of temperature and pressure on rocks, the adiabatic gradient is probably not over a few hundred degrees per 1000 km. If the mantle has never been liquid, then for differentiation to have occurred, the temperature must be above the adiabatic value below 60 km, but below the melting point.

12.8 TEMPERATURE IN THE INTERIOR.

It is now possible to make a reasonable estimate of temperature in Earth's interior. The high near-surface gradients persist until the melting temperature is approached at around 60 km. Occasionally, pockets of rock melt and, where open channels exist, vent thru volcanoes. The temperature at

60 km must lie between 1000° and 2000°C. Below this depth the temperature must lie below the melting temperature. The melting point is estimated from extrapolation of laboratory experiments and theoretical considerations to lie between 4000 and 8000°C at the core boundary for silicates.

The adiabatic gradient would require a temperature increase of only around 500°–1000°C from 60 km to the core boundary. Assuming that Earth has at some time differentiated, the minimum temperature at the core boundary would thus be between 1500 and 3000°C. It must be high enough for the core to be molten, but not high enough to melt the mantle. The melting point of iron at 2900 km is estimated to lie in the range 2300–5000°, with values below 3000° favored. Assuming the inner core and outer core have the same composition, the melting point curve must cross the temperature curve at 5000 km depth. In the outer core the gradient must be close to the adiabatic gradient. This means a rise of some hundreds of degrees in the outer core. The increase in temperature beneath 5000 km would be small, unless there is a change in composition in the inner core, as at these pressures the melting point increases only slowly. It is, therefore, unlikely that the central temperature is over 10,000°C. For an iron core it is not likely to be much over half this and might be below 3000°C. Regardless of what the composition is, it cannot be much under 2000°C. Figure 12.10 shows several estimates of the temperature variation with depth.

12.9 THERMAL HISTORY OF EARTH.

The whole subject of temperature and thermal processes in Earth's interior is fraught with uncertainties. The present temperature can be estimated only within wide limits and the past temperature is even more uncertain. Nor is it possible to calculate with any assurance what must have been the past history to arrive at the present conditions. This is because the processes of differentiation can rearrange the parts. It is not a question merely of changing temperature and decaying heat generation. Redistribution of heat sources with moving layer boundaries and changing mechanisms of heat transfer are involved. As a result of these uncertainties, it is easiest to assume possible starting conditions and check whether they can reasonably be expected to lead to something consistent with present knowledge of Earth's interior. The range of reasonable possibilities is nearly unlimited. Two qualitative examples of such speculations will be given.

The first model is obtained by starting with the slow aggregation of cold material. This matter is assumed to have existed in space for long enough that only long-lived radioactive elements such as are found today in Earth are still generating appreciable heat. Earth was assembled slowly so that very little of the energy of aggradation was retained by the growing body of material. During the whole period of aggradation, Earth remained solid throughout. About four billion years ago Earth is assumed to have reached

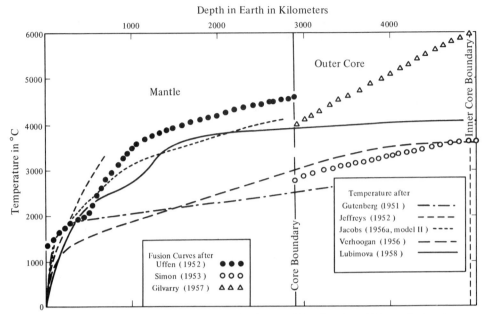

Fig. 12.10 *Estimated temperature variation with depth. (After Gutenberg,* Physics of Earth's Interior. *Copyright Academic Press, Inc.)*

approximately its present size. At this point it was uniform in composition. The temperature at the surface was about what it is today. Temperature increased with depth all the way to the center due to the energy gained by compaction of the interior as the total mass increased. At the center temperature was over 2000°C but not over the melting point of the material.

Radioactive material was distributed throughout the body of Earth, producing an amount of heat about three times the present heat flow at the surface. As a result, all parts of Earth were warming. Near the surface, conduction carried the heat upward, establishing a gradient not greatly different from the present gradient. The depth to which this gradient persisted gradually increased. At first the gradient was very low throughout the rest of Earth, and all parts warmed equally fast. Heat transfer was small except in the near-surface zone where the gradient was steep.

Ultimately the temperature at some point in Earth approached the melting temperature. Let us suppose for simplicity that this process started at the center, although it might start at almost any depth except for the high-gradient zone near the surface. When the melting temperature was approached, two things happened. First, solid convection began, and thereafter restricted the gradient to a low value not very much greater than the adiabatic gradient. Second, differentiation started, with heavy material separating out and being concentrated downward. The radiogenic elements, in spite of their high density, associated with light elements and were carried upward. The volume involved in differentiation and convection gradually grew. As it grew, the thoroughness of differentiation increased. The essential step in the process was the separation from the main body of Earth of two layers: a dense core, from which nearly all the heat-generating material has been removed, and a layer of silicate material which is greatly enriched in radioactive elements and impoverished in volatiles, the present crust. The remainder of Earth, the mantle, was and is intermediate in composition. Differentiation took place continuously in the mantle. At the bottom, metal separated out and was added to the core, which grew in diameter. At the top, the volatiles escaped to the surface and the crust was enriched in the lighter oxides plus rare elements such as uranium and thorium.

The metal-rich core had a lower melting point than the mantle and became liquid. The temperature gradient in the liquid core was less than in the solid mantle. As the mantle thinned, the temperature at the top of the core dropped even though the temperature continued to rise throughout the mantle. This allowed the center of the core to cool, and it froze from the center outward. Throughout the process, the mantle continued to heat. Differentiation and convection proceeded at an ever-accelerating pace, although the rate of rise in temperature declined as a result of the more rapid upward transfer of heat and the steady exhaustion of the radioactive elements.

This model provides us with a continually growing core and crust. Volcanism and igneous activity are the result primarily of the outgassing of the interior. The rate of volcanic activity would have increased throughout the history of Earth as the temperature of the upper mantle rose.

The second model differs in that it is assumed that heat generation was so great at first that all of the mantle and core was once liquid. This permits a rapid initial differentiation, in which Earth separated into two layers—a metallic core and a silicate mantle. The temperature gradient dropped close to the adiabatic gradient everywhere except near the surface, where viscosity was so high that the adiabatic gradient could be greatly exceeded.

As the amount of heat being generated by radioactivity declined, the interior temperature slowly fell. The freezing point was reached first at the center. Gradually, the inner core grew in diameter to nearly its present size. At this time, the freezing point at the bottom of the mantle was reached. Once any part of the mantle froze, further cooling within it was greatly reduced. Nowhere within the mantle is the temperature today more than a few degrees below the melting temperature. The outer core is trapped as liquid between the frozen inner core and mantle.

In the slowly convecting, viscous upper mantle differentiation also occurred with the formation of a light crust depleted in volatiles but enriched in the radioactive elements. This crust was too rigid to convect. Within it the temperature gradient became very high.

Mantle convection was thus gradually confined to a zone of decreasing thickness between the bottom of a growing crust and the top of the solid inner mantle. Eventually the whole mantle froze. Some convection has persisted, however, especially in the upper mantle, where rigidity and viscosity are lowest. Differentiation is still going on, but at an ever-decreasing rate, because of the decreasing heat generation. The rate of convection is just enough to bring the excess heat, beyond what can escape by conduction and radiation, up to the high-gradient zone near Earth's surface. Most of the differentiation occurred early in Earth's history.

Both these models provide for the separation of Earth into layers by differentiation. They differ mainly in the time and rate of this differentiation—whether it is still occurring as rapidly or more rapidly than in the past (Model 1) or whether most of the differentiation occurred early in Earth's history and the situation since then has been a very slow decrease in activity under relatively uniform conditions (Model 2). To decide which is nearer the true situation, it is necessary to examine the manner in which Earth is currently deforming and the history of such deformations in the past. The rest of this book will be devoted largely to these topics.

BIBLIOGRAPHY

Coulomb, J.; G. Jobert, 1963, *The Physical Constitution of The Earth.* Oliver and Boyd. 328 pp.

Eaton, J. P., 1962, *Crustal Structure and Volcanism in Hawaii.* Am. Geoph. Union Monog. 6, pp. 13–29.

Gaskell, T. F., 1967, *The Earth's Mantle.* Academic Press.509 pp.

Gutenberg, B., 1959, *Physics of the Earth's Interior.* Academic Press. 240 pp.

Herbert, D.; F. Bardossi, 1968, *Kilauea: Case History of a Volcano.* Harper and Row, Publ. 191 pp.

Jacobs, J. A.; R. D. Russell; J. T. Wilson, 1959, *Physics and Geology.* McGraw-Hill Book Co., Inc. 424 pp.

Rankama, K.; T. G. Sahama, 1959, *Geochemistry.* University of Chicago Press. 912 pp.

Stacey, F. C., 1969, *Physics of the Earth.* John Wiley and Sons, Inc. 324 pp.

13 EARTHQUAKES

Of all of the sources of information on Earth's interior, the study of earthquake vibrations has provided the most data. To appreciate this, it is necessary not only to observe where and when earthquakes occur, but also to understand in some detail just what happens during an earthquake.

13.1 NATURE OF VIBRATIONS.

On May 18, 1940 a large earthquake occurred at 8:40 P.M. local time in the Imperial Valley of Southern California near the Mexican border. I was a student living in what was known as the "Old Dormitory," a two-story frame structure, on the California Institute of Technology campus. When the initial jar occurred, I thought a large truck had backed into the building. When the shaking persisted, with pulsating intensity, for nearly a minute, rattling windows and pipes, I finally realized that this was an earthquake. I was excited, even elated by the experience, because I had come to California to study earth phenomena.

Most people do not welcome such experiences, and with considerable justification. Earthquakes are one of the most frightening of natural phenomena. Much of our fear of such events, however, is unjustified. As we come to understand them and to prepare better for them, we should be able to reduce their undesirable effects from disaster proportions to a condition of expensive but surmountable annoyance.

The first step is to face them rationally. This is not always easy. Normal experience prepares us poorly for an earthquake. Long before each of us became concerned with such things as gravity, deep down in our subconscious minds our bodies stored the practical wisdom that enables us to walk upright across the land. At first, when we were very small children, we could not do this. In learning to walk a child falls and, in falling, hits the floor. To avoid this unpleasant sensation, the brain learns to direct the limbs so that as one starts to fall a leg is thrust out to maintain balance, sets of muscles tighten or loosen to keep one's center of gravity within the area covered by our feet. Eventually, no conscious effort is required for this. The body has learned that for practical purposes, the ground is a fixed surface of reference, and balance is achieved by holding the body in a certain limited range of positions with respect to the motionless ground.

There are times when the simple rules acquired in infanthood are not enough. As you learn to drive a car, you learn to lean against the car's acceleration in order to maintain your balance. An airplane pilot flying thru a cloud learns that his body cannot tell him which way is down, because the motions of the plane in turning create pulls in other directions as strong as gravity. He learns to depend on instruments which are less easily fooled than his own senses.

When an earthquake occurs, our subconscious inner model of a fixed Earth also fails, for the ground is no longer an unmoving plane of reference, but suddenly becomes a shifting, undependable stranger. Without any action on a person's part, suddenly he is off balance. In spite of his best efforts, he may be unable to stand. The automatic reflexes which normally maintain balance no longer accomplish their task. Those neat and dependable machines, our bodies, are faced with an experience with which they are unprepared to cope. Somewhere deep in the inner recesses of the mind, a hidden operator pushes the red emergency button. The reaction of most people to an earthquake is fear. Sometimes it is panic. Perhaps the adult half-remembers with dread the falls he took as a baby before he developed his unconscious model of the universe. The sensation of unbalance experienced when the ground shifts may well be like the feelings a child has as he learns to walk.

Fear of earthquakes is largely unreasoned as well as unjustified. The discomfort is out of proportion to the danger involved. Death or injury from an earthquake is not nearly so great a hazard as from other, commoner phenomena, such as storms or fires, and certainly is not comparable to highway traffic on a holiday weekend. The chances of death from an earthquake in the United States are less than one in 5,000,000 per year as compared to one accidental death for every 1500 persons annually. The 1906 San Francisco earthquake, one of the worst in the nation's history, killed an estimated 700 people, including those who died in the great fire which followed it. With proper precautions, most of the deaths and destruction which accompanied this earthquake could have been avoided.

Experience with earthquakes overcomes the dread they induce. People living in regions where ground tremors are common, such as some parts of Japan, learn to brace themselves and to wait for the shaking to stop; then they go on with their normal activities. Most earthquakes are small, and, like a gust of wind, soon pass. The majority are so small that they are not felt at all, and are known to exist only because they are recorded by instruments.

A common sequence of sensations during an earthquake begins with a series of jarring motions of the ground, correlating with the arrival of the sound and shear body waves. These may be followed by a rolling sensation corresponding to the surface waves. The motion may be so violent that it is impossible to walk or even crawl. In an automobile, the experience is like driving on a very bumpy road. It may be difficult to steer the car. On a ship at sea, the sensation is like striking a reef. In a metal ship there is sometimes a clearly audible sound as though the hull had been struck with a giant hammer.

The nature of the motions can be better understood by examining a *seismogram*, which is a plot of ground motion against time. In Figure 13.1, time increased from left to right. The small square deflections mark one-minute intervals. The minute mark is wider to indicate the hour. The upper lines in the figure are the records of motion at successive hours preceding the earthquake. The seismogram is made on a drum which rotates in a spiral at a rate of one turn every 59 minutes. The earthquake shown occurred at 20:29:14.5 U.T. (14.5 seconds after 8:29 P.M. universal time) on 23 April 1968 in the Gulf of Alaska.

The vibrations from the earthquake began arriving at State College, Pennsylvania, where the seismogram of Figure 13.1 was made, at 20:37:48. The first motion is the sound wave thru the ground, labeled P on the seismogram. (See Section 10.2 for discussion of velocity of transmission.) The ground alternately expands and contracts with particles vibrating parallel to the direction the energy is flowing (Figure 10.1, p.263). After the first arrival, a series of later pulses continued to keep the ground vibrating until a pulse characteristically well exhibited on the horizontal components of ground motion arrived at 20:44:43, six minutes and fifty-five seconds after the arrival of P. This is the shear wave, labeled S. Here, the ground motion is at right angles to the direction of energy flow. By comparing the time delay of S behind P with Figure 10.6 (p.268), it can be determined that the earthquake occurred 5500 km (3400 miles) from State College. Eight minutes later at 20:53 a third major group of waves began to arrive. This last group is the surface waves. Because they travel thru the low-velocity layers near Earth's surface instead of thru the interior, they take longer to reach the recorder than P and S. The surface waves consist in part of a variety of shear waves called Love waves and in part of a combination of shear and sound waves called Rayleigh waves in which particles move in elliptical orbits. The ground at State College continued to be agitated for several hours, partly by the arrival of surface waves, some of which had travelled the long way around Earth thru the antipodes, and partly by vibrations from aftershocks.

Aftershocks are small earthquakes occurring in the same region as the main earthquake and following the principal event. Almost all large earthquakes are followed by aftershocks. In a typical sequence, the number of aftershocks per day declines hyberbolically with time, but the rate is so varied from one area to another that it is hard to establish general rules. Aftershocks are by definition smaller than the principal shocks which they follow. The strength of the largest aftershocks generally decreases with time, but here again it is hard to establish rules. There is no sure way to distinguish a large aftershock from a subsequent principal shock. Large aftershocks are often followed by their own trains of smaller aftershocks.

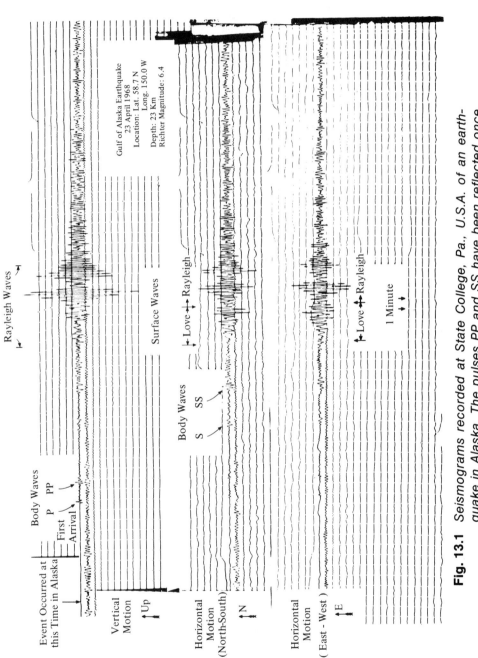

Rayleigh Waves

Event Occurred at this Time in Alaska

Body Waves
P   PP
First
Arrival

Vertical Motion
↕ Up

Gulf of Alaska Earthquake
23 April 1968
Location: Lat. 58.7 N
           Long. 150.0 W
Depth: 23 Km
Richter Magnitude: 6.4

Surface Waves

Body Waves
S   SS

Horizontal Motion (North-South)
↕ N

|← Love →|← Rayleigh

Horizontal Motion (East - West)
↕ E

|← Love →|← Rayleigh

1 Minute
↕

Fig. 13.1 Seismograms recorded at State College, Pa., U.S.A. of an earth-quake in Alaska. The pulses PP and SS have been reflected once from Earth's surface.

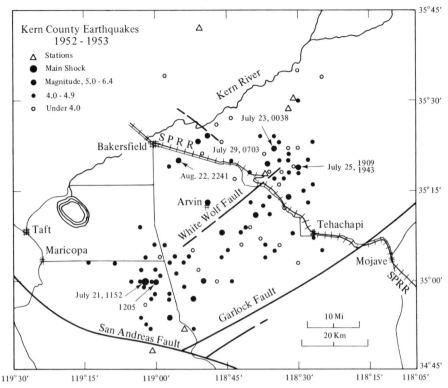

Fig. 13.2 *Locations (epicenters) of the aftershocks of the 1952 Kern County, California earthquake. (Richter,* Calif. Div. Mines Bull. *171, p. 178.)*

Aftershocks do not all originate at exactly the same location as the principal shock in a group, but may scatter over a large neighboring volume (Figure 13.2). As a result, some outlying areas may be shaken harder by an aftershock than by the main shock. The August 22 aftershock of the 1952 Kern County earthquake caused more damage in the city of Bakersfield than the main shock did.

Some earthquakes are preceded by *foreshocks* before the main event. This is common in parts of Japan, and may be useful in such areas as a warning of a larger earthquake soon to come. Generally, however, large earthquakes come without such warnings; and small shocks occur very commonly without being followed by large earthquakes.

In some areas, large earthquakes tend to come in pairs or in swarms. In the Imperial Valley of southern California, damaging earthquakes have been followed many times by a second shock of similar size in the same region within one or two days. Experience has taught the inhabitants to avoid buildings susceptible to damage following on earthquake until the aftershocks have died down. At Helena, Montana, in 1935 there was a sequence of three damaging earthquakes within a period of three weeks.

Sometimes earthquakes at different locations seem to be related to one another. Two earthquakes on July 6 and August 24, 1954 near Fallon, Ne-

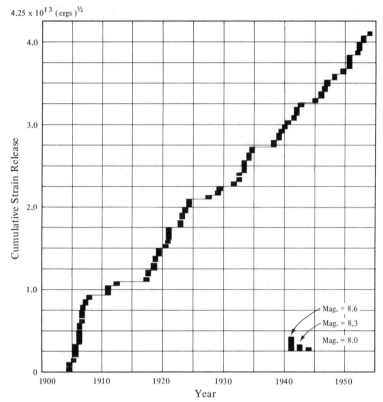

Fig. 13.3 *Cumulative plot of square root of energy released by large earth-quakes. (After Benioff,* Bull. Geol. Soc. Am. Sp. Paper 62.*)*

vada, were followed on December 16 by a third large shock centered about 50 km (30 miles) farther east. There were five very large earthquakes in the year 1906, one in Chile, one in Ecuador, one in California, one in the Aleutian Islands and one in New Guinea. The whole period 1904–1906 was one of exceptionally heavy activity, as were the periods 1917–1924 and 1931–1934 (Figure 13.3).

13.2 MICROSEISMS COMPARED TO EARTHQUAKES.

The ground is never still, but is continually vibrating. Persistent small motions are called *microseisms*. They arise from many sources. Air pressure fluctuations force the ground surface up and down. Wind blows on trees, whose roots pull the ground, and on the sides of hills, which move in response to this pressure. Water waves press on the ocean bottom and are transmitted into the ground (Figure 13.4). Surf breaking on the shore can often be felt for a considerable distance inland. Railroad and highway traffic shakes the ground. Rotating machinery, digging, plowing and the burowing of animals all produce microseisms. In the quietest of regions the ground is likely to be moving up and down, back and forth with ampli-

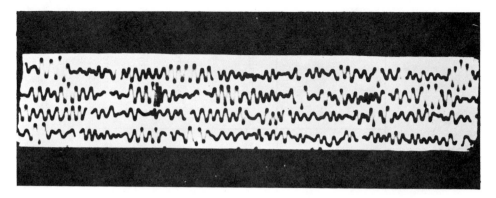

Fig. 13.4 *Seismogram showing typical storm microseisms at State College, Pa., U.S.A.*

tudes of a few microns. (One *micron* equals 10^{-6} meters, .00004 inches.) The smallest earthquakes are so tiny that they constitute part of the microseismic background. Earthquakes differ from microseisms in that they have distinct beginnings and consist of a characteristic sequence of pulses (Figure 13.1).

13.3 EPICENTER AND FOCUS.

The point inside Earth from which energy first starts to radiate is called the *focus*. The point at Earth's surface directly over the focus is called the *epicenter*. The epicenter is usually near but not necessarily in exactly the same place as the most violently shaken part of Earth's surface. The principal reason for this is that the violence of the shaking at the surface depends on the nature of the rocks as well as proximity to the source of energy. The area over which shaking is felt is sometimes loosely called the *epicentral area* or the *meisoseismal area*. At other times these terms are reserved for the narrow region of severe damage. The usage is loose and serves to refer to an area of vague extent surrounding the point of maximum ground vibration.

The epicenter and focus can be located by any one of several means. When an earthquake arrives, the direction of first motion of the ground at a recording site should be either toward or away from the source. Since earthquakes originate within rather than outside of Earth, the choice between toward and away is easily made. An upward first motion is away from the source. A downward first motion is toward it. The direction to the epicenter can be found from the relative strength of the horizontal components of motion. If the ground moves up, north and east in equal amounts, then the epicenter is southwest of the recording site. In Figure 13.5, the north component of motion is about four times stronger than the east, so the epicenter is roughly S22.5°W from State College. Since sound (P) and shear (S) waves travel at different velocities, the time delay between the arrival of P and S is a measure of the distance to the epicenter

Table 13.1. *Earthquake of July 9, 1964*

Observatory	Arrival time of P hour:min:sec	Distance to first trial epicenter	T_p travel-time from assumed epicenter min:sec	Trial origin time $0 = P - T_p$ Average origin time = 16:39:35	$0 - 0_{av}$	Distance error
Baguio, Luzon, Philippines	16:49:16	39.1°	7:40	16:41:36	+2:01	17° farther
Kipapa, Oahu, Hawaii	16:48:32	59.7°	10:09	16:38:23	–1:12	11° nearer
Riverview, N.S.W., Australia	16:44:51	24.45°	5:41	16:39:10	–0:25	5.5° nearer
Charters Towers, Queensland, Australia	16:44:23	11.3°	3:18	16:41:05	+1:30	7.7° farther
Wellington, New Zealand	16:45:16	38.5°	7:36	16:37:40	–1:55	14° nearer

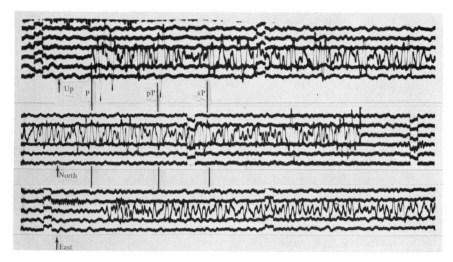

Fig. 13.5 *Correlated motions at the start of the El Salvador earthquake of April 24, 1964 at State College, Pa., U.S.A., were up-north-east, showing that the earthquake occurred southwest of the observatory. The pulses pP and sP are surface-reflected echos of the initial pulse, P.*

(Figure 10.6, p. 268). Knowing distance and direction, the epicenter is easily located. Focal depth is harder to determine, because the energy travels along curved paths (Figure 10.7, p. 269); but earthquakes at greater than 50 km depth of focus are usually followed by one or more surface-reflected echos which can be seen on the seismogram (Figure 13.5). The delay of the time of arrival of each echo behind the direct pulse is a measure of the depth of focus. The precision of this method of epicenter and focus location is poor, because it is difficult to measure accurately the amplitudes of first motion of the ground.

If $S–P$ times are known at three observatories, the epicenter can be accurately located even when its direction from each observatory is not known. Three circles drawn on a globe about the observatories with radii equal to the epicentral distance will cross one another at the epicenter (Figure 13.6). Due to uncertainties in measurements of the $S–P$ time, this method gives results which are accurate for distant earthquakes only to about $\pm$ 100 km (60 miles) except in particularly favorable cases. This degree of uncertainty is hard to avoid because the motion becomes increasingly complicated after the first oscillation begins to arrive because of energy arriving over a variety of paths from the expanding disturbed area. The $S–P$ time interval can be recognized only because S is a strong pulse and the direction that the ground is moving generally changes sharply at the time it arrives.

Greater precision in epicenter location can be obtained by depending only on the more easily determined P arrival times. The method followed in this case is to find, by a series of successive trials, a location and an origin time for the earthquake such that the average difference between the observed and predicted arrival times is a minimum. The method can be understood best by an example. An earthquake was recorded at the times and places listed in Table 13.1. The times of arrival of the earthquake were within one minute of one another at Riverview (near Sydney), Australia; Charters Towers (near Townsville), Australia; and Wellington, New Zealand; and they arrived first at Charters Towers. A trial epicenter was chosen at 10°S, 150°E near the eastern tip of New Guinea, the nearest place to Charters Towers where earthquakes commonly occur (Figure 13.7). The distance from the trial epicenter to each of the observatories was calculated, and the corresponding travel time for P read from Figure 10.6, p. 268). These travel times were then subtracted from the observed arrival times to give five origin times. The average of the five origin times was subtracted from each trial origin time to give an estimated error in each case. The trial origin time at Baguio, Philippines was two minutes, one second later than the average, indicating that the trial epicenter was farther from Baguio than southeast New Guinea by about 17° (Figure 13.7). (The error in distance is again found using the travel-time curve, Figure 10.6, p. 268.) The trial origin time for Wellington is 1 min 55 sec early, indicating that the true epicenter is about 14° nearer this city. Similarly, the true epicenter is shown to be nearer Kipapa and Riverview but farther from Charters Towers.

A second trial epicenter near the western tip of New Caledonia might be tried. Repetition of the procedure described above would soon lead to the true epicenter on the eastern side of the Coral Sea in the New Hebrides Islands at 15.5°S, 167.6°E. This is a region where earthquakes are common.

Epicenter location is complicated by the fact that a large earthquake will radiate energy from a large focal area or volume. The release of energy seems to start at a point, spreading thru the source region at a velocity equal to or less than the speed of transmission of seismic waves. If the initiating disturbance is small and increases in violence as it develops, the first recognizable waves to arrive at a distant observatory may have come from a place other than the one where the earthquake started. When different observatories record first motions from different parts of the focal area, the selection of one exact point as an epicenter is difficult.

A similar argument helps to explain why in some cases the area of maximum shaking at the surface is not right at the epicenter located from arrival times of seismic waves.

13.4 SEISMOMETERS.

In order to observe the motions of the ground, it is necessary to have a place to stand from which to watch the ground move. Since the motions

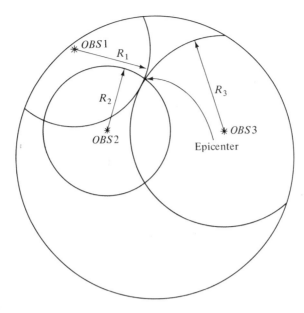

Fig. 13.6 *Knowing the distances from three observatories, the epicenter can be located by drawing three circles with appropriate radii.*

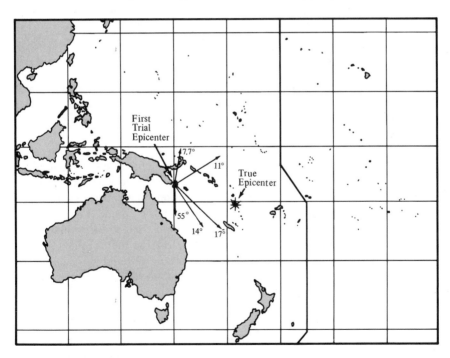

Fig. 13.7 *An epicenter can be located by successive approximations, calculating the errors of a trial epicenter.*

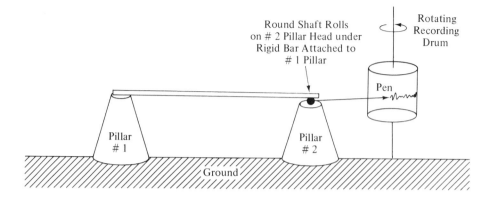

Fig. 13.8 *Principle of the strain seismometer.*

are very small, the observing station cannot be very far away. The basic problem of measuring ground motions is to find a place to stand which does not move along with the point whose motion is being studied.

Devices which measure the vibrations of the ground are called *seismometers*. They are of two general types. *Strain seismometers* measure the change in distance between two points. *Inertial seismometers* measure the motion of the ground compared to a mass which is only loosely attached to it and which tends to stand still as the ground moves.

The principle of the strain seismometer is shown in Figure 13.8. Two pillars are connected by a rod in such a manner that the motion of one pillar with respect to the other is amplified and recorded in some fashion. In the illustration, amplification is obtained by attaching a long pen handle to a thin round shaft which is rotated by the relative motion of the pillars. In practice, sensitive electrical devices are used to provide very large magnification (Figure 13.9).

Very compact instruments can be built by cementing strain-sensitive devices directly onto a rock surface. One type of such a strain gauge consists of a fine grid of wire whose resistance is changed when it is stretched or squeezed. Minute but measurable variations in electrical current thru the wire are proportional to the deformation of the ground (Figure 13.10).

Inertial seismometers are more common than strain seismometers. An inertial seismometer consists of three parts: the inertial mass, the suspension which holds the mass, and the recording system by which the relative motion of the ground and the mass is observed. The suspension system must be strong enough to hold the mass suspended, but so weak that the mass is not dragged along with the ground when the latter moves. Frequently, the suspended mass is a part of a pendulum arranged so that it is able to rotate about an axis. In Figure 13.11, the axis is nearly vertical, so that the seismometer is sensitive to horizontal motion. When the pillar and recording drum move into the page, the mass lags behind them, and the pendulum arm is rotated so that the pen moves across the recording-drum surface. If the ground moves very rapidly, the inertial mass has too little

Fig. 13.9 *The Benioff strain seismometer can detect motions of one part in 10^9 between two pillars 18 meters (60 ft.) apart. (Photo by J. McClanahan. Calif. Inst. Tech. photo.)*

time to follow the drum's motion, and an accurate record of ground movements can be recorded. If the ground moves very slowly, the inertial mass is carried along with the drum, and no motion is recorded. At intermediate speeds, an inaccurate record of motion is made. The ground motion is amplified by the ratio $\dfrac{L_2}{L_1}$, the pen-arm length divided by the distance from the axis to the center of oscillation of the mass. Much greater amplification can be obtained by using electrical devices to detect and record the motions (Figure 13.12).

Seismometers usually operate in sets of three, one each to measure east-west, north-south and vertical motions. The horizontal-component pend-

Fig. 13.10 *A strain gauge mounted on a rock surface.*

ulums rotate about a nearly vertical axis. The axis is tilted just enough that, when the ground is not disturbed, the pendulum always returns to the same rest position because of the pull of gravity. The vertical-component pendulum uses a horizontal axis, and the pendulum is held horizontal by a spring when the ground is at rest (Figure 13.13).

13.5 SEISMOGRAMS.

The record of ground motion made by a seismometer is called a *seismogram* (Figures 13.1 and 8.18, p. 212). Most seismograms are plots on paper of the amplified motion of the ground as a function of time. Seismograms may also be in the form of a magnetized tape, on which the degree of magnetization varies along the length of the tape in proportion to the motion of the ground. Digital seismometers record only a series of numbers representing successive amplitudes of ground motion at fixed time intervals. This last type of recording is particularly convenient for analyzing various features of the ground motion using high-speed computers.

Some seismograms are recorded in such a manner that the plot is proportional to the displacement of the ground. Most seismograms, however, are more nearly a plot of the velocity of ground motion as a function of time. The advantage of this is that velocity is a better measure of the energy of the earthquake than is the ground displacement. Seismic energy is transmitted in two forms: as kinetic energy in the form of the velocity of vibration of the ground with respect to its normal rest position, and as potential energy in the form of the strain the ground undergoes as the seismic

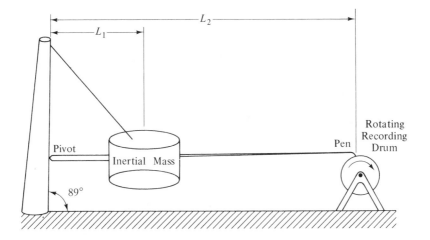

Fig. 13.11 *Principle of the pendulum seismometer.*

Fig. 13.12 *Three of the six seismometers which record continuously at The Pennsylvania State University Geophysical Observatory. The small instruments in front of the large seismometers are galvanometers used to make the recordings.*

waves pass thru it. An inertial seismometer gives information on the kinetic energy. The response of a strain seismometer is more directly related to potential energy. At Earth's surface, where there is little restraint imposed

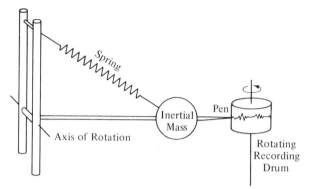

Fig. 13.13 *Principle of the vertical-component seismometer.*

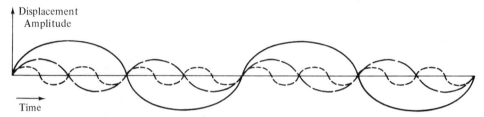

Fig. 13.14 *Three plots of displacement against time. Each represents the same amount of energy because energy is proportional to the square of the product of frequency and displacement amplitude.*

by the overlying air on the motion of the ground, most of the energy is kinetic. The energy per unit mass can be represented by the formula

$$E = \frac{V^2 d}{2} \qquad (13.1)$$

where d is density and V is instantaneous particle velocity. V should not be confused with the velocity at which the energy flows thru the ground. The faster the ground moves, the more energy there is involved (Figure 13.14). At very low frequencies of vibration, the ground can move up and down with a large amplitude of displacement without having at any moment a large velocity. At high frequencies there may be a large amount of energy transmitted by the waves for a very small displacement. The energy is a measure of how much work the waves are capable of doing, and hence of how much damage they can cause. Thus, ground velocity is a quantity of great interest.

13.6 CAUSE OF EARTHQUAKES.

In addition to transient vibrations, earthquakes sometimes produce permanent displacements of the surface of the ground. The Prince William

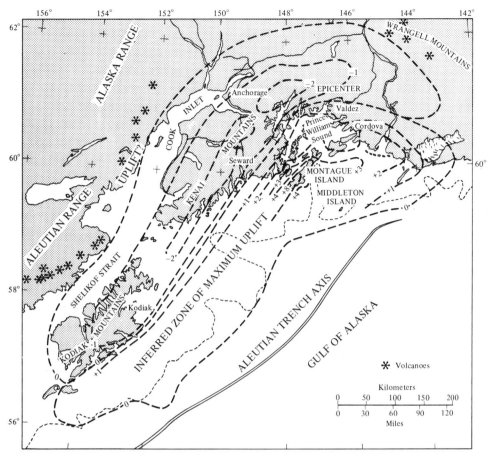

Fig. 13.15 *Vertical ground displacements (dashed contours, in meters) during the 1964 Prince William Sound, Alaska earthquake. The dotted line is the edge of the continental shelf. (After Plafker,* Science *v. 148 p. 1677. Copyright 25 June 1965 by the American Association for the Advancement of Science.)*

Sound, Alaska earthquake of 1964 caused extensive changes in elevation around the Gulf of Alaska (Figure 13.15). During the Kwanto, Japan earthquake of 1923 a large block of ground appears to have shifted generally southeastward compared to seven points roughly in a semicircle surrounding the badly shaken area (Figure 13.16). Beaches were raised as much as 14.4 meters (47.3 feet) in the 1899 Yakutat Bay, Alaska earthquake.

Sometimes the permanent ground displacements seem to be concentrated along individual breaks in the rocks called faults. The 1906 San Francisco earthquake cracked the ground along a 300-kilometer (190-mile) long fault starting at San Juan Bautista in the south and running out to sea at Point Arena on the north. Additional displacements occurred along the California coast in the Cape Mendocino area 160 km (100 miles) further north either on an extension of the same fault or on a parallel fault. Most of the displacement was horizontal, the northeast side of the fault moving

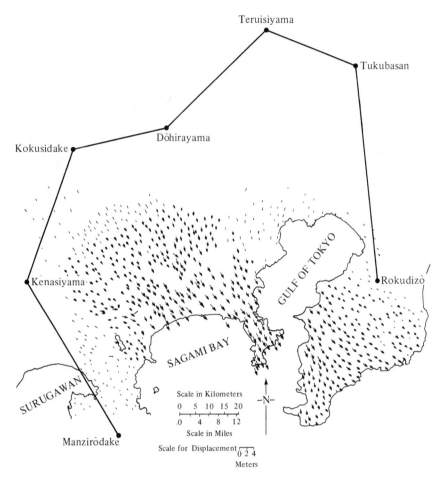

Fig. 13.16 *Ground displacements during the 1923 Kwanto, Japan earthquake in comparison to seven peripheral points assumed to have remained fixed. (Suyehiro,* Engineering Seismology. *Courtesy Am. Soc. Civil Engineers.)*

southeastward relative to the adjoining block (Figure 13.17). Vertical displacements of up to a meter (three feet) also occurred, with sometimes the northeast, sometimes the southwest side being up.

The 1906 displacement occurred on the San Andreas Fault. From San Juan Bautista the fault extends southeastward thru the Coast Ranges to their juncture with the Tehachapi Mountains and the Transverse Ranges, which it crosses. East of Los Angeles it splits into several branches, some lying within the Peninsular Range and others extending thru Imperial Valley to the east into Mexico and the Gulf of California.

Two other earthquakes in historic times have produced displacements on this fault. The section from the Carrizo Plain–Cholame Valley area to a point east of Los Angeles was broken in 1857 (Figure 13.18). A portion from a few kilometers south of the Salton Sea in Imperial Valley to a few

Fig. 13.17 *Road offset by San Andreas Fault during the 1906 San Francisco earthquake. (Courtesy U. S. Geological Survey. Photo by G. K. Gilbert.)*

kilometers south of the United States–Mexican border broke in 1940. The San Andreas Fault is marked by a belt of epicenters along its whole length. Similar well developed earthquake belts associated with frequent fault displacements occur in Turkey, New Zealand and elsewhere.

The general appearance of the fault all along its length shows evidence of displacements within the last few thousand years (Figure 13.19). There are abundant small scarps where one block has been raised relative to an adjoining block. There are streams whose courses have been offset by the horizontal displacement of adjoining blocks. There are pressure ridges where a wedge of material has been squeezed between adjoining blocks, and there are sag ponds formed in down-dropped sections along the fault.

There is reason to believe that displacements may occur underground even when there is little or no surface displacement. A series of en-echelon cracks extended for several kilometers beyond the northern end of the surface displacement produced by the 1940 Imperial Valley earthquake. South of the main part of the break, the line of the fault was marked by *mole tracks*, a characteristic pattern of en-echelon hummocks at the surface, typically cracked across their tops (Figure 13.20). Both cracking and mole tracks represent the response of easily deformed ground to stresses which are too small to produce a continuous fault. It is probable that the relatively easily deformed rocks at Earth's surface yield much more commonly by flexure than the deeper rocks. This may explain the rarity of surface faulting during earthquakes.

The surface trace of a fault in soft ground is rarely straight, but is more typically sinuous (Figure 13.21). Old faults exposed by erosion in firm rocks are generally much more nearly planar. It may be that most faults are plane in firm rocks, and wander only in the unconsolidated surface veneer.

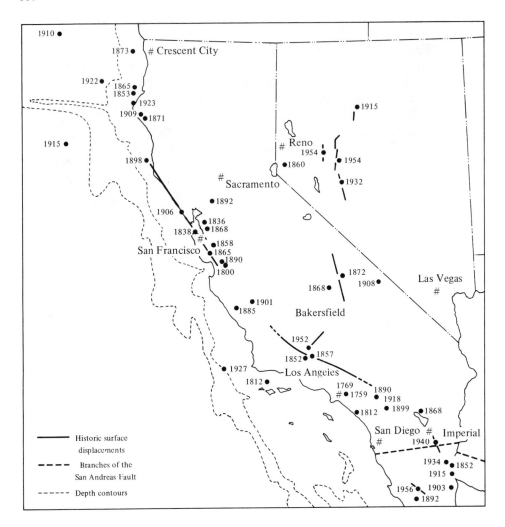

Fig. 13.18 *Larger earthquakes of the California region and lines of accompanying surface faulting. (After Richter,* Engineering and Science, *v. 31, Nr. 2 p. 58.)*

As the surface trace wanders back and forth over the more generally straight trend of the underground fault, alternating sections of predominant compression and tension can be seen (Figure 13.22). Cracks are opened in the tensional areas, and pressure ridges predominate in the compressional areas. This alternation of the pattern of deformation presumably reflects small variations in the mechanical properties of the surface materials. Similar features of compression and tension can be expected on a larger scale. Fundamental patterns of large-scale deformation can easily be obscured by the plethora of such superficial details.

The frequent occurrence of earthquakes along known faults and the occasional surface displacements which accompany earthquakes has led to the hypothesis that most or all earthquakes are the result of sudden fault

Fig. 13.19 *Aerial view of the San Andreas Fault. Note the offset streams. (Courtesy Aero Service Corp. Division of Litton Industries.)*

Fig. 13.20 *Mole tracks occur where soft ground is sheared in an earthquake fault.*

displacements. According to this theory, the only reason that all earthquakes are not accompanied by visible fault displacements is that the foci of the shocks are generally so far below the surface that only buried rocks are displaced.

One version of this hypothesis was developed by H. F. Reid as a result of his studies of the 1906 San Francisco earthquake. It is known as the *elastic-*

Fig. 13.21 *Typically sinuous fault outcrop. Hebgen Lake, Montana earthquake of 18 Aug. 1959. (Courtesy W. B. Hall.)*

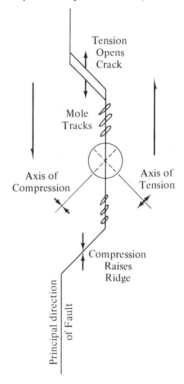

Tension
Opens
Crack

Mole
Tracks

Axis of
Compression

Axis of
Tension

Compression
Raises
Ridge

Principal direction
of Fault

Fig. 13.22

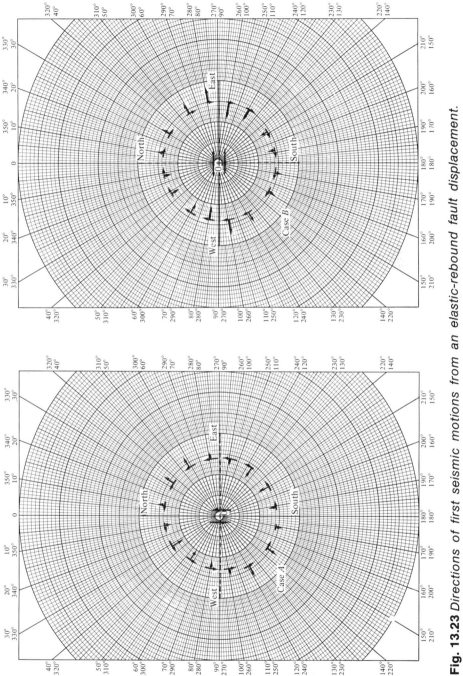

Fig. 13.23 Directions of first seismic motions from an elastic-rebound fault displacement.

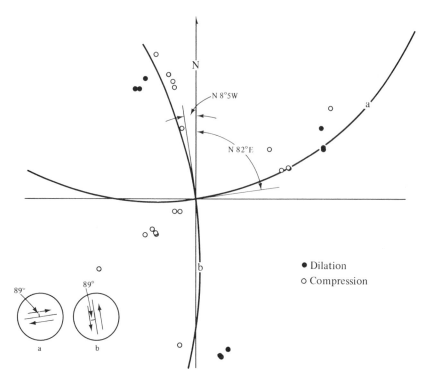

Fig. 13.24 *Fault-plane solution for the Kamchatka, U.S.S.R. earthquake of 4 Nov. 1952. (After Hodgson,* Pub. Dominion Obs. v. 18, p. 235.)

rebound theory. Permanent ground displacements along the San Andreas Fault in 1906 were largely horizontal and varied from zero at the south end to a maximum of 6.5 meters (21 feet) near Tomales Bay. Immediately after the earthquake, a survey was run of the relative positions of carefully located points in California. Comparison of this survey with a previous survey run between 1874 and 1892 showed that points southwest of the fault had moved northwest and points northeast of the fault had moved southeast relative to points further inland (Table 13.2). This is in the same sense of displacement as the fault motion during the earthquake, as would be expected. The amount of displacement decreased with distance from the fault. Furthermore, comparison of the 1874–92 results with an earlier survey in 1851–66 showed that the southwest block had moved northwestward between these two surveys also. The ground motion did not all occur during the earthquake, but involved a persistent drift of the oceanward side northwestward with respect to the continent (or vice versa, since only relative motion could be measured).

Reid explained this by postulating that the earthquake was the result instead of the cause of the observed deformation. The rocks were bent by some unknown subsurface force which dragged the Pacific coast northwest until the rocks were so badly bent they could no longer withstand the strain, and they broke along the San Andreas Fault, snapping to a new, less

Table 13.2 *Displacements of geodetic stations during and preceding the San Francisco earthquake (data from Reid,* California Earthquake of April 18, 1906, *v. 2, Carnegie Inst. Wash. publ. 87.)*

Distance from fault in km	Displacement, in meters	Number of points averaged to get displacement
6.4 northeast	0.58 southeast	1
4.2 northeast	0.86 southeast	3
1.5 northeast	1.54 southeast	10
2.0 southwest	2.95 northwest	12
5.8 southwest	2.38 northwest	7
37.0 southwest	1.78 northwest	1

strained position. The earthquake was the vibration produced by the rocks jumping to their new positions.

This view of the cause of earthquakes can be checked by noting the directions of first ground motion at observatories in different azimuths from the source area. For simplicity, consider a fault with a predominantly horizontal displacement, as was the case for the three large earthquakes on the San Andreas Fault. Suppose also that this fault is in a flat Earth with constant velocities of wave transmission throughout. In the real Earth, the geometry of motions is more complicated, but the same principles apply. If a sudden displacement occurs along a north-south fault at point O as shown in Figure 13.23, Case A, the first motions of the sound waves recorded in the northeast and southwest quadrants will be radially outward, and in the northwest and southeast quadrants will be radially inward. This pattern is observed for many earthquakes (Figure 13.24). Unfortunately, it is impossible to determine from the pattern of first sound wave motions whether the fault runs east-west or north-south, because the displacement shown in Figure 13.23, Case B, will give the same first motions as that of Figure 13.23, Case A. Theoretically, the direction of first motion of the shear waves will distinguish between the two cases, as it is clockwise in 13.23-A and counter-clockwise in 13.23-B. However, the first oscillation of the shear pulse is generally smaller than later oscillations on the seismogram and arrives when the ground is already disturbed by previous arrivals, so that it is rarely possible to determine with assurance what the direction of first motion of the shear pulse is.

The situation is further complicated because a combination of these two types of motion also can occur (Figure 13.25). This double-couple case corresponds to compression of the ground in the northwest-southeast direction and stretching in the northeast-southwest. The first shear-wave motions in this case are out of phase with the first sound-wave motions, going thru a maximum where the sound-wave motions reverse direction and vice versa.

In the majority of cases where the shear-wave evidence is clear, the pattern of first motions fits the double-couple hypothesis. The first-motion evi-

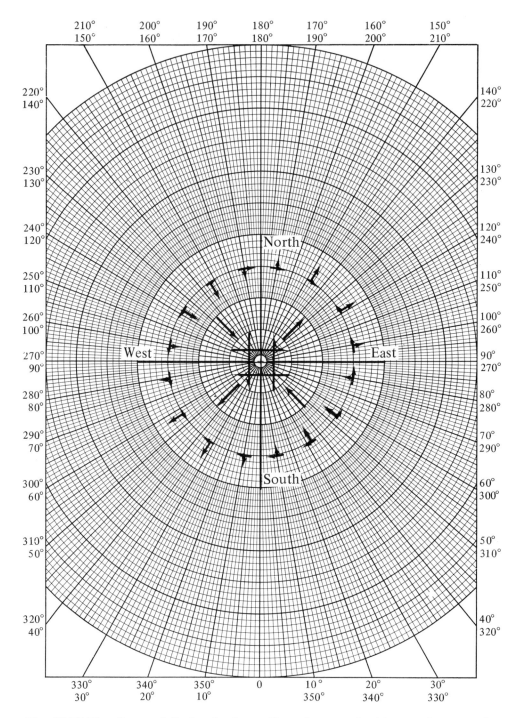

Fig. 13.25 *Directions of first seismic motions from a double-couple ground deformation.*

dence thus seems to be in conflict with the evidence from surface faulting, which suggests the single-couple mechanism. This could be coincidence, since very few earthquakes produce surface faults and equally few earthquakes provide clear first-shear-motion patterns. Or it may mean that the near-surface rocks yield under stress differently from deeply buried rocks. What is needed to resolve the problem is a few cases where first-motion patterns and surface displacements can be compared for the same earthquake.

One of the problems that complicates such studies is that many earthquakes produce surface fractures which are probably of a secondary nature. The 1952 Kern County, California earthquake is an example. This shock produced a small displacement up to 1.3 meters (4 feet) high on the White Wolf Fault. Displacement continued westward into alluvium for about 16 km (10 miles) beyond the point where firm bedrock was broken (Figure 13.26). The small scarp up to one half meter high in the unconsolidated alluvium could mark the continuation of the fault in the underlying rocks, or it could merely be a result of differential settling of loose material due to its being shaken by the vibrations where it abuts a steep mountain front. Alluvium is much too easily deformed ever to accumulate the stress needed to produce an elastic-rebound earthquake.

Similar scarps in unconsolidated rocks are common from earthquakes. Often their pattern is obviously related to local surface conditions rather than to deeply buried rock structures. Figure 13.27 is an example of lurch cracks formed as unconsolidated material sways back and forth. Where the land surface is steep, landslides frequently result, with large displacements along the upper edge of the dropped block. Landsliding and other settling of land associated with earthquakes are responsible for much damage (Figures 13.28, 13.29 and 8.16, p. 210).

Reid's theory remains widely accepted as the best explanation of the cause of most earthquakes because no other equally simple, reasonable process for the sudden release of large amounts of seismic energy has been proposed. The collapse of caverns, the sudden breaking free of entrapped gases, and the stopping of flow of underground lava streams have all been suggested, but shown to be incapable of providing the observed energy. The best alternative proposal is the possibility of a sudden phase change in a large body of rock whereby its volume would be reduced or increased. A body cooled below a phase-inversion temperature could conceivably postpone rearrangement of its atoms until some small disturbance started the change, which might then trigger an explosive shift to the new form. Such a sudden change in phase has never been noted to occur either in the field or in the laboratory. It seems more likely that phase changes progress at a more leisurely pace, taking much longer than the few seconds during which an earthquake occurs.

Unless some as yet unrecognized process is responsible for earthquakes, it seems most likely that they are caused by sudden displacements along faults. At the surface this motion appears to be commonly concentrated

Fig. 13.26 *Scarp formed in alluvial fill by the 1952 Kern County, Calif. Earthquake. (Courtesy Calif. Div. Mines. Compare* Bull. 171 *p. 24. Photo by Lauren A. Wright.)*

Fig. 13.27 *Lurch cracks in an earth dam, 1952 Kern County, Calif. earthquake.*

Fig. 13.28 *The Madison River was dammed by a 100-meter (350 foot) high landslide caused by the 1959 Hebgen Lake, Montana earthquake. (Courtesy W. B. Hall.)*

along a single fault or group of parallel faults. Beneath the surface, the common patterns of first arrival suggest that the faults may consist of networks of intersecting breaks, possibly even involving the crushing of a large volume of rock.

There are a few special cases where other types of events have caused earthquakes. The explosive eruption of Krakatoa Volcano in 1883 must have produced ground vibrations, although no seismometers were in operation in 1883 capable of recording them. The sound was heard in Australia 2000 km (1200 miles) away. Hawaiian volcanic eruptions are preceded by a long series of minor earthquakes whose exact cause is not know, but which may be caused by some process other than faulting of the rocks. The impact of the 1908 Tunguska meteorite produced ground vibrations recorded 893 km (555 miles) away. The sound of the explosion was recorded as an air vibration at many places The barograph at Potsdam, Germany recorded both the direct air wave and the one that travelled the long great-circle path around the far side of Earth.

Any large explosion produces the same types of seismic waves as a natural earthquake. Such artificial earthquakes are so similar to natural earthquakes that it is difficult (though frequently not impossible) to distinguish the one from the other on a seismogram. Blasts generally put a greater proportion of their energy into sound waves as compared to shear waves than

Fig. 13.29 *Aerial view of the Turnagain Heights Landslide at Anchorage, Alaska. It was caused by the 1964 earthquake. (U. S. Coast and Geodetic Survey photo.)*

do natural earthquakes; and because of their shallow depths of focus, blasts put more energy into surface as compared to both sound and shear waves. These distinctions are important because seismograms will be the principal means of checking on clandestine detonations if a treaty to ban underground nuclear explosions is established.

13.7 REPEATABILITY AND PREDICTION.

If Reid's theory is correct, then the gradual deformation of an area preceding an earthquake should be measurable, and might ultimately lead to the possiblity of predicting when the ground is under such extreme strain that it is about to rupture. It would be very useful to be able to predict earthquakes. People could be warned, enabling them to protect themselves and their property.

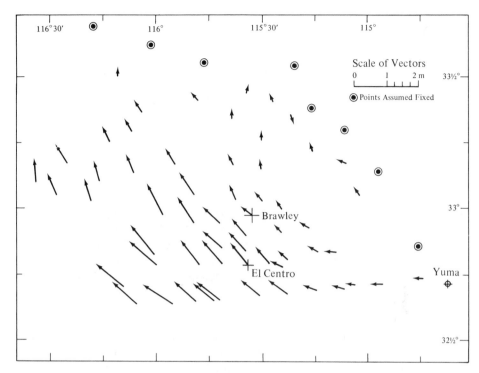

Fig. 13.30 *Ground displacements between 1941 and 1954 in the Imperial Valley, California. (After Whitten,* Trans. Am. Geoph. Union, *v. 37, p. 396.)*

Many experiments are currently being conducted in regions where earthquakes are common with the aim of learning enough about the mechanism to make a warning possible. One type of such studies consists of repeated surveys of the relative positions of surface points (Figure 13.30). Another consists of sensitive measurements of tilts of the ground surface to detect elevation changes. Whether or not these studies ever enable us to make useful predictions, they can certainly be expected in time to reveal a lot more about the deformations of the ground which accompany earthquakes.

Large earthquakes appear to be related to deformations involving most or all of Earth. They sometimes (and perhaps always) produce a change in the path of the axis of figure with respect to the axis of rotation (Section 4.5). It has been suggested but not proven that this change may begin a few days before the earthquake. If it does, this might provide a means of predicting large shocks (Figure 13.31).

Several factors make it unlikely that the prediction problem will be solved easily. One reason is that the ground can deform without breaking. Not only can some rocks deform plastically, as they obviously have done in forming folded mountains (Figures 9.14, p. 240 and 11.11, p. 296), but

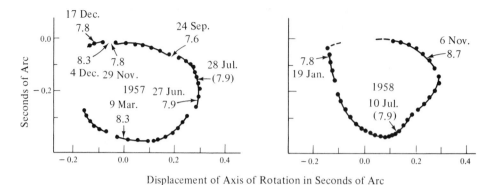

Displacement of Axis of Rotation in Seconds of Arc

Fig. 13.31 *Plots of 1957 and 1958 anomalous Chandler motion (with normal motion removed) showing sudden shifts shortly before large earthquakes. Magnitudes and dates of occurrence of all earthquakes of magnitude 7.5 or larger in 1957 and 1958 are shown. (After Smylie and Mansinha, 1968, Jour. Geophysics Res. v. 73 p. 7663.)*

even along faults, adjoining blocks of rock can move steadily rather than in jerks. Near Hollister, California a section of the San Andreas Fault is undergoing continuous displacement at the rate of 1.3 cm (0.5 inches) per year (Figure 13.32). The motion is sporadic, periods of quiescence alternating with shorter spells of more rapid drift.

Some areas are much more subject to earthquakes than others. Figure 13.33 shows the epicenters of all earthquakes reported by the U. S. Coast and Geodetic Survey for the period 1961–1967. Most of them occurred along the border of the Pacific Ocean. In regions like Alaska, Japan and the Samoa-Tonga area, earthquakes are much more common than elsewhere. The past frequency of earthquakes is some guide to the probability of future disturbances. However, long-term trends are not clear. In Kamchatka in eastern Siberia, the frequency of occurrence of earthquakes rose rapidly after the large earthquake of 1952. On the other hand, large earthquakes can occur in regions which have no history of such disasters. In 1886, an earthquake killed 27 people and did $50,000,000 property damage in Charleston, South Carolina. This large earthquake is unique in the whole history of the east coast of the United States (Figure 13.34).

Although the danger from earthquakes in the United States is usually considered to be much greater in the western cordillera from the Rocky Mountains to the Pacific Coast, possibly the most violent shocks in North America outside of Alaska occurred in southern Missouri in 1811 and 1812. The region was only slightly populated at the time, so that property damage was not great. There were three principal shocks on December 16, 1811 and on January 23 and February 7, 1812, the last being the largest. It was felt as far away as Boston, Massachusetts at 1750 km (1080 miles). A two-meter (six-foot) surface displacement was produced, causing a waterfall where it crossed the Mississippi River. An area over 160 km (100 miles) long and 50 km (30 miles) wide experienced extensive changes in level

Fig. 13.32 *A gradual slippage of 1.3 cm/year is slowly offsetting this ditch south of Hollister, Calif. (E.S.S.A. Photo.)*

causing ponding of lakes, forming of new swamps, tilting of stream beds and raising of low domes. Although this region has experienced many small earthquakes before and since the 1811–12 shocks, none of a comparable size has reoccurred. Clearly, a long history of low activity is no guarantee of future immunity.

Another problem is the infrequency with which large earthquakes occur with the same epicenter as a preceding earthquake. In Yugoslavia there is an historic record extending back for over two thousand years. In 1963 an earthquake occurred which caused widespread damage in the city of Skopje, the largest city in the province of Macedonia. When historical records were studied it was found that large earthquakes had occurred previously in Macedonia in 518, 1555, 1904 and 1921. But in none of these cases had the area of heaviest shaking, the epicentral area, been identical. Although earthquakes occur again and again in some regions, exact repetition rarely if ever occurs. This makes it hard to predict where or when the next big one will strike.

13.8 MAGNITUDE AND INTENSITY.

The amount of energy released by an earthquake can be calculated from the amplitude of the vibrations it produces. Suppose that an earthquake occurs right at Earth's surface and radiates a series of seismic waves in such a fashion that, 2.5 minutes after the earthquake has occurred, the ground is

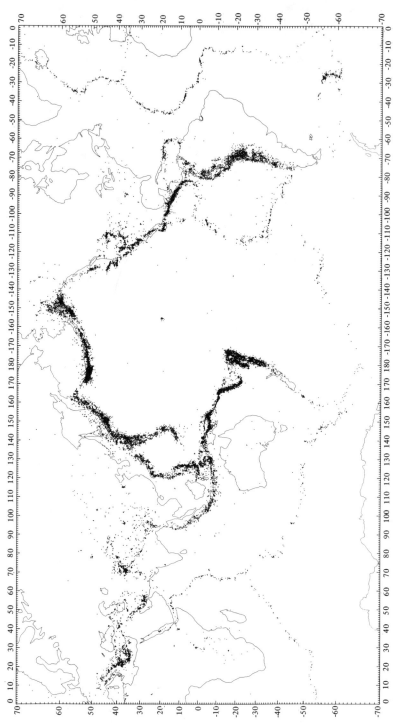

Fig. 13.33 *Epicenters of earthquakes reported by U. S. Coast and Geodetic Survey for the period 1961–1967. (Courtesy James Dorman.)*

vibrating at an average velocity of 0.1 centimeter per second over the distance range from 20 km to 120 km from the epicenter (Figure 13.35). The kinetic energy is

$$E_K = \frac{1}{2} M V^2 \tag{13.2}$$

where V is the velocity at which the ground is moving (0.1 cm/sec, or 10^{-3} m/sec) and M is the mass of the ground which is in motion. The volume of rock involved is a hemisphere of outer radius $R_o = 120$ km (1.2×10^5 m) and inner radius $R_i = 20$ km (2×10^4 m). Its volume is

$$V_R = \frac{2}{3} \pi (R_o^3 - R_i^3) \tag{13.3}$$

If the average density, d, of the rocks involved is 3 gm/cc (3000 kg/m³) then

$$
\begin{aligned}
E_K &= \frac{1}{2} V_R d V^2 \\[2mm]
&= \frac{\pi}{3} (R_o^3 - R_i^3)\, dV^2 \\[2mm]
&= \frac{\pi(12^3 - 2^3)\, 10^{12} \times 3000 \times 10^{-6}}{3} = 5.4 \times 10^{12} \text{ joules} \tag{13.4}
\end{aligned}
$$

The energy at any instant will be divided between kinetic and potential energy, so our hypothetical disturbance will actually involve about twice this much energy, or a little over 10^{13} joules. This is a big enough earthquake to be widely felt and to do considerable damage in the epicentral area. The 1906 San Francisco earthquake, one of the largest earthquakes of the 20th century, released about 2×10^{17} joules, or 20,000 times as much energy. An atom bomb of the Hiroshima type releases roughly 8×10^{13} joules, or less than one thousandth the energy of a large earthquake.

The sizes of earthquakes can conveniently be compared by noting the relative amplitudes of the ground motions they produce at a standard distance from the epicenter. The most commonly used measure is called the *magnitude*. It is found by measuring the largest amplitude of motion on the seismogram, calculating from this the instantaneous ground velocity, taking the logarithm of this number and adding a correction for the effect of distance from the epicenter:

$$\text{Magnitude} = \text{Distance Correction} + \log_{10} \frac{\text{Seismogram amplitude}}{\text{Seismometer sensitivity}} \tag{13.5}$$

There are several varieties of magnitude in common use, each based on a different method of measuring seismogram amplitude. When the largest

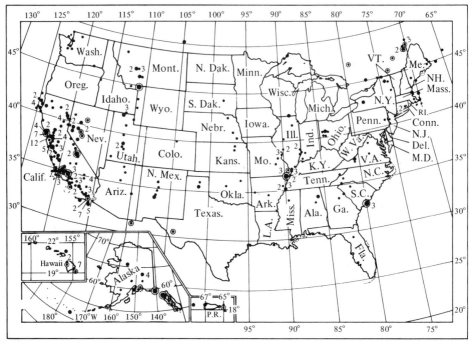

Legend

- Intensity VII-VIII or 25,000+ sq. mi. felt area
- Intensity VIII-IX or 150,000+ sq. mi. felt area
- ◉ Intensity IX-X or 500,000+ sq. mi. felt area
- ◎ Intensity X-XII or 1,000,000+ sq. mi. felt area

Fig. 13.34 *Epicenters of damaging United States earthquakes thru 1967. (E.S.S.A.)*

amplitude of a sound-wave pulse is used, the magnitude is called *body-wave magnitude*, M_B. When the largest amplitude of a surface-wave pulse is used, it is called *surface-wave magnitude*, M_S. The two are roughly related empirically by

$$M_B = 0.63 M_S + 2.5 \qquad (13.6)$$

The scale is designed so that the smallest earthquakes recorded have magnitudes, M_S, of about zero ($M_B = 2.5$). The M_B of the largest earthquake recorded is 8.2 (M_S was 8.9). Since the scale is logarithmic, one step in magnitude corresponds to an increase in the amplitude of ground motion of a factor of ten, or an increase in peak energy density of a factor of one hundred (compare Equation 13.2).

Earthquakes of magnitude, M_B, of 8 or more are rare, occurring about once every ten years (Figure 13.36). If the annual frequency graph is projected linearly to larger magnitudes it predicts one earthquake of $M_B = 8.2$ every 20 years and one of $M_B = 8.4$ every 50 years. Such is not the case. The largest earthquake observed between 1905 and 1971 was the 1906 Ecuador earthquake of magnitude 8.2. Earth seems unable to accumulate and release more energy than this at any one time.

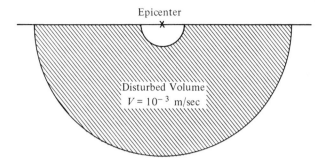

Fig. 13.35

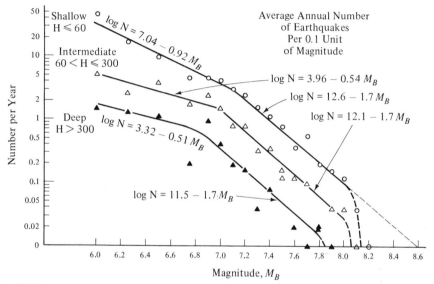

Fig. 13.36 *Annual frequency of earthquakes of different magnitudes. (After Gutenberg,* Quart. Jour. Geol. Soc. London, *62, p. 8.)*

The amount of energy released per earthquake increases with magnitude faster than the annual number of earthquakes of that size decreases. As a result, most of the energy released by earthquakes must come from the largest events (Figure 13.37). The total amount of energy released per year is about 10^{18} joules.

To a person experiencing an earthquake, the important feature is not how much energy is released, but what is happening in his immediate neighborhood. Earthquakes can also be rated as to the intensity of the damage they do. Table 13.3 is a summary of an intensity scale often used in the United States. Investigators interrogate residents of the shaken area and observe the effects of the shock. On the basis of these studies, maps are drawn showing the area within which the degree of shaking exceeds each level of the intensity scale. Such maps are called *isoseismal maps* (Figure 13.38). An earthquake is said to have had an intensity of nine if the highest degree of shaking was level nine on the scale.

Table 13.3 *The Modified Mercali Intensity Scale of 1931 (abridged).*

Intensity	Persons	Weak Structures	Strong Structures	Other effects	Typical body-wave magnitude if shallow
I	Not felt except by few under favorable circumstances.				
II	Felt by few at rest.				
III	Felt noticeably indoors. Standing cars may rock.			Delicately suspended objects swing. Duration estimated.	3.8
IV	Felt generally indoors. People awakened.			Cars rocked. Windows etc. rattled.	4.3
V	Felt generally.	Some plaster falls.		Dishes, windows broken. Pendulum clocks stop.	4.8
VI	Felt by all, many frightened.	Chimneys, plaster damaged.		Furniture moved, objects upset.	5.3
VII	Everyone runs outdoors. Felt in moving cars.	Moderate damage.		Waves seen on ponds, steep stream banks collapse.	5.8
VIII	General alarm	Very destructive and general damage.	Slight damage.	Monuments, walls down. Furniture overturned, sand and mud ejected, changes in well-water levels.	6.3

Table 13.3 (*Cont.*)

Inten-sity	Persons	Weak Structures	Strong Structures	Other effects	Typical body-wave magnitude if shallow
IX	PANIC	Total destruction.	Considerable damage.	Foundations damaged. Underground pipes broken. Ground fissured and cracked.	6.8
X	PANIC	Total destruction.	Masonry and frame structures commonly destroyed. Best buildings survive. Foundations ruined.	Ground badly cracked. Rails bent. Water slopped over banks. Large landslides.	7.3
XI	PANIC	Total destruction.	Few buildings survive.	Broad fissures. Fault scraps. Underground pipes out of service.	7.8
XII	PANIC	Total destruction.	Total destruction.	Acceleration exceeds gravity. Waves seen in ground. Lines of sight and level distorted. Objects thrown in air.	

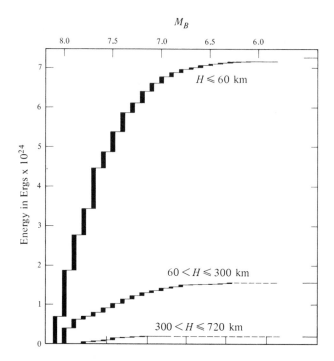

Fig. 13.37 *Cumulative average annual energy released by earthquakes of different magnitudes, M_B. H = depth of focus. (After Gutenberg, Quart,* Journal Geol. Soc. London, v. 112, p. 9.)

Usually there is a rough correlation between maximum intensity and magnitude levels. For earthquakes with a shallow depth of focus, the empirical relation is

$$\text{Maximum intensity} = 2M_B - 4.6 \qquad \textbf{(13.6)}$$

This relationship is a very rough approximation, because magnitude and intensity are affected by different factors. Deep-focus earthquakes generally have substantially lower intensities for the same magnitude than shallow shocks, but the shaken area is likely to be larger for the same maximum intensity.

Where there is surface faulting, the intensities along the fault will generally be level IX–XII. Elsewhere, the intensity depends greatly on the nature of the ground. Soft unconsolidated ground tends to vibrate more than consolidated rock. In Figure 13.38, note the expansion of the intensity contours along the sediment-filled central valley of California from Red Bluff to Hanford compared to their constrictions to the northeast along the axis of the igneous mass of the Sierra Nevada mountains. Intensification of shaking is particularly strong during the surface-wave portion of the ground motion. It is believed to be a resonance phenomenon similar to seiches in lakes. As wave motions strike the boundary between two media,

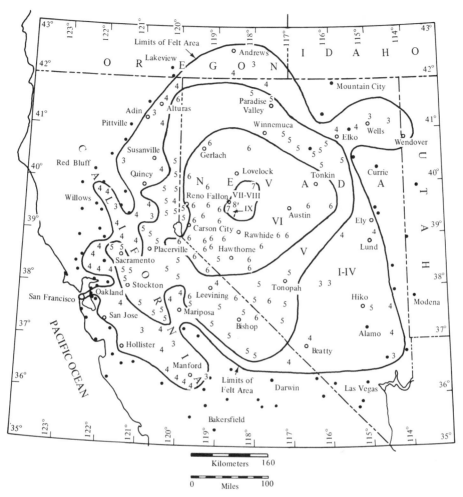

Fig. 13.38 *Isoseismal map of the Fallon, Nevada earthquake of 6 July 1954. (After Cloud,* Bull. Seis. Soc. Am. *v. 46, p. 34.)*

part of the energy is passed thru and part is reflected back into the first medium. Large superficial bodies of unconsolidated material tend to trap waves which reflect back and forth within them, building up in amplitude. An enhancement of amplitude by a factor of three or more is common.

Building damage also results, in part, from similar resonances. All structures have natural free periods at which they vibrate most easily. If the frequencies of ground vibration match the building's free resonances, then motions are larger and there is a greater danger of damage than otherwise. In regions where earthquakes are common, it is good practice to construct all buildings so that free oscillations will be damped out as quickly as possible or are at frequencies above those likely to be stimulated by earthquakes.

The magnitude of an earthquake calculated at different observatories is also subject to the effects of the ground. As the energy of a distant earthquake flows upward toward the surface under the observatory, it has to pass thru the layers of the crust. The top of each layer reflects back part of the energy of each pulse, and the bottom redirects part of the back-travelling energy surfaceward again. The complete ground motion observed at the surface is composed of a series of reflected and rereflected pulses formed in this fashion. The particular pattern of oscillations of the wave train formed in this way is called the *transfer function* of the ground. The result is that every seismogram consists of a complicated series of motions which have inherent in them much more information on Earth's structure than has anywhere as yet been completely unravelled.

Nearly every large earthquake teaches engineers something new about safe design. When the 1952 Kern County, California earthquake occurred, there were a number of recently built skyscrapers in Los Angeles which had never been shaken by a large disturbance. In several of these, offices were lit by heavy fluorescent fixtures hung from the ceiling on rods. These fixtures were set into oscillation like pendulums, and many broke loose and fell (Figure 13.39). Fortunately this occurred before 5 A.M. when the offices were unoccupied, so no injuries resulted. Weakly attached objects are particularly vulnerable to earthquake damage and should be avoided in construction.

Many states have special provisions in their building codes regarding protection against damage by earth vibrations. Masonry structures in particular require extra care in their design. Foundations tilt, causing whole buildings to sway back and forth and concentrating stresses wherever there is yielding. Masonry walls and decorative pieces which are not firmly attached have a tendency to fall (Figure 13.40). In a large city, especially in the downtown area where there are many multi-story buildings, masonry falling in the streets is likely to be more of a danger than the collapse of roofs (Figure 13.41). In older communities, especially where masonry without steel reinforcing has been used, it is wise to get out-of-doors as quickly as possible in an earthquake, and to stay out until aftershocks cease. Even modern buildings are not safe if they are not properly built (Figure 13.42).

Fire is one of the most serious hazards from earthquakes (Figures 13.41 and 13.43). Such fires may start from cooking fires, gas leaks which are ignited by sparks, or short-circuited electrical wiring. Their extinction is delayed by the general confusion following a large earthquake, and because many fires start all at once, streets are blocked by debris and water pressure fails where mains are broken or where tanks supplying water have collapsed (Figure 13.44). In San Francisco in 1906, fires got out of control in large part because the water supply to the city was broken at several places (Figure 13.45). It is desirable, in areas where earthquakes occur, to have small reservoirs scattered throughout cities for such emergencies and not to assume that the regular supply will be everywhere available at all times.

The amplitudes of ground motion at the surface during an earthquake are usually very small. Although it is obvious that motions of several me-

Fig. 13.39 *Fallen fluorescent-light fixtures in Los Angeles following the 1952 Kern County earthquake. (From Karl V. Steinbrugge collection.)*

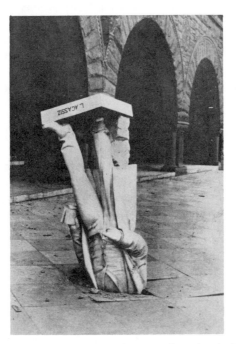

Fig. 13.40 *This statue fell from a wall niche at Stanford, California during the 1906 earthquake. (From Lawson, et al.,* The California Earthquake of 18 April 1906. *Courtesy Carnegie Institution.)*

Fig. 13.41 *Well-built buildings stand, but fallen masonry clogs the streets. San Francisco in 1906. (From Karl V. Steinbrugge collection.)*

ters sometimes occur, because fault displacements of this size have been produced, relative ground displacements of nearby points rarely can exceed a few millimeters or destruction from earthquakes would be much more widespread than it actually is. Only in the central epicentral areas of very large earthquakes is there more than superficial damage to well built structures. In spite of the obvious limitations on the amplitudes of ground motion demonstrated by the extent of resulting damage, following a large earthquake there are frequently claims that waves have been seen in the ground. Yet buildings reported to have been subjected to these waves show no damage. In part these claims may be the result of optical illusions induced by the unfamiliar motions to which the observer is subjected. A man's body acts like the inertial mass of a seismometer. It constitutes an inverted pendulum, easily caused to sway by the movements of the ground underfoot. The apparent motion of surrounding objects is, according to this theory, the result of uncompensated shifts in the frame of reference of the observer.

Sometimes waves may be suggested by the swaying of trees or grain as the seismic waves pass, although these motions are more likely to occur at the slow natural periods of oscillation of the plants than at the rapid rates of passage of the seismic waves. Seismic waves ordinarily travel at velocities of several kilometers per second. Some reports of observed motions of the ground during earthquakes may be purely the figments of excitable imaginations.

There are a few cases where observers have reported seeing cracks open and close in the ground as wave crests pass. In soft ground, such cracks

Fig. 13.42 *The Saada Hotel before and after the 1960 Agadir, Morocco earthquake. (Courtesy American Iron and Steel Institute.)*

would be expected from theory. However, most ground cracking is the result of lurching or settling (Figure 13.27, p. 358). Large cracks are almost invariably of this nature. Stories of fissures opening and swallowing people are common but are almost never verifiable. A story has often been told of a cow which fell into a fissure during the San Francisco earthquake. The fissure then closed, leaving only the tail exposed. The truth of the story is that a cow did fall into an earthquake-produced fissure, was injured by the fall and had to be killed by its owner, who buried her where she lay. In retelling, a little shoveling of earth developed into a natural closing of the ground. Thus do tales grow.

The only authenticated case of a person being killed by a fissure which opened and closed during an earthquake is from the 1948 Fukui, Japan

Fig. 13.43 *San Francisco on fire after the 1906 earthquake. (From Karl V. Steinbrugge collection.)*

Fig. 13.44 *This water tank near Bakersfield, California collapsed during the 1952 Kern County, Calif. earthquake. (Karl V. Steinbrugge photo.)*

earthquake. Witnesses saw a four-foot wide crack form, a woman fall into it, and the crack close upon her. Her crushed body was later recovered.

Although cracking of the ground, a sign of tension, is more commonly reported for earthquakes than is evidence of compression, the latter does

Fig. 13.45 *During the 1906 San Francisco earthquake, water mains were broken by the faulting, depriving the city of water to fight fires. (From Lawson, et al.,* The California Earthquake of April 18, 1906. *Courtesy Carnegie Institution.)*

frequently occur (Figure 13.46). One of the most striking results of such compression is the production of natural fountains. These are most commonly reported in areas of soft ground. Water or mud is spurted into the air to heights as great as a few meters, sometimes steadily, less often pulsating. The play may be brief and coincident with the ground vibrations, or it may continue after the shaking has subsided. Sand and mud craters found following many earthquakes attest to the prevalence of this phenomenon (Figure 13.47).

The phenomenon has not been thoroughly studied, so its cause is partly speculative. A likely explanation is that buried water-saturated rocks are squeezed. The water escapes to the surface thru a few cracks which permit a large and forceful flow to develop. This explanation is supported by observed fluctuations in some water wells during earthquakes. Floats are used in many large wells to indicate water depth. These sometimes show oscillations of a meter or more (several feet), indicating that water has been forced into the well and sucked out as the earthquake waves passed. Water-level gauges in wells make good strain seismometers where water-saturated rock is sensitive to ground pressure.

Fig. 13.46 *Compression of the ground during the 1948 Fukui, Japan earthquake buckled this railroad track. (Courtesy U. S. Geol. Survey, photo by Mainichi Graphic.)*

13.9 PSEUDOSEISMS.

Air vibrations can easily be mistaken for earthquakes, especially if the sounds are of too low frequency to be heard by the ear. The passage of low-frequency air waves past the body gives a physical sensation very similar to ground motion. Vibrations of this sort are called *pseudoseisms*.

Pseudoseisms can be produced by distant explosions. Sound waves in air are generally bent upward by a decreasing velocity of transmission with elevation (Figure 6.12, p. 149). This occurs because sound velocity in air is proportional to the square root of the temperature, which normally decreases upward throughout the troposphere. Occasionally, however, a temperature inversion occurs with warm air overlying cold air. Under these conditions, sounds may be bent back to the surface without having been observable over an intervening interval. The sound is usually distorted, and low frequencies may be emphasized over high frequencies. This increases the likelihood of sound being mistaken for earth vibrations.

Sometimes the atmosphere has a focusing effect which greatly increases the strength of the sounds at some particular distance from their source. Blasting and the detonation of large guns have been observed at distances as great as several hundred kilometers as a result of such concentrations of energy. There can even be some danger of minor damage from the air vibrations due to the focusing of the sound. For this reason it is wise to postpone large surface blasts on days when an air-temperature inversion exists.

A jet aircraft going faster than sound produces pressure waves in the atmosphere which are easily mistaken for the noise of a blast. At large distances, when the more easily heard higher frequencies have been filtered out, such sounds may also become pseudoseisms.

Fig. 13.47 *Craters formed by fountaining of water during an earthquake. (Courtesy U. S. Geological Survey. Photo by W. R. Hansen.)*

13.10 EARTHQUAKES AS A MEANS OF ENERGY TRANSFER.

The energy involved in the earthquake process is substantially less than the heat flux thru Earth's surface, which is about 2.4×10^{17} kg-calories per year or 10^{21} joules (Section 12.5). The amount of energy released annually by earthquakes is only about 10^{18} joules, or one thousandth of this. Earthquakes could thus be a minor consequence of the much larger heat flux.

The distribution of earthquakes, however, is highly irregular. Most of the energy is released in a few narrow belts (Figure 13.33). This is particularly true for the intermediate and deep shocks. Even within these belts, however, the total energy involved probably does not exceed that represented by the heat flux. Their distribution does imply, however, that stresses are not uniformly generated or relieved in the crust or mantle. The distribution of earthquakes is a clue to processes which are going on in Earth.

Seismic waves carry energy with them from their focal regions, and distribute it throughout Earth. Studies of the variation of the amplitudes of seismic pulses with distance from the source indicate that the lower mantle and core are probably much more nearly perfectly elastic than the crust and upper mantle. As a result, almost all of the energy from seismic waves is absorbed in the same layers where it is generated. Earthquakes may help to spread energy horizontally, but they probably carry very little energy downward.

All of these arguments are based on the supposition that the history of seismic activity since the development of widespread instrumental recording around 1905 is typical of past geologic time, which may not be true.

BIBLIOGRAPHY

Adams, W. M., 1964, *Earthquakes*, D. C. Heath and Co. 122 pp.

Davison, C., 1936, *Great Earthquakes*, Thomas Murby and Co. 286 pp.

Freeman, J. R., 1932, *Earthquake Damage and Earthquake Insurance*. McGraw-Hill Book Co., Inc. 904 pp.

Gutenberg, B., 1959, *Physics of the Earth's Interior*. Academic Press, 240 pp.

Gutenberg, B., C. F. Richter, 1954, *Seismicity of the Earth*. Princeton Univ. Press. 310 pp.

Heck, N. H., 1945, *Earthquakes*. Hafner Publ. Co. 222 pp.

Hodgson, J. H., 1964, *Earthquakes and Earth Structure*. Prentice Hall, Inc. 166 pp.

Howell, B. F., Jr., 1959, *Introduction to Geophysics* McGraw-Hill Book Co., Inc. 399 pp.

Iacopi, R., 1964, *Earthquake Country*. Lane Book Co. 192 pp.

Kirkland, W. G., 1962, *The Agadir, Morocco Earthquake*. American Iron and Steel Institute, 96 pp.

Lawson, A. C., H. F. Reid, and others, 1908, *The California Earthquake of April 18, 1906*. Carnegie Institution of Washington v. 1, 451 pp., v. 2 (1910), 192 pp.

Richter, C. F., 1958, *Elementary Seismology*. W. H. Freeman and Co. 768 pp.

14 TECTONIC PATTERNS

14.1 VOLCANISM.

One of the most obvious conclusions that can be drawn from study of any part of the universe is that change is going on everywhere. Such change normally involves a transfer of energy. Just as the galaxies are spreading in space, so is energy being redistributed more evenly throughout the universe. On the whole, Earth is losing energy. It radiates away more heat than it receives. It gives up rotational energy to Moon. Hydrogen escapes from its upper atmosphere.

In the case of Earth's interior, the most effective means of energy transfer appears to be thermal. More energy flows thru the surface as heat than as any other form. It might seem that a study of the detailed pattern of heat flow would be the best way to discover the fundamental pattern of changes in Earth. This, however, is not the case. Heat flow is difficult to measure. Its variations from place to place are relatively small; and it is strongly affected by local conditions near the surface (see Chapter 12). With the data currently available, it is not possible even to say for sure whether the heat flux is larger under the oceans or on land; although it appears probable that the oceanic flux is the greater. Although the heat flux is exceptionally large at many places along the oceanic ridges, the flux field has not been mapped anywhere in sufficient detail to determine whether this is a general rule or not, or even to say whether such areas are the places of greatest heat flow.

An alternative method of studying the pattern of heat loss is to examine the distribution of active volcanoes over Earth's surface (Figure 9.10, p. 235). It might reasonably be expected that where the heat flow is greatest, the temperature gradient will be highest, the melting point of rocks will be reached nearest the surface, and volcanoes will be most likely to form. Shown in Figure 9.10 are 460 volcanoes known to have had eruptions within historic time. The coverage is probably more complete in some places (e.g. Europe and the Near East) than in others, because written records go back farther. It is probably very incomplete in ocean-covered areas, because volcanic events there will be noticed only if they are violent enough to break thru to the water surface, and even then only if there is a ship nearby when the eruption occurs.

The distribution of volcanoes is strikingly limited. Most are found in a narrow belt surrounding the Pacific Ocean. This includes the 49 active volcanoes of the United States exclusive of Hawaii (Table 14.1. Figure 14.1). Sunset Crater, Arizona is of special interest. Indian legends tell of its eruption. The date has been established by the measurement of tree-ring patterns in logs buried by the ash of the eruption. Comparing these with ring patterns of still living trees and ancient logs studied in nearby regions shows the year when the buried trees ceased to grow as 1060, presumably the year the volcanic eruption killed them. There are other equally fresh-looking craters in the western part of the Great Basin of the United States which may have erupted as recently as Sunset Crater. Parts of Inyo and Mono Counties, California and adjoining areas in Nevada show extensive evidence of recent volcanism (Figures 14.2, 14.3). This region can be presumed to be still active, and further eruptions similar to those which buried thousands of square kilometers under lava and ash can be anticipated in the future.

Along the east side of Asia, the same volcanic belt of which California and southern Alaska are parts extends southward from Kamchatka along the Kurile Islands to the Japanese Island of Honshu. Here it splits into two trends. One branch extends southwest thru the Ryukyu Islands to Taiwan, the Philippines and Celebes in Indonesia. The other runs south along the Bonin, Volcano and Marianas Islands.

At Celebes the main volcanic belt again splits. One part runs eastward past the Bismarck Archipelago and the Solomon Islands to Hunter Island in the New Hebrides. The other follows the southern arc of Indonesia from Timor thru Java, Sumatra and the Andaman Islands to Burma.

Starting northeast of the New Hebrides, a string of volcanoes continues the circumpacific trend from Samoa south thru the Tonga and the Kermadec Islands to Ruapehu on the north island of New Zealand. Farther south, there are six volcanoes in the Antarctic area. The list may be incomplete because of the lack of observations on this continent and in the seas along its coast. There are several active volcanoes in the South Sandwich Islands.

The belt continues northward along the west coast of South America to Ecuador, where it again branches. The main branch follows the Pacific coast thru southern Mexico and Baja, California to complete the circumpacific loop, sometimes called the *ring of fire*. The other branch crosses Venezuela to form the chain of volcanoes of the Lesser Antilles, which separate the Caribbean Sea and the Atlantic Ocean.

In addition to this principal belt of volcanoes and its branches, volcanoes are found at a number of places along the Mid-Atlantic Ridge, notably Iceland, Jan Mayen Island, and Tristan da Cunha, and in the Cape Verde, Canary and Azores Islands east of the Ridge.

In the Mediterranean area, there are ten active volcanoes in or near Italy and six more between Sicily and the Aegean area plus Nimrod Volcano in Armenia.

There are twelve volcanoes scattered along the rift system extending south from the Red Sea to Tanzania.

Fig. 14.1 *Lassen Peak, California. This volcano last erupted in 1917.*

Table 14.1 *Active Volcanoes of the United States*

Name	Location	Latest Eruption
Lassen	California	1917 Lava
Shasta	California	1876 Ash
Baker	Oregon	1870 Explosive
Cinder Cone	California	1851
St. Helens	Washington	1841–3 Lava and Ash
Sunset Crater	Arizona	1060 Ash
Redoubt	Alaska	1966
Pavlof	Alaska	1964 Ash
St. Augustine	Alaska	1964 Ash, lava
Trident	Alaska	1963 Ash, lava
Katmai	Alaska	1962
Martin	Alaska	1960
Spurr	Alaska	1953 Explosive
Torbert	Alaska	1953
Iliamna	Alaska	1947
Mageik	Alaska	1946
Veniaminof	Alaska	1944
Aniakchak	Alaska	1931 Ash, lava
Chiginagak	Alaska	1929
Novarupta	Alaska	1912 Ash
Atlin Lake	Alaska	1898 Ash
Onnimah-Strasse	Alaska	1856 Submarine
Mt. Wrangel	Alaska	1819
Peulik	Alaska	1814 Explosive

Table 14.1 (cont.)

Name	Location	Latest Eruption
Kiska	Aleutian Islands	1969 Explosive, lava
Pinnacle Island	Pribilof Islands	1964 Explosive, lava
Pogromni	Unimak Island	1964 Explosive, lava
Shishaldin	Unimak Island	1963 Explosive
Amukta	Aleutian Islands	1963
Akutan	Aleutian Islands	1952
Great Sitkin	Aleutian Islands	1947 Lava dome
Okmok	Umnak Island	1945 Lava, ash
Mt. Cleveland	Chaginadak Island	1944 Explosive, lava
Makushin	Unalaska Island	1938 Explosive
Yunaska	Aleutian Islands	1937 Explosive
Kanaga	Aleutian Islands	1933
Bogoslav	Aleutian Islands	1931 Lava dome
Gareloi	Aleutian Islands	1930 Lava
Kagamil	Aleutian Islands	1929
Seguam	Aleutian Islands	1927
Tanaga	Aleutian Islands	1914 Lava
Recheschnoi	Umnak Island	1878 Explosive
Semisopochnoi	Aleutian Islands	1873
Isanotski	Unimak Island	1843
Carlisle	Aleutian Islands	1838
Little Sitkin	Aleutian Islands	1828
Fisher	Aleutian Islands	1826 Explosive
Sarichef	Atka Island	1812
Pavlof Sister	Aleutian Islands	1786
Kilauea	Hawaii	1969 Lava
Mauna Loa	Hawaii	1950 Lava
Hualalai	Hawaii	1801 Lava
Haleakala	Maui	1750 Lava

Isolated volcanoes occur in the Cameroons (West Africa), on Grand Comoro and Reunion Islands in the Indian Ocean, in Manchuria and in the South China Sea. There are three volcanoes on the Galapagos Islands off the coast of Ecuador, and four in the Hawaiian Islands.

Every few years another volcano is added to the list of active vents. Surtsey off the south coast of Iceland first erupted in 1963 (Figure 14.4). Paricutin in Mexico first erupted in 1943 (Figure 14.5). Furthermore, there are a large number of inactive or dormant volcanoes which appear from the rocks which surround them to have erupted within the past few thousand years. These vents are scattered along the belts of active volcanoes. (A *dormant volcano* is one which erupted in prehistoric times and is going to erupt again. It is often impossible to tell until it erupts whether a volcano is finally inactive or only dormant.)

About 98 percent of the active volcanoes occupy three groupings: the ring of fire surrounding the Pacific and the branches of this ring, the Atlantic Ocean belt including the volcanoes in the Azores, Canary and Cape

Fig. 14.2 *Pre-historic but fresh-looking lavas cover parts of Inyo County, California.*

Fig. 14.3 *Warm water from hot springs in Mono County, California, provides a good use for volcanic heat.*

Fig. 14.4 *Surtsey Volcano, a new island forming south of Iceland. This photo-graph was taken November 26, 1963. (Courtesy S. Thorarinson.)*

Verde Islands, and the group extending from the central Mediterranean east and then south along the Red Sea to Tanzania and out into the Indian Ocean. The three groups are not continuous. They may really consist of more than three groups, since there are long gaps between sub-groupings in each main group.

The fact that nearly all of the volcanoes of the world fall along so small a number of lines in contrast to the vast areas in which there is no evidence of volcanic activity for millions and sometimes hundreds of millions of years points to a concentration of some kind of exceptional thermal activity in these special regions.

14.2 SEISMICITY.

The distribution of earthquakes is remarkably similar to that of volcanoes (Figure 13.33, p.364) except that earthquakes occur more often than volcanic eruptions, so that the earthquake belts are more continuously marked than the volcanic ones. Earthquakes are common along all the coasts facing the Pacific except Antarctica. The belt starts just north of the Palmer Peninsula in Antarctica, follows the Falkland Island Ridge thru the South Sandwich Islands, returns to Tierra del Fuego in Southern Chile, continues unbroken to Alaska, south again thru Kamchatka, Japan and New Guinea to the New Hebrides, deflects (just as the volcanic belt does) to Samoa, and reaches south to Macquarie Island beyond New Zealand.

Fig. 14.5 *Lava from the new volcano, Paricutin, in Mexico covered the town in the foreground in 1944. (Photo by F. O. Jones. Courtesy U. S. Geological Survey.)*

Sections which had few earthquakes between 1961 and 1967 (Figure 13.33) are filled in by shocks in other years. Unlike the volcanic pattern, the earthquake belt does not continue thru Antarctica, where earthquakes of all sizes are extremely rare.

The earthquake belt branches in Venezuela to loop around the Lesser Antilles and Hispaniola. It splits in Japan to the Marianas and Palau Islands; and it divides in Indonesia. The Indonesian branch is particularly well developed, and extends thru Burma into Central China, where it spreads out over a broad area reaching from the Himalayas on the south at least to the Altai Mountains along the Russian border on the north. Westward the belt includes Afghanistan, Persia, Turkey, Greece, Yugoslavia, Italy, Algeria and Morocco.

The Atlantic, Indian and Antarctic oceanic ridges are frequently marked by earthquake epicenters. This is true also of the East Pacific Rise, but not of the central Pacific ridge. There are no earthquakes in the central Pacific area except around Hawaii, where they commonly precede volcanic eruptions.

The African rift system from the Red Sea to Lake Nyasa is also marked by a belt of epicenters. Other minor belts such as the St. Lawrence River zone are revealed by plotting events occurring over a long period of time (Figure 13.34, p. 365).

Although occasional small earthquakes can occur anywhere, their absence in the centers of the oceanic basins and in the continental-shield areas points to the stability of these places. Whatever the processes of change are that act near Earth's surface, the energy release, both seismic and thermal, is concentrated in three systems of belts: the circumpacific belt plus its branch thru central Asia to the Mediterranean, the oceanic ridge system and the African rift system.

14.3 MOUNTAINS.

One of the effects of volcanoes and earthquakes is to build up new land masses. Over the whole Earth, roughly a cubic kilometer of lava and ash is emitted from volcanoes in an average year. After many successive eruptions, the summit of a volcano is raised high above its surroundings. Its mass is visibly added on top of previously existing layers of the crust. Similarly, large blocks of land are raised (and dropped) as much as ten meters or more in large earthquakes. Although the frequency of repetition of large shocks in individual areas is too low to tell how fast large mountain ranges have been formed, it is only necessary to suppose that the largest observed uplifts are repeated once every few thousand years, continuing for a few tens of millions of years, to raise the highest mountains seen today. Geologic evidence of large changes in elevation is abundant (e.g., Figures 9.15, p. 241 and 13.15, p. 348).

Mountain ridges are obviously temporary features. Weathering and erosion eventually wear away the peaks on land and fill in the valleys of the oceans. Sharp topographic relief is thus proof of recent disturbance.

The location of recently formed mountains is closely associated with the belts of volcanic and seismic activity. Figure 14.6 shows the distribution of mountain chains which, from the ages of the rocks of which they are composed, are known to have been formed during the last 65 million years. These belts include most of the high mountains of the world: the Andean–Rocky Mountain–Pacific Coast cordillera of the Americas and the Himalayan–Caucasus–Alpine system of Eurasia and northwestern Africa. They also include most of the chains of islands along the western margin of the Pacific from the Aleutians on the north, thru Japan and the Philippines to the New Hebrides on the south, plus the West Indies and Falkland Island ridges in America. The whole Indonesian archipelago is a chain of recently formed mountains.

Examined in detail, the mountains of the continents reveal other characteristic patterns. One of these is the thickness of the crust as revealed by seismic refraction surveys. In general, the higher the average surface elevation in any area, the thicker the crust is underlying that area. This would be expected from isostasy (Section 9.7). Assuming five kilometers as the normal value for the ocean depth, and five kilometers as the normal thickness of oceanic crust, then for isostasy to hold, the length of the root extending into the mantle beyond ten kilometers under any part of the continents will be proportional to the height of the continent above the sea bottom according to the following formula (compare Figure 14.7):

$$d_w H_w + d_c H_{oc} + d_m H_r = d_c (E + H_w + H_{oc} + H_r) \qquad \textbf{(14.1)}$$

where d_w = density of ocean water = 1.028 gm/cc

d_c = density of crustal rocks, here taken as 2.7 gm/cc

d_m = density of mantle rocks, gm/cc

H_w = ocean-water depth, here taken as 5 km

H_{oc} = thickness of oceanic crust, here taken as 5 km

H_r = length root of continent extends into mantle beneath oceanic crust in kilometers

E = height of continent above sea level in kilometers

Solving for H_r by rearranging terms

$$d_m H_r - d_c H_r = d_c E + d_c H_w - d_w H_w \qquad \textbf{(14.2)}$$

$$H_r = \frac{d_c E + (d_c - d_w)H_w}{d_m - d_c} \qquad \textbf{(14.3)}$$

and substituting in numbers

$$H_r = \frac{2.7E - 8.36}{d_m - d_c} \qquad \textbf{(14.4)}$$

Values of root length, H_r, for different elevations and different values of density contrast, $d_m - d_c$, are given in Table 14.2. Note that Equation 14.4 is derived in such a manner that it is not necessary to require that density be constant throughout the crust. The denominator, $d_m - d_c$, is the difference in density at any depth. Both d_m and d_c would be expected to increase with depth due to pressure.

Since we know from seismic refraction studies that the crust is generally 30 to 40 km thick (root of 20–30 km) over large parts of the continents where surface elevations average less than 1000 meters, the density contrast must be between 0.3 and 0.4 gm/cc. This is consistent with the maximum measured crustal thickness of 72 km where the surface elevation is around 5000 meters in the Academy of Sciences Mountains of the Soviet Union.

What this means is that mountain ranges involve the whole thickness of the crust, not just the upper layers. The structure and history of the roots is as important to an understanding of mountain formation as is the better-known superficial appearance at the surface.

The roots of current mountains are not available for direct study, being deeply buried by other rocks. Erosion of older, once lofty mountains, how-

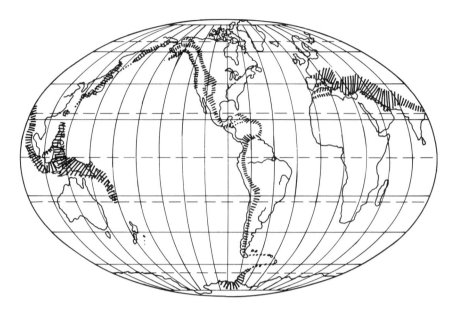

Fig. 14.6 *The most recently formed mountain belts.*

ever, has exposed the internal structure of ranges. From studies of such regions, geologists have discovered a number of features which seem to be characteristic of the type of old mountain belts found on continents.

There are six additional principal characteristics of continental mountains which seem to be observable in almost all ranges. First they occur in long narrow belts (Figure 14.6). The width of the belts may vary. They may split into several branches, either parallel or diverging. The branches often separate and then merge again. Chains of different ages often are parallel, less often overlap or intersect. In any case, each chain or group of chains occupies only a small part of the total surface of Earth. No such marked restriction of mountain formation to such a limited distribution at any one time and to other localities at other times could have resulted from a uniform, world-wide set of forces. Mountains are built by forces which were concentrated in the local areas where the mountains are found.

The second characteristic is that a considerable part or all of the area currently occupied by the mountain system was formerly covered by a shallow sea whose basin gradually sank as sediments accumulated in it. Such a basin of deposition is called a *geosyncline*. The axis of the geosyncline and the axis of the mountains are generally nearly parallel and may coincide. The sedimentary rocks exposed in the mountains are largely those which were deposited in the geosyncline. The types of sediments involved are similar to those being deposited along the continental margins today in places like the Gulf of Mexico and the Atlantic Shelf (Figure 9.13, p. 239).

Table 14.2 *Crustal-root length in kilometers for different values of density difference,* d_m-d_c, *between mantle and crust and different elevation, E, of continents above sea level.*

Density difference, $d_m - d_c$, in gm/cc

E	0.2	0.3	0.4	0.5	0.6
0	41.8	27.9	20.9	16.7	13.9
1,000m	55.3	36.9	27.7	22.2	18.4
2,000m	68.8	45.9	34.4	27.5	22.9
5,000m	109.4	72.9	54.7	43.7	36.4
10,000m	176.8	117.9	86.4	70.7	58.9

Third, rocks originally deposited nearly horizontally in the geosyncline have been compressed as a part of the mountain-building process. If individual formations were flattened out so that each fold was returned to a horizontal position, the edges of the geosyncline would be moved apart by a few to tens of kilometers. Mountains represent a shortening of the crust. The actual amount of compression is hard to measure because there is a tendency for individual layers of rock to be stretched parallel to their surfaces even as the whole bundle of layers is squeezed. The total amount of the compression can best be estimated by noting the increase in crustal thickness from 30–40 kilometers to 40–70 kilometers.

Fourth, the direction of the shortening was at right angles to the axis of the geosyncline. The force must have acted largely in this direction also, because mountains are long, narrow structures. Since the length of individual mountain systems is great, we are dealing with a system of forces, not a single force.

Fifth, for millions of years after mountains were first formed, the whole folded belt continued to rise, exposing by erosion rocks which had formerly been deeply buried. This must mean that the forces producing the original compression ceased their action, or moved elsewhere.

Three stages of the mountain building process can be recognized: first, the accumulation of sediments in the geosyncline; second, the compression forming the mountains and their root; and third, the removal of the ridge by erosion. Presumably the root shrinks also, rising to replace the material eroded away at the surface, and possibly also either spreading sideways or being absorbed in the underlying rocks.

The final principal characteristic is that extensive erosion typically exposes a core of igneous rocks at the center of the range. Very large bodies of igneous rock of this sort are called *batholiths*. The average chemical composition of batholiths is very close to the average composition of geosynclinal sediments, even though the mineralogical composition is different. At places, large masses of partially metamorphosed sedimentary rocks grade into igneous rocks (Figure 14.9). Much if not all of the magma which solidified to form these batholiths must have been derived from the

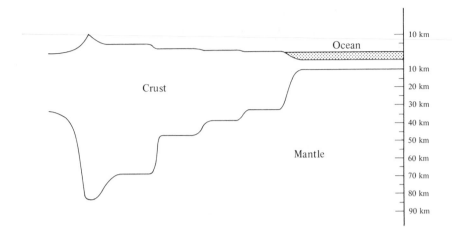

Fig. 14.7 *The higher the surface elevation the greater the expected crustal thickness.*

sediments which were folded to form the mountain range. The batholiths were formed after the folding, and clearly take the place of invaded rocks (Figure 14.8). Their composition is in marked contrast to that of basaltic rocks which commonly intruded the sediments during or after their deposition and compression or were extruded thru them onto the surface.

Batholiths are of great horizontal cross-section, and increase in size as one goes deeper beneath the surface. They are elongated parallel to the mountains. They appear to have been emplaced over a period of time, involving several paroxysms of successive intrusion with the composition of the intruded rocks changing with time. The sequence of composition of the intrusions is consistent with a tendency for the heavier mineral crystals to settle downward in the molten mass, and lighter silicon-aluminum-rich and associated compounds to concentrate upwards. Studies in many areas indicate that the history of solidification of deep-seated igneous bodies involves the upward concentration of low-density minerals at the expense of heavier constituents. The average surface rocks of the continents are chemically more like the average of the exposed tops of batholiths than the average of lavas escaping from volcanoes on the continents. This is not to say that there are not sialic (granitic) volcanoes. Such volcanoes do occur on the continents; but there are also basaltic volcanoes here. In the oceanic areas there are only basaltic volcanoes. This suggests that basalt comes from the lower crust or mantle of Earth and sial is a derivative product separated from it in the melted roots of mountains on the continents.

The oceanic ridges, while not topographically so prominent as the continental mountains, also contrast sharply in elevation to the surrounding ocean floors. Their lower but still relatively great heights are also marked by belts of earthquakes and volcanoes. The degree to which the characteristics listed above for continental mountains also apply to oceanic mountains is unknown, because the oceanic ranges are not accessible for direct observation. Since oceanic mountains do not commonly protrude above

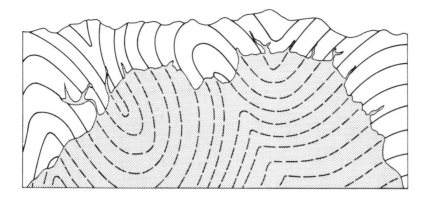

Fig. 14.8 *A batholith (shaded) replaces folded rocks in the root of a mountain range.*

sea level, and where exposed are more simatic in composition than average continental rocks, it is likely that they differ also in other ways.

Lastly, the rift systems of Africa are flanked by this continent's highest mountains, from Mt. Sinai at the north end of the Red Sea thru Ethiopia and Tanzania to Mt. Mlanje in Mozambique. The towering Drakensberg of Natal lies along this trend to the south. These mountains differ from the circumpacific belt in a number of ways. Although their component rocks are generally similar in composition, they appear to have reached their present elevation more by vertical displacement on faults than by the compressive processes typical of continental mountains. They resemble the oceanic ridges in that they have a central trough and lack intermediate and deep earthquakes.

14.4 TRENCHES.

The deepest depressions on Earth's surface are found close to and parallel with mountain belts (Figure 9.10, p. 235). Narrow depressions lying below normal ocean-bottom level are called *deeps*. Where they lie in front of arcuate ridges fronting continents, they are called *foredeeps*. In some cases they afford greater topographic relief than the highest mountains above sea level (Table 14.3). The deepest known is the Challenger Deep southeast of the Marianas Islands, reaching 10,863 meters (35,640 feet) beneath sea level.

Near land such troughs quickly fill with sedimentary debris. A deep deposit-filled basin lies along the south edge of the Himalayas. The Persian Gulf may be a similar structure. In Western North America a remarkable chain of valleys begins south of Mount Saint Elias in Alaska, forms the inside passage behind Vancouver Island, Canada, and extends south along Puget Sound and the Willamette Valley. It is interrupted by the Klamath Mountains in Oregon, but continues as the Sacramento and San Joaquin Valleys in California. The Tehachapi Mountains and the Transverse

Table 14.3 *The Greatest Depths in the Oceans*

Name	Location	Maximum depth in meters
Challenger deep	South of Marianas Is.	10,863
Tuscarora deep	East of Kurile Is.	10,540
Emden deep	East of Philippines	10,500
Ramapo deep	East of Japan	10,374
Penguin deep	East of Tonga-Kermadek	10,035
Planet deep	East of New Britain	9,410
Milwaukee deep	North of Puerto Rico	8,750
Bonin trench	East of Bonin Is.	8,660
Byrd deep	20° South of New Zealand	8,590
New Hebrides trench	West of New Hebrides	8,320
South Sandwich trench	East of Falkland Is. Ridge	8,264
Atakama trench	West of Peru	8,050
Aleutian trench	South of Aleutian Is.	7,680
Ryukyu trench	East of Ryukyu Is.	7,480
Sunda trench	South of Indonesia	7,455
Romanche trench	South of Mid-Atlantic Ridge along equator	7,300
Cayman trough	North edge of Caribbean Sea	7,200

Ranges separate this area from the Ventura and Los Angeles basins to the south. A continuous valley connects the 14,000 meter (45,000 ft.) deep Los Angeles basin with the Salton Sink and the Gulf of California to the south.

14.5 GRAVITY ANOMALIES.

In addition to such obvious evidences of disturbance as mountain ranges and ocean deeps, subtler evidence comes from measurements of gravity. After correcting for the more obvious features of mass distribution, anomalies in gravity remain (compare Section 9.7). The largest of such variations amount only to around 2/10,000 of the normal pull of gravity, but this is still more than 20 times as large as the average variation. Like the other tectonic features discussed already, they tend to be found in narrow belts parallel to the edges of the continents (Figure 9.10, p. 235).

Large gravity deficiencies are commonly found between island arcs and the foredeeps which parallel them. The best known such belt lies along the south edge of Indonesia, looping north between Ceram and New Guinea and extending along the east edge of the Philippines (Figure 14.10). It follows the intermittent chain of islands off the coast of the larger islands. The axis of the anomaly passes thru the islands of Soemba, Timor and Tanimber. Less pronounced areas of positive gravity anomaly lie on either side of the minimum.

Fig. 14.9 *This light-colored granite, which has intruded older, dark-colored metamorphic rocks, is typical of the core of the Grand Teton Mountains, Wyoming, U.S.A.*

Similar belts of anomalies lie along the foot of the Himalaya Mountains in India and along their extensions south thru Burma, west of the Japan deep, at places along the west cost of South America, along the Russian coast on the northeast side of the Black Sea, along the north edge of the Carpathian Mountains in central Europe, in the Adriatic Sea between Italy and Greece and along the the Atlantic side of the West Indies. There are less clear patches of large anomaly along the west coast of the United

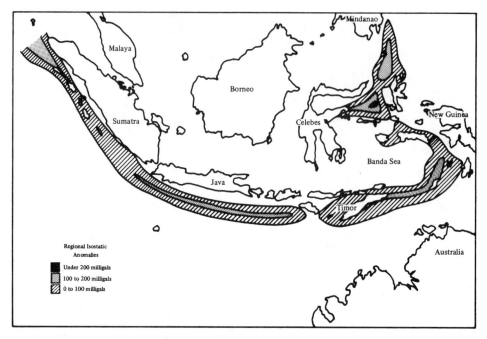

Fig. 14.10 *Pattern of gravity anomalies in Indonesia. (After Heiskanen and Vening Meinesz,* The Earth and Its Gravity Field, *p. 388. Courtesy McGraw-Hill Book Co., Inc.)*

States, of which the large anomaly near Seattle is the most prominent example (Figure 9.25, p. 255). Many others may exist. Earth's gravity field has not yet been thoroughly mapped.

These gravity lows represent deficiencies in mass compared to the normal distribution. They must mean that low density material extends to greater depths in these areas than on the average. Unless they are compensated by equal amounts of exceptionally dense rocks at greater depths, they show that in the anomalous areas the rocks are not in isostatic equilibrium. This suggests that forces are currently stressing these areas, or that they acted so recently that there has been too little time for isostatic equilibrium to have been reestablished. Large gravity anomalies are, therefore, another mark of recent or current deformation.

14.6 MAGNETIC ANOMALIES.

Magnetic anomalies are even less thoroughly mapped. Patterns of rock magnetism were discussed in Section 7.4. In the oceans these occur as bands, usually paralleling the oceanic ridges (Figure 7.17, p. 183). Although only a small fraction of the oceans has as yet been mapped and the belts at some places are displaced horizontally by faults, showing that blocks of the ocean floor have been offset horizontally (Figure 15.16, p. 428), the parallelism of the magnetic anomalies to the other tectonic fea-

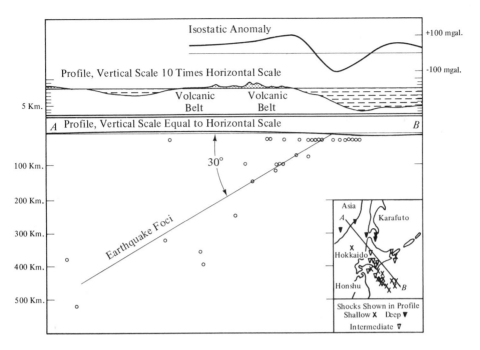

Fig. 14.11 *Tectonic profile, northern Japanese region, showing earthquake foci, relief and gravity anomalies. (After Gutenberg and Richter, Bull. Geol. Am. Sp. Paper 34.)*

tures provides a means of mapping the tectonic pattern nearly everywhere. On land, the magnetic field is more complicated, and no simple widespread pattern has been detected.

14.7 TYPICAL TECTONIC PROFILE.

Seen in cross-section, the different tectonic features of the circumpacific belt are revealed to have a systematic arrangement. This is shown diagrammatically in Figure 14.11 using a southeast-northwest profile across Hokkaido, the north island of Japan. The lower cross-section is to correct scale. The upper one shows the crust with the vertical scale exaggerated. Proceeding from the deep ocean toward the land, the profile first crosses the foredeep, which off Japan reaches a maximum depth of 10,374 meters (34,035 ft). From the axis of the trench continentward, shallow earthquakes are common, especially along the east coast and beneath the steep continental slope. Low values of gravity are encountered from the oceanic edge of the trench nearly to land. The volcanoes of Japan lie largely in the central and western parts of the islands except for Mount Fuji, which, however, may be the northernmost member of the branching belt running south thru the Bonin Islands. The active volcanoes fall roughly onto two parallel lines. The eastern group is more numerous, generally younger and more active than the western group. From the eastern edge of the volcanic

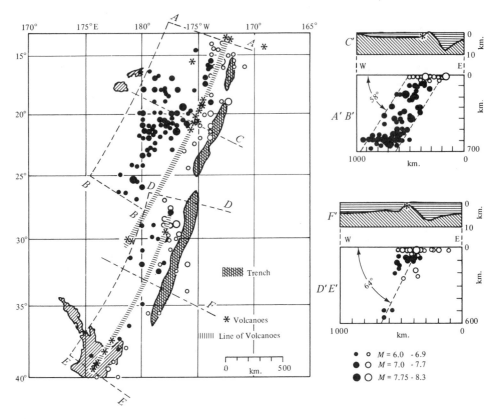

Fig. 14.12 *Tectonic profile Tonga-Kermadec region. (After Benioff,* Geol. Soc. Am. Sp. Paper 62, *p. 68.)*

belt continentward, the epicenters of the largest earthquakes are usually deeper than normal. Behind the island arc lies the shallow Japan Sea, and beyond this the partially eroded stumps of old mountain ranges in Korea and Siberia. The deepest earthquake foci lie beneath these older mountains. The Japanese archipelago constitutes a first fringe of young mountains; the Sikhote Alin and Korean mountains is a second fringe; and beyond these in the interior of the continent lie still older ranges, eroded until only their roots remain, revealed by study of exposed rock formations.

The earthquake foci lie in a zone dipping beneath the continents. The dip varies from place to place, being shallow where the tectonic belt lies close to a continental margin, as in Japan, and steep where the belt lies out to sea, as in the Tonga-Kermadec area (Figure 14.12). North of Japan, in the Kurile-Kamchatka region, the zone seems to steepen going from the intermediate to the deep-focus shocks (Figure 14.13).

Not all of the typical features listed above are found in every cross-section. Deep-focus earthquakes are much more limited in distribution

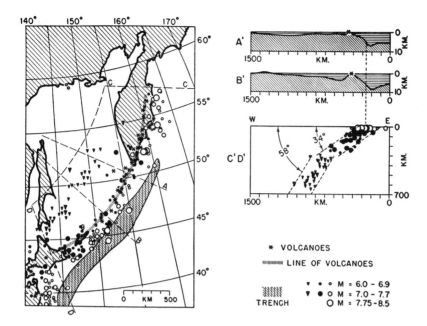

Fig. 14.13 *Tectonic profile, Kamchatka-Kurile Islands region. (After Benioff, Geol. Soc. Am. Sp. Paper 62, p. 69.)*

than are the shallow ones, and the width of the shallow-earthquake belt is very variable. The development of basins behind the fringing ranges is sometimes confused by their being filled by deposits or by there being several closely spaced mountain ridges, as in South America.

The oceanic ridges and the African rift belt differ greatly from the circumpacific system and its branches. Intermediate depth earthquakes are rare in these belts, and only one marginal trough has been observed, the Romanche deep in the Atlantic. Instead, there tends to be a central trough in the middle of the ridge. This trough never reaches exceptional depths (Section 9.1, Figure 9.3, p. 228). The oceanic ridges are broader and gentler in proportion to their height than are the mountain ranges which fringe the continents. This may be in part because there are no thick sections of sediments available in the middle of the oceans. The continental borders are marked by shelves of such deposits. Folded, faulted and upraised, such rocks constitute an important fraction of continental mountains. On the other hand, the oceanic ridges may be of a fundamentally different nature from the continental mountains. The African rift belt more closely resembles the oceanic ridges than it does typical continental mountains such as the circumpacific belt or the Himalayan-Alpine system.

Belts of differing type can intersect as in the case of the East Pacific Rise, which meanders northward across the Pacific Ocean between the 100th and 120th west meridians of longitude until it strikes into the coast of Mexico (Figure 14.14). Examined in detail, the belt of earthquakes lying

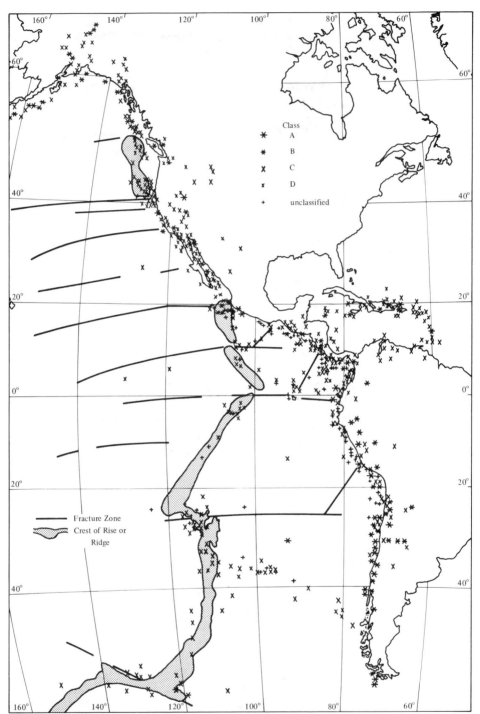

Fig. 14.14 *Earthquake epicenters along the East Pacific Rise and adjoining areas. (H. W. Menard,* Marine Geology of the Pacific, *p. 130. Courtesy McGraw-Hill Book Co., Inc.)*

along the San Andreas Fault in California more nearly trends into this oceanic line than southeastward along the coast of Central America.

The explanation of the complex interrelations of the two types of tectonic belts presumably lies in the nature of the forces which deform Earth's crust. The possible origin of these structures will be the subject of the next chapter.

BIBLIOGRAPHY

Gutenberg, B.; C. F. Richter, 1954, *Seismicity of the Earth*. Princeton Univ. Press. 310 pp.

Hart, P. J.; V. V. Beloussov (Ed.), 1969, *The Earth's Crust and Upper Mantle*. Am. Geoph. Union Monograph 13. 735 pp.

Howell, B. F., Jr., 1959, *Introduction to Geophysics*. McGraw-Hill Book Co., Inc. 399 pp.

Menard, H. W., 1964, *Marine Geology of the Pacific*. McGraw-Hill Book Co., Inc. 271 pp.

Poldervaart, A. (Ed.), 1955, *Crust of the Earth*. Geol. Soc. Am. Sp. Paper 62. 762 pp.

Umbgrove, J. H. F., 1947, *The Pulse of the Earth*. Martinus Nijhoff. 358 pp.

15 ORIGIN OF MOUNTAINS AND CONTINENTS

The continents would long since have been worn away by weathering and erosion if some process for their regeneration were not going on today. Therefore, they must still be forming. In the last chapter, we saw that the current tectonic deformation of Earth was concentrated in long narrow belts associated with the highest mountain ranges. The origin of mountains and the origin of the continents are most easily explained as being a single process.

Continents are anomalous rather than the usual condition. In Chapter 9 it was pointed out that the surface of Earth can be divided into two principal areas: the ocean basins, occupying about 55 percent of the surface, and the continental platforms, occupying less than 35 percent. The small remainder is transitional or in one of the special categories of extremely high or low elevations. The differences between the special features of the continents and the more prevalent though less well studied features of the ocean basins need to be explained.

15.1 FOUR DISTINCTIVE CHARACTERISTICS OF CONTINENTS

There are four features of the continents which must be explained by any theory of their origin. These are: (1) their height above the ocean floor, (2) their peculiar arrangement largely in one hemisphere, (3) their tendency to a triangular shape and (4) the roughly concentric, mosaic arrangement of rocks of different geologic ages.

Elevation is the most obvious distinction between continents and ocean basins. Actually, however, elevation is only a secondary characteristic. It results from the difference in density between the rocks constituting the continents and those underlying the oceans. Continents stand high because they float isostatically in denser simatic rocks. The simatic material is denser than the sialic continents because of differences in chemical and mineralogical composition.

This difference in composition is the most fundamental difference between the two areas. The column labeled *sial* in Table 15.1 is an average of over 5000 continental igneous rocks. Sial is relatively rich in silica

Table 15.1. *Percent composition by weight of different types of rock.*

Compound	Sial[1]	Basalt[2]	Tektites[3]	Add to basalt to produce sial
SiO_2	59.12	48.80	73.1	69.44
Al_2O_3	15.34	13.98	12.2	16.70
$Fe_2O_3 + FeO$	6.88	13.37	5.1	0.39
MgO	3.49	6.70	2.3	0.28
CaO	5.08	9.38	2.8	0.78
Na_2O	3.84	2.59	1.2	5.09
K_2O	3.13	0.69	2.4	5.57
TiO_2	1.05	2.18	0.8	–0.08
P_2O_5	0.30	0.33	——	0.27
MnO	0.12	0.17	0.1	0.07
H_2O	1.15	1.80	——	0.50
Other	0.50	——	——	1.00

[1] Average crust of Clark and Washington, 1924, U.S.G.S. Prof. Paper 127.
[2] Average of 43 basalts. R. A. Daly, 1942, Geol. Soc. Am. Sp. Paper 36.
[3] Average of five columns, Table 4, p. 101, O'Keefe et al., 1963, Tektites.

(SiO_2), alumina (Al_2O_3), soda (Na_2O) and potash (K_2O). Basalt, the type of rock found underlying the ocean basins, is richer in iron, magnesium and calcium oxides. Individual rocks, particularly in the continents, differ greatly from these averages. The water content in all rocks is hard to determine accurately, and the figures given for this component may not be truly typical. The overall contrast between the varying iron-magnesium-calcium content as compared to the silicon-aluminum-sodium content of the two sets of rocks leads to more compact and hence denser mineral forms for basalt than sial. Any theory of the origin of the continents must explain this contrast in composition.

A second peculiarity of the continents is their restricted distribution on the surface of Earth. Hold a globe so that a point at 180° longitude and 38°S latitude (between New Zealand and the Kermadec Islands) is nearest you. Note that the only large land masses on the side of the globe you can see are Australia and Antarctica (Figure 15.1). Now turn the globe thru 180° so that the point nearest you is 0° longitude, 38°N latitude (between Spain and the Balearic Islands). Most of the continents are now on your side of the globe. Earth can be divided into two hemispheres, one largely water covered and the other containing most of the land (Figure 15.2). The water hemisphere is 89% water covered, 11% land. The land hemisphere is also largely water covered, 53% to 47%. Over 80 percent of all land is in the land hemisphere. The contrast would have been even more striking if the areas had been measured to the boundaries of the continents, because most of the water-covered shelf areas of the continents are in the land hemisphere.

Such an uneven distribution of sial is not likely the result of chance. If the land masses were randomly distributed over Earth's surface, the ratio

Fig. 15.1 *One half of Earth is largely water covered, the other contains most of the land.*

of land to water would be nearly the same in every hemisphere. There must be something in the process which produces continents which has had more effect in one half of Earth than in the other. Any reasonable theory of the origin of the continents must explain this distribution.

The third distinctive feature is the triangular shape of the continents. North and South America and Africa show this most clearly; but the Eurasian mass also has three south-directed points in Arabia, India and Malaya. Only Australia and Antarctica lack this rough triangular tendency.

Furthermore, there is a tendency for the continental edges to be smooth in shape. Most of the irregularity of shorelines shown on any map is the result of the flooding of shelf areas. Hudson's Bay, the North and Baltic Seas and the Yellow Sea between China and Korea are extreme examples of the irregularity which results from flooding of the continental margin. The outer edges of the shelf are much smoother than the shoreline itself.

The transition from continent to ocean is relatively abrupt. At most places it takes place in a distance which is a small fraction of the width of either area (Figure 9.3 p. 228). Around most of the Pacific, deep water lies within a few kilometers of the land. Elsewhere, from the edge of the continental shelf to the ocean basins, the bottom drops off at the angle of repose of sediments or more steeply. Whatever it is that produces the continents builds them at certain limited locations with the process ending sharply at an edge.

Continent building rarely forms incomplete continents. There are few if any partially developed continents whose crests have not yet reached sea level. There are only a few small continental masses like New Zealand lo-

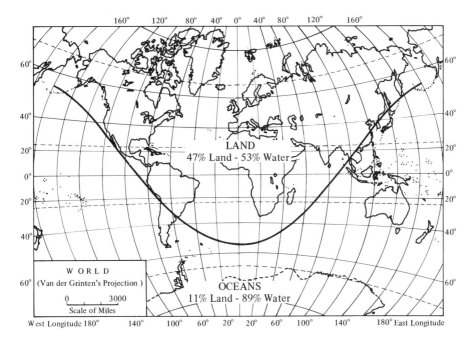

Fig. 15.2 *Boundary between the land and water hemispheres. (Base map Nr. DD98. Courtesy A. J. Nystrom and Co., Chicago.)*

cated in a large oceanic area. Sial typically comes in large masses, usually triangular. This must be a consequence of the nature of the originative process.

Finally, each continental mass consists of a mosaic of strips of rocks of varying ages, with a tendency for the oldest rocks to be toward the center. Large parts of the central portions of each continent are underlain by exposed or shallowly covered crystalline shields (Figure 9.10, p.235). These shield areas are fringed with mountain belts. Furthermore, the ages of formation of rocks of which the shield areas themselves are composed show a tendency to increase toward the center, as though each continent had been built by the addition of successive mountain ranges along its edges.

North America illustrates this rough pattern (Figure 15.3). Some of the oldest rocks found anywhere on Earth crop out northwest of Lake Superior. They were formed 2.8 to 3.5 billion years ago. A large belt of rocks between this region and Hudson's Bay is all over two billion years old. Southeast of this area is the Grenville province, whose rocks were formed 800–1100 million years ago. To the northwest in the Yellowknife and Great Bear provinces are rocks of a similarly intermediate age. Outside of this central region are rocks such as those of the Appalachians and the mountains of Labrador, Greenland and Ellesmereland, ranging in age from 200 to 600 million years. Mountain ridges of similar age occur at several places buried beneath the sediments of the central United States; and the remains of such mountains are found in the Great Basin between the Rockies and the Pacific coastal ranges.

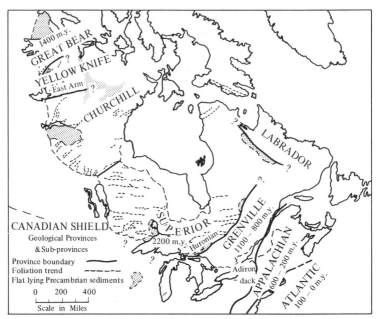

Fig. 15.3 *The structural pattern of North America consists of roughly concentric belts of rocks. (J. T. Wilson,* Trans. Am. Geoph. Union *v. 31, p. 104.)*

In the eastern and southern United States, great wedges of undeformed sediments deposited in the last 200 million years flank the continents (Figure 9.13, p. 239). In the west, a series of chains of mountains extends from the Rockies to the Pacific coast, the westernmost ranges still being in the process of formation (Figure 9.15, p. 241).

On other continents, a similar pattern can be seen of peripheral mountains surrounding central plains of covered and exposed shield areas (Figure 9.10, p. 235). Most of central Africa is a great shield with young mountains at the very south and in Morocco at the north. Central Australia is a shield, with mountains at its east end. Europe and Asia are two mountain-fringed plains separated by the Urals.

The concentric pattern is not as clear as the other characteristic features. This may be because dating and mapping of the older ranges are far from complete; or it may be because the concentric character is itself not so essential a characteristic of the continents.

In the sections which follow, different theories of the origin of the continents will be tested by examining the ability of each to explain these four principal characteristics of continents: (1) their elevation, which is due to their low density, which is due to their sialic composition; (2) their concentration largely in one hemisphere; (3) their tendency toward a triangular shape with abrupt edges; and (4) their mosaic arrangement of mountain belts with those in the center tending to be the oldest.

15.2 CLASSIFICATION OF THEORIES OF CONTINENTAL ORIGIN

All respectable theories of the origin of continents have one feature in common. They attempt to explain how masses of light sial or heavy sima moved from some former distribution to their present locations. This makes it possible to classify all theories of continental origin according to the path the material took. Table 15.2 presents this classification.

There are five ways the sial can move. It can have reached the crust from below (from the mantle) or from above (from space). Alternatively, it can have been in the crust "originally," that is from the end of some earlier stage in Earth's evolution, and then moved up, down or horizontally.

The only other possibility is that simatic rather than sialic material moved. There could thus be five corresponding simatic types of movement. However, since the sialic material is the lighter, it would be expected to concentrate in any sorting process above the simatic material. As a result, only the possibility of horizontal simatic movement leads to an independent, reasonable theory of the origin of the continents.

One example of each of these six possibilities will be presented, and its reasonableness tested by seeing how well it explains the major features of the continents discussed above. The theories are not mutually exclusive, and the truth may be a combination of two or more individual possibilities.

15.3 DIFFERENTIATION-IN-PLACE THEORY

Detailed geologic examination of igneous rocks has shown that, as a magma cools, it tends to form successively rocks of a variety of chemical and mineralogical compositions. This process is called magmatic differentiation (see Section 12.3). The principal process involved in magmatic differentiation is fractional crystallization. When a magma first starts to solidify, the chemical composition of the crystals formed is different from the average composition of the magma itself. Generally these first crystals are enriched in iron and magnesium. They are normally denser than the magma both because of their iron content and because atoms pack more tightly together in solids than in liquids. Therefore, they have a tendency to sink to the bottom of the liquid body. In this way, the upper part of the magma becomes more sialic, and the lower part more simatic. When the magma finally cools so far that it becomes entirely solid, there is a gradation in composition from the top to the bottom (Figure 15.4).

The corresponding process of early sialic crystallization is theoretically possible, but is rarely observed. If crystals of less density than the magma were to form, they would be expected to float to the top. They do not form because the solid phase is normally more dense than the liquid phase. (Ice is one of the few exceptions.) Therefore, to float, a crystal must be substan-

Table 15.2. *Classification of theories of the origin of continents.*

Direction material moves	Example of theory
Mantle to crust	Differentiation in place
Crust to mantle	Reabsorption
Crust to space	Rotational fission
Space to crust	Meteoritic
Horizontally in crust (sialic elements move)	Convection
Horizontally in crust (simatic elements move)	Sedimentary differentiation

tially enriched in the lightweight components of the magma from which it forms. The reverse order of crystallization is expectable. More energy is required to keep a molecule composed of heavy atoms sufficiently agitated to be liquid than is required for one composed of light atoms. Consequently, simatic compounds tend to solidify at higher temperatures than sialic compounds, and come out first as the magma cools.

When a magma is partially solidified, there will be a stage when it consists of an open aggregation of crystals with liquid filling the pores. If the magma is squeezed at this stage, and there is some path along which the liquid can escape, it will do so, leaving the crystals behind. The liquid at this stage will be sial-enriched. Because it is lighter than the solid crystals, the liquid will tend to escape more often upward toward the free surface of Earth than downward. Thus, sialic material has a tendency to move upward wherever magma is solidifying. This process is called filter pressing.

When a magma solidifies, the liquid part gradually becomes enriched in volatile components such as water, carbon dioxide and sulfur dioxide, which are gases at surface pressures. If the magma is at a shallow depth so that pressure is not too great, bubbles of these gases will form in the liquid magma. The formation of bubbles is accompanied by expansion. This reduces the density of the liquid-bubble mixture. If there is an open escape route to the surface, the bubbly magma will expand in this direction. That part of the material moving up the vent toward the surface experiences decreasing pressure as the surface is approached. This allows the bubbles to expand further, and may result in the process of *gas streaming* whereby the volatile materials flow out of the magma as a steady stream, often sweeping other materials with them. They may appear at the surface as hot springs or geysers (Figure 12.6, p.308). In extreme cases, they result in violent volcanic explosions, in which the outrushing gases carry lava and ash high into the air (Figure 12.5, p.306).

These processes of magmatic differentiation are well established by studies of rock bodies. Collectively, they are fully adequate in their nature and rate of action to have formed the whole volume of continental rocks found on Earth today. They have been demonstrably involved in the formation of most or all igneous rock bodies.

Fig. 15.4 *A gradation in composition is often observed in igneous rocks.
(Courtesy U. S. Geol. Survey. Photo by P. K. Sims.)*

The question then arises, are these processes by themselves enough to explain the origin of the continents? To decide this one must ask whether magmatic differentiation provides explanation for the four principal characteristics of the continents described in Section 15.1.

The answer to this question is "no." Magmatic differentiation explains neither why the continents are largely in one hemisphere nor why they are triangular in shape with sharp borders. Magmatic differentiation would be expected to produce sialic material everywhere on Earth, though not necessarily equal amounts at each place. The distribution of sial should be more random. Nor is the mosaic pattern of concentric mountain ranges and old interior shields explained. Magmatic differentiation is an igneous process, and typical mountain ranges involve the compression of geosynclinal sediments.

Even the high elevation of the continents is not necessarily explained, although the presence of sialic rocks at the top is an expected result. The differentiation-in-place theory would require that the process of separation into light and heavy components result in a column of rock longer than an undifferentiated or less thoroughly differentiated column. There is no reason to expect that this would be the case. It would be equally reasonable to expect that the formation of a dense fraction would result in a decreased volume of the heavy part of the column which exactly counterbalanced the increased volume of the light fraction. Indeed, the loss of volatile gases from the rock during magmatic differentiation might lead to the expectation that the differentiated column would be the shorter one.

The obvious conclusion is that the continents cannot be explained solely by differentiation in place. Nevertheless, igneous rocks show such general evidence of magmatic differentiation that this process is called upon as a preliminary step in most of the other theories of the origin of continents.

15.4 THE SEDIMENTARY-DIFFERENTIATION THEORY

If one examines the processes by which the rocks of the continents are weathered and eroded away, a process whereby simatic components are concentrated into the ocean basins is revealed. When an igneous rock is exposed to air, moisture, organic materials and daily and seasonal temperature fluctuations at Earth's surface, most minerals in igneous rocks decompose, forming a new suite of substances more stable in the new environment of temperature, pressure and contact with surrounding materials (Section 9.3). The most stable common component is the relatively light mineral, quartz (SiO_2). Much of the alumina (Al_2O_3) is converted into clays, which are also very stable in the surface environment. Lime, magnesia and iron oxides are relatively easily dissolved, and are carried by streams to the oceans, where eventually they are precipitated, often by organic action, forming coral reefs or layers of ooze on the ocean bottom. There is, therefore, a tendency for the processes of weathering and erosion to carry the simatic suite of oxides preferentially to the ocean basins leaving sialic material behind.

If one postulates further that volcanism and the consequent creation of new land tend to occur preferentially in the places where land has already begun to form, then it is easier to explain the continents than by calling on igneous differentiation alone. A continent is postulated to produce a hot spot (due to enrichment in radioactive elements) stimulating magmatic differentiation beneath itself, with greatest activity along its edges where the sialic layer is thin. In this way, large centers of differentiation once started tend to grow more rapidly than small centers. Eurasia is the largest such center. Africa, the Americas, Australia and Antarctica are smaller but still relatively large. Some of the large islands like New Guinea, New Zealand, Iceland and even Hawaii are presumed to represent the beginnings of continents which have not yet had time to develop fully. In the last two cases, very little sedimentary sorting has yet occurred, and these islands are almost completely simatic.

This theory has the advantage that it comes nearer than does magmatic differentiation alone to explaining the distribution of the continents in one hemisphere. Treating Eurasia as one, there are only six large continents. It would be equally reasonable to suppose that Eurasia was originally several continents which grew together to form one large mass. With so few centers it is not unreasonable that they are grouped as they are. The distribution is numerically nearly in balance, four in one hemisphere and two in the other. Remember also, that the division into land and water hemispheres in Figure 15.2 was deliberately made to emphasize the contrast.

The theory also explains the rough concentricity of the continents by supposing that volcanism is a peripheral process occurring mainly along the continental margins. Why this should be so is not clear; but then the concentric pattern is only a rough one, and not all volcanism is peripheral, as in the case of the rift volcanoes of Africa. According to this theory, erosion of lavas produces the sedimentary rocks which become involved in the formation of mountain ranges and which are melted or metamorphosed to produce the granites found in the shield areas.

A desirable feature of the sedimentary segregation theory is that it explains the difference in height of the oceanic and continental columns. There are real differences in overall composition between them, resulting from the horizontal transfer of dense sediments. Coupled with isostatic adjustment, this makes the ocean basins sink and the continents rise.

The theory can be tested by inquiring whether the volume of volcanic rocks is sufficient to have produced the continental rocks by their decomposition. Table 15.3 shows the area of the continents and the volume of continental material rising above the ocean floors, which are assumed to start at a depth of 5000 meters. The volume of this material is 11.2×10^8 cubic kilometers. Assuming that the continents are underlain by a root five times as large as the part exposed above the ocean floor, then the total volume of continental material is about 6.7×10^9 cubic kilometers. The rate of new lava extruded at Earth's surface annually is about one cubic kilometer. At least as much probably solidifies without reaching the surface. The amount of new rock which would have been formed in the roughly 5×10^9 years of Earth's history if volcanism had occurred on the average at its present rate is thus fully adequate to produce all the sialic material now found in the continents.

The difficulty with this theory is: where did the simatic material go? The total volume of simatic sediments sitting on the ocean floor is many orders of magnitude too small to be the material which would have been carried off the continents. Unless this sediment has somehow been converted back into crystalline rock and forms a part of the basement found by seismic surveys beneath all the oceans, the theory must be considered untenable. There is no known reason to suppose that temperatures and pressures one to two thousand meters beneath the ocean floor are anywhere near large enough to produce such a conversion. Furthermore, if there were such a change, it should produce a gradual transition from the veneer of deep-

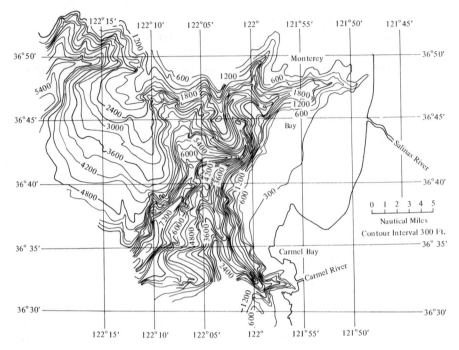

Fig. 15.5 *Monterey submarine canyon, California. (F. P. Shepard,* Calif. Jour. Mines and Geology *v. 34, p. 300.)*

ocean sediments to the crystalline rocks. Instead, their boundary is a sharp transition easily mapped by seismic methods.

This problem is one of several which led earth scientists to propose the MOHOLE, a hole to be drilled thru the five kilometers of crust at some point in the deep ocean where it would pass thru the Mohorovicic discontinuity and enter the mantle. Cores from such a hole, or even a shorter one a few kilometers long, would probably show whether there is some process going on which metamorphoses sediments into crystalline rocks beneath the sea.

Another objection to the sedimentary segregation theory is that the triangularity of the continents is left just as unexplained as in the case of igneous differentiation.

15.5 THE REABSORPTION THEORY

One of the principal features of the sedimentary differentiation theory is that it requires that there be a transformation of surface sediments into denser crystalline material beneath the ocean. If such a reabsorption can dispose of a few thousand meters of sediments, what are the chances that a normal thickness of continental crust could have disappeared in the five-billion-year life of Earth? Could Earth once have had a universal sialic crust, and could this crust have recombined with mantle rocks over limited areas to form the ocean basins?

Table 15.3

	Area of continental Materials in		Average height above −5000m	Volume	Mass if density is 2.75 gm/cc
	% of surface	Km²	in m	Km	Kg
Subaerial	29.2	148,900,000	5,840	8.7×10^8	24×10^{20}
Shelves	5.4	27,500,000	4,900	1.3×10^8	3.6×10^{20}
Slopes	9.8	50,000,000	2,400	1.2×10^8	3.3×10^{20}
Total	44.4	226,400,000		11.2×10^8	31×10^{20}

This would require that the ocean floors have sunk as the crust was absorbed. There is abundant evidence that the floors have sunk. Guyots and coral atolls appear to have formed in this way (Figures 9.7 and 9.8, p.232). There is also an abundance of marine sediment covering much of the surface of the continents, as though sea level had formerly stood higher than at present, submerging vastly greater areas of continental shelf.

Finally, there is the evidence of submarine canyons cut into the continental shelves (Figures 15.5 and 8.4, p. 196). These have profiles and branch into arms very like stream valleys on land. They are so similar to land valleys that for many years scientists found it hard to believe that they could be eroded by mud flows or thinner density currents on the sea floor; and they thought it more reasonable that they formed originally above sea level and have since sunk. This might be what would be observed if surface rocks were being absorbed into a gradually expanding ocean floor.

This view of forming the ocean basins by reabsorption of the sial into the mantle explains the triangular shape of the continents by supposing the points to be produced where expanding oceanic basins overlap. The concentration of continents in one hemisphere is the result of the very large size of the Pacific cell, which has wiped out all land over nearly half of Earth's surface. Of the four principal characteristics of the continents, the theory fails to explain only the mosaic structure of the continents.

In spite of this theory's seeming ability to explain the origin of continents and ocean basins, hardly any scientists favor it. The reason is the total lack of any evidence for processes of absorption of sediments into the mantle under ocean-bottom conditions. This is a negative argument and hence not a strong one. Nevertheless, geochemists believe that they understand the behavior of rocks and minerals well enough to say that reabsorption has not occurred in this fashion.

15.6 ROTATIONAL FISSION THEORY

The converse of reabsorption is for part of the crust to have been lost to space. Considering what a difficult time men have launching rockets weighing at most a few tons, it is hard to see how ocean-sized patches of crust could be removed from Earth's surface. Indeed, under present conditions this would not occur. However, many changes have taken place in the lifetime of Earth, and five billion years ago dismemberment might have taken place more easily.

The least unreasonable version of this theory postulates that Earth was once larger than it is now and rotated in about 2.65 present-time hours. The rate of rotation was gradually increasing to conserve angular momentum as dense material differentiated out of the mantle to form the growing core. As the spin rate increased, the equatorial bulge expanded. At a sufficiently large spin rate, one equatorial diamater can grow more rapidly than the others, and the body takes the form of a cigar-shaped ellipsoid (Figure 15.6). Continued rotational acceleration makes such a body unstable, and a piece or pieces break off the ends of the ellipsoid. In the case of Earth the largest fragment became Moon. Any smaller fragments either escaped into space or were later reabsorbed into Earth and Moon. Tidal friction then gradually transferred momentum from Earth to Moon, letting the latter move away and the former rotate more slowly. It is postulated that Moon broke from what is now the Pacific basin, and that the continents have slipped part way into the hole thus formed, opening up the Atlantic and Indian oceans between them.

Moon has a density of 3.34 gm/cc, about right for a mixture of 80% upper mantle (density 3.5 gm/cc) and 20% crust (density 2.75 gm/cc). Its size is equal to a layer averaging 65 km thick over two thirds of Earth's area or 130 km thick over a third of Earth. This would correspond to a sialic layer originally 13 or 26 km thick. The larger of these figures is close to the present thickness of sialic crust, which is found over an area slightly greater than one third of Earth's surface.

This theory explains composition, distribution and shape of the continents reasonably well, but not their mosaic structure. Its greatest weakness is the improbability that the scar produced by Moon's loss would still be recognizable. If Earth were fluid enough to deform as postulated, probably it would be fluid enough for all evidence of the former surface conditions to have been obliterated. Even the infalling of the smaller fragments of disruption would be expected to cover so much of Earth's surface that the original hole from which Moon came would no longer be recognizable, just as Moon's surface is marked by meteorite impact scars today. Also, calculation of the shape of Earth at disruption indicates a probable ratio of axial lengths of 8:10:23. With the longest axis nearly three times the shortest axis, one would expect to find much evidence of the change which has brought Earth to its present shape. No such evidence is observed.

Fission itself presents serious mechanical problems. The combined angular momentum of the present Earth and Moon would provide a rotational period for the proto-earth of about four hours. A lot more angular momentum is needed to reduce the period to 2.65 hours, especially for a cigar-shaped body, whose moment of inertia is greater in proportion to its mass than that of a sphere. The theory is unlikely for this reason to serve even as an explanation of Moon's origin.

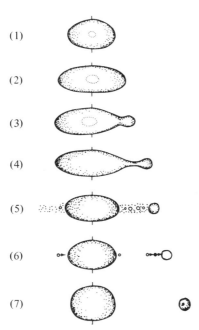

Fig. 15.6 *Disruption of Earth by rotational fission. (After D. U. Wise,* Jour. Geoph. Res. *v. 68, p. 1548.)*

15.7 METEORITIC ORIGIN.

Another wild hypothesis is that the continental material may have reached Earth as one or more giant meteorites. It is known from studying Moon and Mars that large meteorite impacts are common in the solar system. One may picture a body somewhat smaller than Moon but larger than any known planetoid landing near the north pole and splattering southward to form six large continental lumps (Figure 15.7). One may postulate that throughout the life of Earth, these have gradually spread by erosion and by plastic flow, mixed with underlying material as simatic volcanism broke thru them, and been deformed by mountain-building cycles along their spreading edges. Repeated metamorphism of the mixture of intrusives and original meteoritic material gradually transformed the mixture into today's sialic crust. The concentration of material in one hemisphere is explained by the meteorite having struck near the north pole with a drift toward what is now 0° longitude. The concentric structure is supposedly the result of spreading of the originally more concentrated material. The tapering shape results from the decrease in the amount of original material with distance from the impact crater.

If it is assumed that the present volume of 6.7×10^9 cubic kilometers of sial is an equal mixture of meteorite and basaltic material, a body one sixth

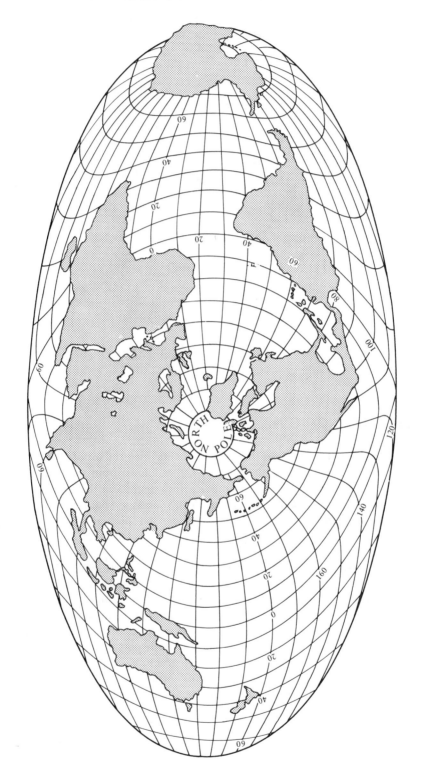

Fig. 15.7 *The continents are clustered about the water of the Arctic Ocean. (Courtesy Athelstan Spilhaus.)*

the size of Moon is required, and its composition would have been that given by the right hand column of Table 15.1. This is remarkably similar to the composition of tektites (next to right column, Table 15.1), the lightest of the known meteorites. The most popular theory of the origin of tektites is that they are fragments of sialic rocks thrown clear of Earth or Moon by meteorite collisions; but there is no certain evidence as to their origin. A large tektite planetoid remains a possibility.

There is no serious support for the meteorite theory of the origin of the continents, largely because of one very serious flaw. The energy of a large infalling meteorite is so great that it is difficult to see how the body could escape complete disruption, scattering pieces all over Earth's surface. A body falling to Earth from far away gains energy as it falls thru Earth's gravitational field. Falling from an infinite distance, if nothing interferes with it, at impact it will be moving with a velocity of 11 km/sec. Each kilogram of the meteorite will have kinetic energy of 6.3×10^7 joules, which is enough to heat itself from 0° Kelvin to well above its melting point and to melt it several times over. A small meteorite loses most of its energy to the air as it passes thru the atmosphere. A large meteorite has so large a ratio of volume to surface area that very little of its energy is lost in this way. Even if two-thirds of its energy were given to the rocks surrounding the impact site, still the meteorite probably would retain enough energy to vaporize itself. If it evaporates, it is reasonable to suppose that whatever part of the meteorite did not escape to space would condense as a world-wide layer of continental material.

If the meteorite theory has any virtue at all, it is to remind us that a theory which explains most of the observed facts is not enough. It should also lead to other unpredicted events, not to inconsistencies.

Meteorites may still play one role in the origin of continents. If Earth formed by the gathering together of previously widely scattered material, as is commonly postulated, the late stage of this assembly could have been marked by the infall of meteorites of a variety of compositions. The original Earth must have been an unsorted mass of such meteoritic material. Because meteorites of various compositions are known to strike Earth today, it seems reasonable that the range of compositions of the early meteorites was at least as great. Let us suppose that the proto-earth near the end of its assembly was fluid except for a thin crust. If some of the meteorites which struck it then were large tektites, might they not have embedded themselves in the still fluid but highly viscous mantle? If such is the case, present patterns of volcanism and other tectonic features may be controlled at least in part by compositional variations in the mantle resulting from irregularities in the distribution of material added as Earth formed. It is a common habit to think of Earth's mantle as being more uniform than the surface rocks. There is no evidence for this. Calculated rates of present-day convective flow in plastic rocks are so slow that, if the whole mantle is involved, it can have been overturned only a few tens of times in five billion years. (The circumference of a circle 2700 km in diameter is 8500 km. At a rate of 1.7 cm/year, this large a circle would rotate ten times in 5 billion years.) This is not enough to have done a thorough job of mixing a mass of originally varied composition. Perhaps compositional differences as well as temperature differences help to determine convective patterns in the mantle.

15.8 THE CONVECTION HYPOTHESIS.

Nearly all theories of the origin of the continents begin with the differentiation of sialic material from the interior of Earth. Many mechanisms have been proposed whereby this material, once it has separated out at the surface, can be gathered together in a few principal areas. It has been suggested, for example, that gravitative sliding or erosion and sedimentation may have gathered a once universal crust into depressions on a surface made irregular by unequal shrinking or expanding of the interior. Another proposal is that the oceans were excavated by meteorite impacts. This differs from the meteorite theory of the last section by having the continent thrown aside from an originally universal crust instead of being composed of infalling material.

By far the most popular theory, however, is that the continents are floating on a plastic, convecting mantle, and are collecting over the sinking parts of the cells (Figure 15.8). Beneath each of the oceans there is presumed to be a rising current of hot mantle material, spreading apart on either side of the oceanic ridges. Sediments deposited in the ocean basins are carried continentward by the upper surface of these cells.

Where the flow turns down, most of the sialic material, especially the part which accumulates largely along the continental margins, is compressed to form mountain ranges, uplifting and rejuvenating the continents. The remainder of the sialic material is carried under the continents where it is added to the continental roots and stimulates continental volcanism. The simatic sedimentary deposits which once capped the oceanic rises and ridges are also carried beneath the continents; but, being denser than the sialic sediments, they tend to join the descending material and are recycled into the mantle. This explains why very few examples of deep-sea oozes and other central-ocean deposits are found among old sediments on the continents.

This theory explains the four basic characteristics of the continents outlined in Section 15.1 better than any of the other theories discussed above. It provides for the concentration of the sialic material in a few areas, largely in one hemisphere by postulating that the Pacific Ocean is the site of the largest of several convective cells. The sharp borders of the continents are explained by the sweeping action of the convection. Their sharp, triangular corners result from the presence of overlapping suboceanic cells. The roughly concentric mosaic of continental mountain ranges results from the repeated compression of the sediments which tend to accumulate along the continental margins.

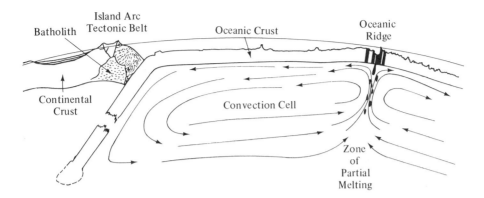

Fig. 15.8 *The convection current hypothesis of the origin of continents and mountain ranges.*

The convection hypothesis has the further advantage that it explains reasonably well nearly all the characteristic features of the typical tectonic plan and profile presented in Chapter 14. Its consistencies with these patterns will be described first, and then some problems presented by the theory will be examined. The oceanic ridges are postulated to be upraised because they are underlain by the hot, and hence expanded, column of the rising parts of convection cells. This explains the unusually high values of heat flow found at many places along the ridges. It produces the oceanic volcanoes such as those in Iceland and Tristan da Cunha on the Mid-Atlantic ridge. The outward flow splits the crust apart to form the central trough which is characteristic of the ridge (Figure 9.3, p. 228). Horizontal drift of the ocean floor carries the central volcanoes shoreward. As they leave the central area, they become inactive, their tops are eroded off, and the ocean floor drops, carrying the summits below sea level to become guyots or the bases of coral atolls. Belts of new crust solidify along the mid-ocean ridges, magnetized in bands of alternating polarity, as found in the Reykjanes area and elsewhere (Figure 7.17, p. 183).

Seismic activity at the sites of upwelling material has a consistent pattern. Along the oceanic ridges, the belt of epicenters closely follows the central trough and the segments of tear faults which offset it (Figure 15.9). Along the trough sections, the patterns of earthquake first motions are most easily explained by vertical fault displacements. Along the tear faults, the first motions are more consistent with horizontal block displacements. The direction of these displacements is always away from the median trough; and epicenters occur almost always between the sections of oceanic ridges, not on the structural lineations which continue beyond the ridge crest into the ocean basins. This type of faulting is called transform faulting. It can occur only where there is new crust being created along the ridge crest in contrast to the displacement of the ridge crest along a fault (Figure 15.10). The motion can occur without the offset sections of ridge crest moving apart because it is newly formed crust which moves, not the location of the ridge.

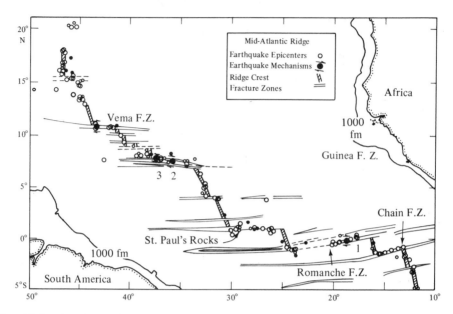

Fig. 15.9 *Earthquake epicenters between 1955 and 1965 on a part of the mid-Atlantic ridge and the most probable shear direction as determined by first-motion studies of seismograms of four earthquakes. (Sykes, 1967,* Jour. Geoph. Res. *v. 72, p. 2137.)*

The Atlantic Ocean basin is postulated to have been formed entirely by the lateral displacement of the Americas relative to Europe and Asia. This continental drift appears to have occurred so gently that the borders of the continents still retain their shapes, and can be fitted together like pieces of a jigsaw puzzle (Figure 15.11). Similarities in the geologic history of eastern Canada, Scotland, Greenland and Scandinavia confirm that they were once close together. Similar correlations exist between Africa and South America. The ages of the rocks found on volcanic islands in the Atlantic increase with distance from the central ridge.

The creation of new crust must somewhere be compensated by the destruction of old crust (unless Earth is expanding). The Tonga Islands north of New Zealand are an example of what happens where crust is being destroyed. Because there is no nearby land mass providing sediment to bury the ocean floor, what goes on is more clearly illustrated here than at most places. Approaching Tonga from the east, the sea floor dips down into the deep Tonga trench. That the sea floor really has been bent is demonstrated by the tilt of Capricorn Guyot along the edge of the trench (Figure 15.12).

The path of the displaced crust is marked by the zone of seismic activity dipping westward under the islands (Figure 15.13 and also Figure 14.12, p. 398). Shallow earthquakes occur over a wide area with a variety of first-motion patterns, particularly those characteristic of thrusting. A study of precise focal locations for 1965 found that all the intermediate and deep

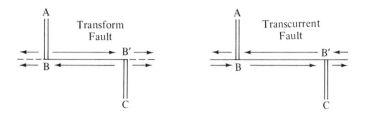

Fig. 15.10   *On a transform fault, new material rises along the lines AB and BC. Right-lateral displacement occurs on the fault only between B and B', and no displacement outside of this. On a transcurrent fault (right), B and B' move apart, and left-lateral motion extends along the whole fault length.*

earthquakes occurred in a zone less than 20 km thick. First motions were generally of the double-couple type (Section 13.6, Figure 13.25, p. 358). For deep earthquakes, the compressive axis was generally parallel to the dip of the seismic belt. For intermediate-depth earthquakes, sometimes the compressive and at other times the tensional axis was parallel to the dip.

This is what might be expected if a 20-km thick slab of the crust were moving downward into the mantle along this zone. A reasonable explanation is as follows. Near the surface, the type of deformation depends on local rock types. Shear-thrust displacements are common where blocks of oceanic crust are underriding adjoining blocks. At intermediate depths, the descending slab is being stretched, causing tension. Below 300 km, the earthquakes are the result of compression of the slab, whose further descent is somehow impeded from below.

Studies of the absorption of seismic waves in the area suggest that the upper mantle is most absorptive (least viscous) under the Tonga ridge and is stiffer in the down-thrust slab. This stiffness contrast explains the restriction of earthquake foci to this narrow zone.

In an area like Tonga where there is little sediment on the sea floor, a low-lying island arc is formed where one block of crust is tilted up as another passes under it. Only the summits of a few volcanoes, formed where mantle material escapes thru cracks in the crust, rise above sea level. In a few hundred million years, when earthquakes have ceased and the volcanoes have become inactive, when the summits are eroded off by the sea and the site of deformation has moved elsewhere, there will be little to distinguish this area from other parts of the ocean basins.

Where the convection cell turns down under a continental area, on the other hand, the result will be different. Here part of the crust will be light material not easily carried downward. The ridge raised in front of the oceanic deep will be larger, and may consist of one or more ranges of folded mountains (Figure 15.8). Sediments deposited along the continental margin will be brought to the site of downthrusting and compressed to form new mountains as the denser underlayer of crust sinks beneath them.

Fig. 15.11 *Sea-floor spreading is postulated to have torn the Americas loose from Africa and Europe. Blocks have been fitted together for minimum overlap (black areas) and gaps (dotted areas) at the 915 meter (3000 foot) contour. (After Bullard, Everett and Smith,* Phil. Trans. Royal Soc. London *v. A258 p. 41.)*

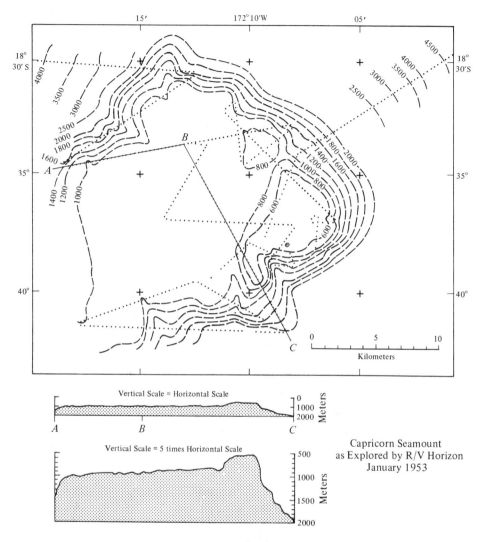

Fig. 15.12 *Capricorn Guyot tilts from the ocean basin into the Tonga Trench. (Raitt, Fisher and Mason,* Geol. Soc. Am. Sp. Pap. 62, p. 244.*)*

Where the crust has been thickened, the unusual downward extent of sialic rocks will produce the gravity anomalies typical of tectonic belts. Behind the frontal range, the back-tilting of the crust will favor the formation of a basin, in which more sediments will tend to accumulate. As the basin fills, its floor will sink isostatically, making room for more deposits. The basin will become a geosyncline.

The deepening and filling of the geosyncline may be the cause of its own destruction. The crust bends under the load of sediment. Its lower portion will increase in temperature. This weakens it. The force of the spreading ocean floor will tend to crack the crust at this weak spot, making the

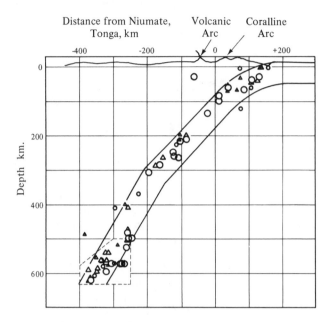

Fig. 15.13 *Locations of earthquakes during 1965 in the southern half of the Tonga area and the focal mechanism for structures of this type predicted from first-motion studies. (After Isacks et al.,* Jour. Geoph. Res. *v. 73, p. 5871 and 5874.)*

geosyncline the most likely site for the start of a new descending slab, with accompanying compression of the surface rocks.

By the time the locus of compression moves on to a new site, the lower part of the thickened wedge of crust will have increased in temperature, because sialic rocks are poor heat conductors and are rich in radioactive elements. The root of the mountain range will melt, becoming a batholith. Only after many thousands of meters of material have been eroded from the raised summit of the range will this resolidify to form a part of the stiff crystalline shield of the continent. By this time, several parallel ranges may have been added to the belt of peripheral mountains forming the continent.

The whole Earth is postulated currently to contain a system of interrelated convection cells of this type. Some of these are growing to cover expanding areas of the surface. Others are shrinking correspondingly. The edges of the cells can be found by noting the locations of zones of compression and tension. Between the tensional belts, which mark loci of rising material, and the compressional zones, which mark descending material, the whole crust and underlying top of the mantle drift horizontally with little or no surface disturbance. The crust can be visualized as a number of blocks, each typically generated along one edge (an oceanic ridge) and disappearing down into the mantle at its opposite side or overriding an adjoining block along a tectonic belt. Figure 15.14 is one attempt to outline these blocks. Yielding is complicated by the curvature of Earth's surface.

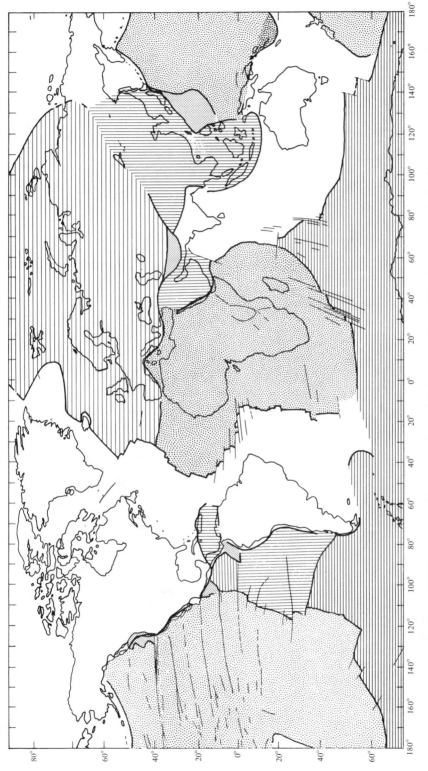

Fig. 15.14 *The crust can be divided into seventeen blocks, generated along oceanic ridges and consumed along tectonic zones. (Morgan, Jour. Geophys. Res. v. 73, p. 1960.)*

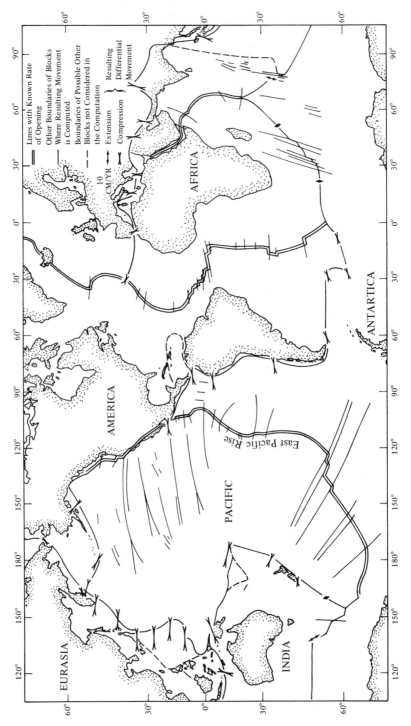

Fig. 15.15 *Differential movement in the principal compressional areas of Earth. (LePichon, Jour. Geoph. Res. v. 73, p. 3675.)*

Places of predominantly strike-slip displacement are required as at the Samoan Islands, the Falkland Islands and the west end of the Aleutian Islands (Figure 15.15).

The number of blocks may well be greater than is shown in Figure 15.14. The African rift zone, extending from the Red Sea south to Mozambique, could be the beginning of another spreading zone.

15.9 PROBLEMS WITH THE CONVECTION HYPOTHESIS.

Although the convection hypothesis seems to explain remarkably well a large number of observed phenomena, there are tectonic features which are not entirely consistent with the theory. The Pacific Ocean in some ways looks like the site of a large convection cell with tectonic belts surrounding its borders except in Antarctica. Yet the North Pacific lacks an active central ridge, there being earthquake epicenters and active volcanoes only in Hawaii. To fit the theory, it is necessary to suppose that most of the floor of the Pacific was generated in some old cycle of convection, and that spreading is now restricted to the East Pacific Rise.

The relationship between oceanic tectonic belts and continental border structures is particularly confused in western North America. The East Pacific Rise, a typical oceanic ridge (Figure 15.15), joins the American Cordillera, a typical continental tectonic belt except for the lack of deeper-than-normal earthquakes, on the west coast of Mexico. Is the Gulf of Baja California a frontal deep along the continental margin, or is it a crestal trough on an oceanic-type ridge? Is the San Andreas fault of California a giant transform fault along the continuation of this ridge, or does the zone of tension extend northward under the continent in the Great Basin between the Rocky Mountains and the Pacific Coast Ranges? This high plateau has many of the features of a zone of tension, including block-faulted mountains and high heat flow.

The patterns of magnetic anomalies off the Oregon, Washington and British Columbia coasts have been interpreted as marking the course of the crest of the convection cell (Figure 15.16). However, the pattern is far more complex than that of the supposedly typical Reykjanes Ridge area in the Atlantic (Figure 7.17, p. 183).

One interpretation is that the whole of North America is being shoved by the active Atlantic convection cell over the north end of the East Pacific cell. However, this presents a problem in the horizontal transfer of stress. In a classic experiment in which physical properties such as viscosity of the crust and mantle were carefully and properly scaled in proportion to the reduction in size from kilometers to centimeters and in time from thousands of years to minutes, David Griggs showed that a force exerting on the crust from below by moving mantle material could easily compress the crust into a thickened wedge, but that the compressive force is quickly dissipated (Figure 15.17). At a distance of a few times the crustal

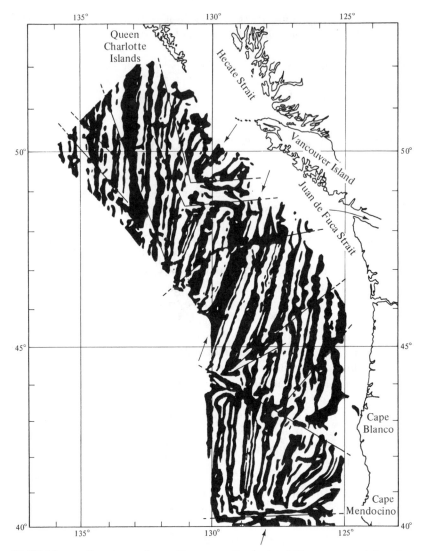

Fig. 15.16 *Magnetic anomaly pattern southwest of Vancouver Island, north-west Pacific Ocean. Supposed oceanic ridge segments (marked by arrows) are offset by frequent faults (located by topographic evidence and shown by lines.). (Raff and Mason, Bull. Geol. Soc. Am. v. 154, p. 1106. Copyright 16 December 1966, American Association for the Advancement of Science.)*

thickness, the horizontal stress no longer persists. The conclusion to be drawn from this is that mountains must form where the forces act upon the crustal rocks. North America cannot be shoved westward beyond the end of the horizontal part of the convective flow. Where then does the cell spreading from the Atlantic turn down? Does it extend all across North America, descending finally under the Great Basin along the projection of the supposedly upwelling East Pacific Rise?

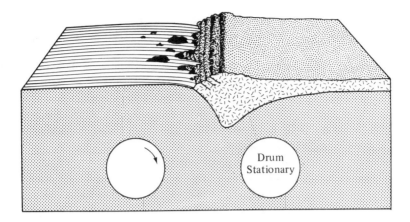

Fig. 15.17 *Crustal material forced to one side by drag acting on it from below piles up where the force stops without disturbing more distant crust. (After Griggs, Am. Jour. Sci. v. 237, p. 643.)*

Heat flow also is a problem. Although many areas of higher-than-average heat flow are formed along the oceanic ridges, there is very little if any more heat reaching the surface under these areas than anywhere else. If the oceanic ridges are truly the sites of upwelling convection cells, then there should be thermal evidence of a steady drop in temperature from the site of upwelling to the locus of descending flow.

A related problem is the difference in radioactive-element content of the oceanic as compared to the continental rocks. Sial is much more radioactive than sima. The thermal gradient in continental areas would, therefore, be expected to be greater than in oceanic areas. This should make the mantle hotter under the continents than under the oceans, and the continents should be the sites of upwelling. Only if the overall heat loss from the mantle is so great or so deep-seated that it can override the blanketing effect of continental rocks is it reasonable that the oceans should be the sites of upwelling.

One possibility is that convection cycles start by upwelling under continental masses but continue long past the point when the continents have been swept off the crests of the cells. This is consistent with the theory that the Pacific Ocean is an old cell and also with the idea that mantle upwelling is now starting under the rift zone of Africa and under the Great Basin of the United States. According to this view, the present pattern of oceanic upwelling is already starting to be replaced by a new pattern of continental disruption in these two areas.

The depth to which convection cells extend is another uncertainty. Their thickness must be limited to a layer of nearly constant equivalent density, such as would result from a single chemical composition. (Two rocks have the same equivalent density if both would have the same density if they were at the same temperature and pressure.) Preferably, the cell should be limited to a layer of constant mineralogical composition, as the energy consumed or released in going from one mineral phase to another will be a barrier to flow. Suppose that under normal conditions a phase change occurs at a depth H from Earth's surface (Figure 15.18). When convection is occurring, this surface is dragged upward in the rising current and carried downward in the descending current. The deflection of the transformation surface produces a density difference between the two columns opposite to the density difference driving the convection cell. If the step in density at the phase transformation is Δd, and the two surfaces are ΔH different in depth in the rising and descending columns, then a pressure $\Delta d \Delta H$ opposes the convection at depth H.

Convection can cross a phase-change level only if the temperature difference in the two columns is large enough to create a pressure difference greater than $\Delta d \Delta H$ between the two columns. Thus a larger temperature gradient is required for convection in the presence of phase changes than in a mineralogically uniform layer, and what convection can occur will take place more slowly.

As the material moves, the phase transformation will progress thru the material. The amount of resistance that the phase transformation offers to convection depends critically on how fast the phase transformation progresses thru the flowing material. Once the cell gets under way, there will be a temperature difference between the two columns. High temperatures have a tendency to cause transformation to low-density forms. Consequently, the high temperature in the rising column tends to make the phase transformation progress to greater depths here; and the lower temperatures in the descending column make it rise to shallower depths there. If this progression equals the rate of convective flow, then the phase transformation has no effect on the rate of convective overturn. If it proceeds faster than the convection, then flow accelerates. If it proceeds slower, then the motion is opposed.

The reasons convection cells start and stop may be related to this process. If the overturning of a cell occurs fast enough, then hot rock may be moved from the bottom to the top of the cell in less time than it takes the roots to be rewarmed to the temperature required for an overturning. Under these circumstances, convection will be spasmodic rather than continuous.

The rate of convection will also be affected by any differentiation which occurs at the top or bottom of the cell. Removal of light constituents at the top or heavy constituents at the bottom will enhance the density difference between the rising and descending members.

The limitation of earthquakes to the upper 700 km of Earth may be a clue to the vertical extent of convection in the mantle. Earthquakes imply deformation. Does this mean that flow extends only to 700 km? Or is the lack of deeper shocks the result of lower yield strength of the deeper rocks? It would be easier to understand how the convection cells can rise largely in oceanic areas if they extended more deeply into the mantle than 700 km, because the greater the depth to which they extend the less influence the heat-blanket effect of the continents would have on their locations.

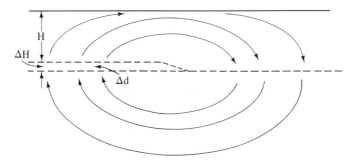

Fig. 15.18 *In a convection cell, a phase transition may take place at different depths in the rising and falling sides.*

A convection cell may occur in a layer only a few tens of kilometers thick, even where its horizontal dimensions extend for thousands of kilometers. Similar wide flat cells are found in both the atmosphere and the oceans. If such a cell is to drag most or all of the crust as floating blocks on its surface, it seems reasonable to expect it to extend to a depth of twice the crustal thickness or more. There is plenty of room for such a cell system between the bottom of the crust and the level of minimum seismic velocity at 150 km. This is the region of the largest predicted thermal gradients.

Another way of viewing convection is to suppose that it is driven by the sinking of cold dense blocks of crustal rock into the hot underlying mantle. This requires that the suboceanic crust have the same composition or very nearly the same composition as the mantle. It may differ at most by the loss of light components, especially those which are volatile, where new rock is formed along the oceanic ridges. The cold crust, according to this view, is denser than the underlying mantle. Breaks occur in the crustal slab most commonly (but not always) along the continental margins, which tend to be zones of weakness. Where the crust has cracked, the whole slab beneath the adjoining ocean tends to slide down into the mantle, opening a crack in the center of the ocean and driving warmer, underlying mantle material to the surface. According to this view, temperature plays a less important role than differentiation in providing the density differences which cause parts of the mantle to overturn. The loss of light (volatile and sialic) elements leaves a residuum of denser-than-normal material at the top of the convection cell. Similarly, a downward segregation of iron-rich compounds at the bottom may provide a layer of lighter-than-normal material below the dense surface layer.

The rates of displacement are determined from the widths of bands of alternating magnetic polarity (Figures 7.17, p. 183 and 15.16) and the duration of the epochs of magnetic reversal. They are a few centimeters per year (Figure 15.15). This is about the same rate that deformation occurs on faults like the San Andreas in California. Convective flow at this rate could occur in a mantle with a viscosity of 10^{22} poises assuming a temperature gradient around 10°/100 km above the adiabatic gradient. This is the viscosity required for the observed uplift of Scandinavia (Figure 9.29, p. 258). This temperature gradient is within the range of reasonable temperatures in the upper part of the mantle (Section 12.8). Motions of one to ten centimeters per year can accumulate to large drifts of blocks of crustal material with respect to one another in a few million years, which is a short time geologically. The observed current rate of gradual displacement of 1.3 cm/year on the San Andreas Fault near Hollister amounts to 1.3 meters per century, or 13 km per million years. There is some geologic evidence that this fault has existed for several tens of millions of years, providing (on the uncertain assumption that the current rate is typical) some hundreds of kilometers of continental drift of the part of California southwest of the San Andreas Fault with respect to the rest of North America.

Another difficulty with the convection-current hypothesis is the distribution of volcanism. Surprisingly few volcanoes have been observed along the oceanic ridges considering that they are presumably the sites of upwelling (Fig. 9.10, p. 235); even though it must be recognized that large numbers of submarine eruptions may occur without being observed at the surface. Most volcanoes are found in the tectonic belts at the continental borders. It is difficult to understand why material should have a tendency to work its way to the surface in the middle of a belt which is supposedly the focus of compression. Possibly continental volcanism somehow results from the metamorphism of water-rich sediments carried under the continental edges. As new crystalline rocks are formed from this material where the convection cells turn down, a volatile-rich fraction escapes upward.

The proof of what has actually occurred is being sought as this book is being written. The magnetization of the ocean floor in bands of alternating polarity presumably provides a detailed history of crustal deformation back for many millions of years. Once the ocean floor is fully mapped in sufficient detail and the magnetic evidence analyzed, the history of seafloor motions will become clearer. The extension of this history back thru the complete geologic record will provide a challenge to earth scientists for many years to come.

15.10 CONTRACTION HYPOTHESIS.

The probably important role of convection in tectonic deformation was only recently recognized. Previous to the realization of the possibility that rocks could deform plastically even under near-surface conditions, other theories of the origin of mountains were popular. Isaac Newton was one of the first to propose that mountains might be explained as a result of contraction. Before the discovery of radioactivity, it was assumed that Earth must be cooling; and that, as a consequence, its volume must be decreasing. Surface temperature is controlled by the incidence of solar radiation, which cannot have varied greatly for billions of years, because geologic processes and life forms similar to those observed today have been common throughout much of Earth's history. The surface rocks, according to this view, would have been fixed in area. If the interior shrinks, the surface layers will possess an area greater than is needed to cover the underlying material (Figure 15.19). The crust is postulated to have adjusted to a shrinking interior by wrinkling like an apple skin as the fruit dehydrates. Mountain ranges are these wrinkles. Compression is predominant to an uncertain depth possibly as great as 300 km, beneath which tension is more general to about 700 km. The reversal from compression to tension is used as an explanation of the minimum of earthquake activity at 300 km depth (Figure 11.10, p. 294). Beneath 700 km, Earth is assumed to have been so little affected as yet by the surface heat loss that earthquakes are rare.

The amount of wrinkling is a measure of the amount of shrinking of the interior. The change in circumference of Earth, ΔC, is related to the decrease in radius, ΔR, by the formula

$$\Delta C = 2\pi \Delta R \qquad\qquad (15.1)$$

Estimates of the amount of compression represented by individual mountain belts range from a few to a few hundreds of kilometers. From the fact that the crust is almost never thickened by more than a factor of two under mountains, it can be estimated that the shortening of ranges is a fraction of their width. But even 6 kilometers of shortening implies a decrease in radius of nearly a kilometer. Common igneous rocks contract about one part in 100,000 on cooling 1°C, so a kilometer of shrinkage requires 10^5 degree-km of cooling (e.g. 1000°C over a column 100 km long, or 2000°C over a column 50 km long). The shrinkage is postulated to be going on in the depth range 300–700 km, the zone of deep-focus earthquakes.

The shield areas of the continents are laced with old mountain ranges. If each requires a mere six km of shrinking, then the interior must have cooled many tens of thousands of degrees throughout its history. Since there is considerable doubt that it is cooling at all (Section 12.6), decrease in temperature of the interior seems to be entirely inadequate as a cause of contraction.

An alternative proposal is that Earth shrank due to phase transformations as the interior differentiated. The postulate here is that Earth began as a mass of material of uniform composition melted or very close to melting, and that it has been separating into layers by upward concentration of the lighter compounds and sinking of heavy components to the core throughout its life. In this case, the limit on shrinkage is set by the amount of density difference between the original components and the present Earth. If the total change in density were a factor of two from 2.75 to 5.5 gm/cc, then Earth's volume would be halved. The original radius would have been $\sqrt[3]{2}$ times the present radius, or about 8000 km. This would provide for a shrinkage in radius of 1630 km. This figure is probably several times that actually available, because solid-solid phase changes of the type expected in Earth are usually accompanied by volume changes of only a few percent. This change in radius is, however, more nearly of the amount needed.

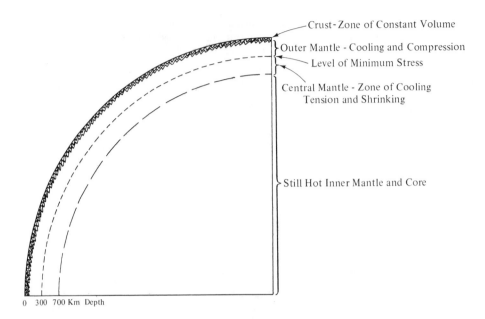

Crust - Zone of Constant Volume

Outer Mantle - Cooling and Compression

Level of Minimum Stress

Central Mantle - Zone of Cooling
Tension and Shrinking

Still Hot Inner Mantle and Core

0   300   700 Km   Depth

Fig. 15.19 *According to the contraction hypothesis, the crust must wrinkle forming mountains as it adjusts to a shrinking interior.*

A third possibility is that Earth has been shrinking due to loss of material from its interior. Hydrogen is by far the commonest element in the universe. If one supposes that the primordial Earth was composed half by volume of hydrogen, and that this has been working its way slowly to the surface, where it escapes to the atmosphere and then into space, the requisite shrinking is obtained. At Earth's surface, this hydrogen is combined with oxygen to form water. Water is known to be a fluxing agent, reducing the melting temperature of magmas and making them more fluid. According to this concept, the crust, especially in the continents, is the dry remainder left after the hydrogen has escaped. The sima of the mantle could still be hydrogen-rich. Volcanism results wherever differentiation of the mantle brings an excess of hydrogen-rich material to the surface. This could explain why water is given off so abundantly in volcanic processes.

One of the requirements of the contraction hypothesis is that the crust be strong. A rigid surface layer would have to slide over a less rigid, presumably plastic interior. For mountains to form in one or a few long narrow belts at any one time in geologic history, the compression must be transmitted from a large area of surface to the yielding zone. Current mountain building is concentrated on a few intersecting zones, such as the circumpacific and Indonesian-Himalayan-Alpine zones.

To explain the tendency for compression to be concentrated most commonly along the edge of the continents, it is supposed that the thick accumulations of sediments, which occur most often here, depress the surface layers of rock and form a weak belt. The shallow part of the earthquake zone of Chapter 14 is visualized as a shear in which blocks of the rigid crust are riding over one another (Figure 15.20). Gravity anomalies are formed by the thickening of light crustal rocks while ocean trenches are formed where the ocean bottom is slipping under the continents. The back-tilting of the raised blocks forms the shallow basins behind island arcs. Magma invades the crust along cracks formed by tension in the tilted block as the arcuate range is raised.

Once a major yielding starts, it proceeds thru a typical series of stages. The first stage is the downbending of the original geosynclinal flexure, with coincident accumulation of a thick section of sedimentary deposits. When the section becomes thick enough, the underlying crust collapses, buckling and faulting in such a manner that the crust is thickened over a broad zone. The rock layers of the geosyncline are either pinched between converging blocks of the collapsing crust or slide off the upraised ridge.

Once the principal spasm of compression has occurred, erosion starts to wear away the upraised welt, depositing it nearby to reinitiate the cycle along another trough. The depressed root melts either because of the heat generated by its high proportion of radioactive elements or because of invasion by material differentiating from the mantle. The invading material itself may be presumed to be enriched in radioactive elements. Chemical studies of the processes of differentiation predict that the radioactive elements will tend to be concentrated in such rising solutions. Later, the removal of overlying rocks brings the melted batholith so near the surface that its heat escapes. Finally, erosion exposes the now solidified batholith itself.

The principal objection to this theory comes from the limited distribution of mountains being formed at any one time. As was pointed out above, such a sharp limitation of their distribution implies that the forces acting were locally much greater than those acting elsewhere. Contraction must put the whole crust under stress. It is difficult to see why there is not relief of this stress in many areas simultaneously, not necessarily equally developed, but certainly more widespread than is the case at present, this expectation is reinforced by Grigg's scale-model experiments, which show that stresses cannot be transmitted large distances horizontally thru layers as thin as the crust. Contractive compression cannot be transmitted around Earth to concentrate in a few areas of deformation. Mountains must be produced by local forces originating in the areas where the mountains are found.

15.11 EXPANDING-EARTH HYPOTHESIS.

An even more questionable hypothesis is that the ocean basins are the result of expansion of Earth. It is postulated that the universal constant of gravitation is not a constant, but relaxes with time, allowing all bodies (including the universe) to expand. According to this view, Earth's interior is swelling as compact crystal phases transform to less dense ones. The continents originally covered the whole Earth. The ocean basins have opened

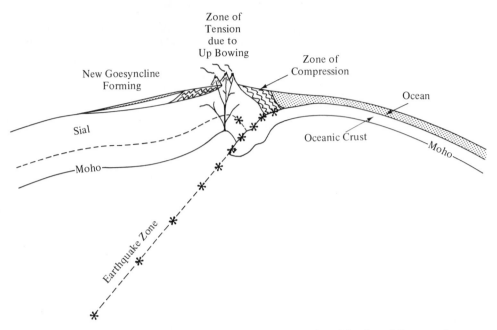

Fig. 15.20 *The contraction hypothesis postulates that blocks of the crust override one another to produce mountain ranges.*

up as the core and mantle expanded. New crust is still forming today, according to this view, along the crests of the oceanic ridges. This theory fails completely to explain the origin of mountain ranges and the rejuvenation of the continents. Its chief value is to remind us that every postulate is subject to question. Everything else in the universe seems to be undergoing change. Why should the universal "constant" of gravitation be an exception?

Regardless of which one or which combination of the above theories is correct, it is obvious that Earth is still evolving. The origin of the continents is only one aspect of the overall evolution of the whole Earth, a still progressing process of change.

BIBLIOGRAPHY

Daly, R. A., 1938, *Architecture of the Earth*. D. Appleton-Century Co., Inc. 211 pp.

Heirtzler, J. R., 1968, Sea-Floor Spreading. *Scientific American*, v. 218 (21), pp. 60–70.

Howell, B. F., Jr., 1959, *Introduction to Geophysics*. McGraw-Hill Book Co., Inc., 399 pp.

Hurley, P. H., 1968, The Confirmation of Continental Drift. *Scientific American*, v. 218(4), pp. 53–64.

Isaacs, B; J. Oliver, L. R. Sykes, 1968, Seismology and the New Global Tectonics. *Jour. Geophys. Res.* 73:5855–5899.

Jacobs, J. A.; R. D. Russell; J. T. Wilson, 1959, *Physics and Geology*. McGraw-Hill Book Co., Inc. 424 pp.

Stacey, F. D., 1969, *Physics of the Earth*. John Wiley and Sons, Inc. 324 pp.

Takeuchi, H., 1967, *Debate About the Earth*. Freeman, Cooper and Co. 253 pp.

Umbgrove, J. H. F., 1947, *The Pulse of the Earth*. Martinus Nijhoff. 358 pp.

Wise, D. U., 1963, On Origin of the Moon by Rotational Fission during Formation of the Earth's Core. *Jour. Geophys. Res.* 68:1547–1554.

16   EVOLUTION OF EARTH

Previous to Copernicus and Galileo, man was accustomed to thinking of Earth as the center of the universe. Still self-centered, modern man finds it difficult not to think of Earth as something special. Yet the origin of Earth is most likely just a typical, minor event associated with the origin of that very ordinary star, Sun, a member of one of the common spiral varieties of galaxy, Milky Way. There are probably billions of other planets like Earth in the universe.

The basis of speculations on events which may have preceded the emergence of Earth as a distinct member of the universe was presented in Chapter 2. The discussion there was factual, with the emphasis on evidence. Unfortunately, the evidence is sketchy and not conducive to firm conclusions. It places limits on what can have happened, no more. The discussion here will be largely speculative. One reasonable theory of the possible development of the present situation will be presented. There are alternatives all along the way. The reader is invited to construct his own model of reality, using whatever evidence he can find in this text or elsewhere. Select the best evidence you can find. Evaluate each "fact" for accuracy. Put together a better picture if you can. Each man who adds one piece to the puzzle belongs to the family of Johannes Kepler, Isaac Newton and Albert Einstein.

16.1 ORIGIN OF THE ELEMENTS.

In the beginning, everything was very concentrated. All the mass and energy seen in the universe today and much more were gathered into a volume much smaller than a galaxy. The concentration of matter was so great that distinctions between mass and energy which are fundamental to conventional thinking were not meaningful. Matter was a soup of sub-fundamental particles so agitated and mutually interfering with one another that it is indescribable in terms of what can now be identified. How the universe got into this state is beyond our determining. Perhaps the universe is oscillatory, as some believe, alternately expanding, as at present, and contracting. Its condition at the beginning was an extremely unstable one, so unstable that it simply exploded, scattering fragments of many sizes all over space. Whether this beginning was merely the start of the present cycle, a transformation from some previous, different but unidentified state, or the creation of matter from nothing is unknown and perhaps cannot be known. For our purposes this time is simply the beginning of those events for which we have even the faintest evidence.

As the tightly packed material separated, sub-atomic particles as we know them condensed out of the protomatter which preceded them. At first the soup was too hot for any two particles to stick to one another at all. As expansion occurred, the energy level dropped, and new organizations of matter developed. Protons, neutrons, electrons, etc. formed; and these, in turn, gathered themselves into units of various sizes from individual sub-atomic particles to groupings far larger than any atoms known today. A wide variety of nuclear reactions occurred. While the packing was still tight, some small nuclei collided with one another to form larger ones. Others collided and were broken apart. The final varieties which formed were determined by the speeds of reaction of different nuclear processes in the rapidly changing environment of decreasing pressure and temperature. All of the elements and their isotopes which we know today were created, as were many others so unstable they have long since disintegrated into more stable forms.

Eventually, the chaotic impulse of the initial explosion left all this matter scattered in a great, diffuse cloud, rapidly expanding in space. Most of the mass was in the form of hydrogen atoms. Other atoms occurred in abundances decreasing with the size of the nucleus (Figure 16.1). When the density became low enough, most of the initial energy was radiated away or consumed in giving relative motion to the parts. The average temperature of the gas cloud dropped to a low value.

16.2 ORIGIN OF GALAXIES.

The distribution of mass in the cloud would have been uneven. If there were not already some centers of concentration at this stage, they would soon have begun to develop. Gravitative forces would have become of importance in controlling motions. Wherever the average density of the cloud was even a little greater than elsewhere, matter from surrounding areas would be deflected toward this center. The universal spreading of all such centers away from one another continued, as it does today, but the single large cloud broke up into galactic fragments with open spaces between them.

Generally, the matter in each galactic mass possessed a certain amount of angular momentum. This allowed it to contract rapidly in one direction, flattening into a rough disk. Parallel to the disk, contraction occurred relatively slowly. Any parts of the cloud not moving parallel to the disk tended to collect toward the center, while fast-moving fractions kept their distance. Thus most galaxies developed a spiral form, although a few shrank to form ellipses (Figures 2.7, p. 24 and 2.9, p. 26).

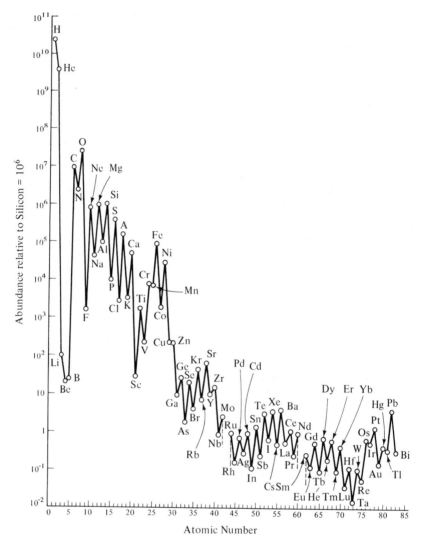

Fig. 16.1 *Relative cosmic abundance of the elements. (After Ahrens,* Distribution of the Elements in our Planet, *p. 14. Courtesy McGraw-Hill Book Co., Inc.)*

16.3 ORIGIN OF STARS.

Each galaxy broke up similarly into sub-groupings; and, within each of these sub-groupings, individual centers of mass concentration developed. Most groupings were small centers of crystallization, where a few atoms clustered to form loose cometary lumps. Here and there throughout the galaxy some centers would become so strongly developed that they would draw together large amounts of the material. The larger a given center of concentration became, the more efficiently it drew together the small bodies in its neighborhood. Bigger and bigger assemblages developed and eventually dominated the mass distribution.

The gravitative energy gained by the infalling of assembling material raised its temperature. Where enough material had collected to raise the internal pressure to the point where atomic structures began to break down, nuclear energy was released and glowing stars were formed. Today, most matter in the universe finds itself in stars.

Stars are still forming today from the remnants of the original clouds of dispersed matter. Others have lived their lives, converted their hydrogen into helium, and either exploded or collapsed. Throughout the present galaxies, the raw material of stars is partly original material from the initial dispersion and partly the debris from nova and super-nova explosions.

Most of the assemblages of material which formed were small. The vast majority in total number, but not in mass, were small. For some reason, the process of condensation tends to form groups of stars within galaxies; and the stars themselves are surrounded by systems of smaller bodies: planets, moons, planetoids, meteoroids and comets, all too small to be seen from outside the star's vicinity. Occasionally, the largest of a star's companion bodies is itself big enough to glow, and a double star results.

16.4 ORIGIN OF EARTH.

In one of Universe's myriad spiral galaxies is a very ordinary G-type star, surrounded by a cloud of satellite bodies, including nine planets, 31 moons associated with these planets, thousands of minor planets moving among the orbits of the larger planets, and billions of less condensed comets mostly in orbits beyond the outermost larger planet. The third planet is called Earth.

All these bodies condensed from the same cloud of material as Sun; but, because of differences in density, pressure, temperature and velocity of rotation with radius from the center, different elements accumulated at different rates in each body. Condensation presumably began as soon as temperature in the original gas cloud dropped low enough to allow molecules to crystallize when atoms touched one another. The loose crystalline masses so formed attracted one another gravitationally. Relatively large aggregations developed wherever the density exceeded some critical value. Thereafter, material from a large region was accelerated toward this center, and the more material moved toward it, the more strongly gravitational forces accelerated the convergence.

As material spiraled inward toward the body which was to become Sun, eddies formed secondary centers of condensation. At first there were not any great differences in composition between the accumulating bodies, but as Sun grew large enough to radiate great amounts of electromagnetic energy, the differences in temperature with distance became important. At some stage, temperatures nearer Sun than Jupiter became so high that the lighter gases could no longer condense on bodies as small as Earth. From this point on, the evolution of Earth consisted of a loss of light material while at the same time Earth drew in to itself whatever small centers of condensation passed close enough to it.

The number of bodies moving about the inner part of the solar system at this stage is uncertain. There may have been fewer or more than at present. The smallest bodies were gradually swept up by the larger ones just as meteoroids are collected by Earth today. Large bodies may have existed which have since been deflected out of the solar system or, more likely, been broken up by collisions with one another. A planet between Mars and Jupiter was one of the last of these to be destroyed; and the impacts of its fragments mark the surfaces of Moon, Mars and even Earth today. The planetoids are fragments of this or possibly other former planets or their satellites.

Earth, then, began its existence as a secondary center of condensation in the shrinking solar cloud. Like Sun, it presumably went thru a critical stage when density passed some limit which led to instreaming of material. As the mass accumulated, the pressure rose; and what was initially a loose aggregation of particles became so tightly packed that phase changes began to occur. From the start, these phase changes may have favored the preservation of crystals composed of the elements making up the present bulk of Earth. Hydrogen, nitrogen and the inert gases (helium, argon, etc.) may have entered relatively little into these compounds, and tended to be concentrated upward.

In the deep interior, temperature rose until it was high enough to melt the material. This favored further differentiation. Two layers formed. An excess of metallic elements, with iron the commonest one, settled to the center. The remainder consisted mostly of oxygen and silicon with lesser quantities of such elements as aluminum, carbon, sulfur and magnesium in proportions not greatly different from those found today in Sun (Table 2.2, p. 34). Lesser amounts of all the elements occurred throughout this mixture. The whole was very close to melting almost from the start due to the heat gained from impact as new material arrived. From an early stage, then, Earth had a metallic core and a siliceous mantle. The core was at first very small, but throughout Earth's history it has grown as heavy metallic elements settled out of the mantle and light elements moved upward.

This means that mantle composition has changed progressively also, becoming poorer both in iron and other core elements and in the volatiles, which escaped to form the atmosphere and oceans. In the mantle, differentiation was controlled by the temperature. Whenever temperature was near or above the melting point, convection was relatively rapid, transferring volatile-rich material upward and volatile-poor material downward. This suggests also that temperature may have gradually increased in Earth's interior. The richer a material is in water, the lower its melting temperature, and hence the easier for convection to act, holding temperature down. As a result of the slow upward migration of the volatiles, particularly hydrogen in the form of water, a gradual warming of the interior is required to maintain convection.

Convection has been particularly slow near the surface, where the rocks are firmer than at great depth. This has allowed differentiation to occur here largely in association with volcanic processes. Mid-ocean volcanism consists largely of the release of the gaseous elements, leaving behind basaltic crust. This is carried horizontally to the continental areas across the upper surfaces of convection cells.

Beneath the continents a much more complicated sequence of chemical steps occurs. The dominant process in the upper mantle is the slow convection by which heat, water and sialic elements are removed from the deep interior. The heat transfer powers the convection cells which sweep the ocean basins clean of sediments, compress and thereby uplift the continents, and produce the typical tectonic profile of mountain building so well developed around the Pacific Ocean.

The feature of this profile most difficult to explain is the presence of volcanism at the sites of the down-turning convection cells. For some reason, an exceptionally fluid magma is generated here, even though this is the site of descending cool material. One theory which has been offered to explain this is the *serpentinization theory* of H. H. Hess. Serpentine is a low-pressure iron-magnesium silicate that contains water as one of its component parts. Hess suggests that serpentine forms at the top of the mantle where the water-rich rising column of the convection cell drops to a temperature of about 500°C. Much of the water remains frozen in the crustal layer until it is carried downward at the edge of the continents. As the former surface layer descends, it warms. A phase transformation occurs during which water is released. This produces a zone of water-rich rocks. Pockets of this material, caught at the base of the belts of compression, eventually work their way to the surface thru volcanoes.

This could explain the relative uniformity of the lavas of mid-ocean volcanoes in contrast to the more varied composition found on continents. Where sialic rocks abound, the mixing of such materials with simatic melts gives rise to a variety of volcanic rocks. Compositions range from basalts not very different from oceanic basalts to material whose components are close to the average of all continental rocks.

One of the implications of the convection hypothesis is that deep-sea sediments are carried beneath the continents. They are almost never found among the formations which constitute mountain ranges. This must mean that the process of convective descent is efficient in carrying them downward. It could be that this part of the suboceanic material becomes mixed with the descending material and, being more simatic than average crust, helps to drag the descending current down. If this is the case, however, it is hard to see what causes the volcanism along the continental borders. It is more likely that some of the volatiles involved in this volcanism are derived from water-saturated marine deposits carried under the continents by the convection currents.

Understanding of the nature of the differentiation process at the continental roots may hold the key to many geologic problems. The location of most types of mineral deposits is controlled by the geochemical reactions which concentrate different elements and compounds. Copper, lead and zinc separate from magmatic solutions and are deposited from the escaping volatiles. Aluminum ore, coal and oil are formed from sedimentary rocks as they change over long periods of time. Local geologic conditions are not, however, the only factors controlling the occurrence of many types of ores. Tin, for instance, is found in abundance in only a few places in the world (Figure 16.2). In these areas tin is produced from vein deposits where magmatic water has deposited the tin as it escaped from the parent bodies and from sedimentary rocks formed from erosional debris derived from vein deposits. Similar geologic conditions exist at many places on Earth, yet tin is found in the crust in sufficient quantities to make it an ore only in southeast Asia and a few other places. Why? Could there be a place in the mantle beneath Malaya which is the source of this tin? If this is the case, then the mantle is not of uniform composition, either due to its being originally composed of a variety of different substances or due to a tendency for certain elements to concentrate in limited regions at some stage of Earth's evolution.

Convection as we see it occurring today presumably does two things. First, it breaks up previously existing patterns, mixing materials which may have existed as concentrations in only certain parts of the mantle. In the process of this stirring of the mantle it may bring batches of these ancient concentrates to the crust to form metallogenic provinces where some one mineral resource is unusually abundant.

Second, convection is itself a process of differentiation involving vertical transport of different elements. Iron and certain other elements such as nickel which behave in a similar chemical fashion to iron are moved downward to the growing core. Hydrogen, in the form of water, moves upward to ground water, to the oceans and to the atmosphere, bringing with it other elements in solution. These tend to separate out as ore deposits as temperature and pressure drop, decreasing the ability of the water to carry compounds in solution.

Heat, much of it derived from radioactive disintegration, supplies the energy which keeps this process going. It is the steady outflow of heat from Earth's interior which moves the convection cells and thereby renews the continents. It is heat which drives vapors from the magma, producing valuable concentrations of essential ores. Without this running down of energy, conditions on Earth would become stagnant, and life as we know it would be suppressed.

Over the 6000-year span of human history, the principal features of Earth have remained fixed. Careful measurement of the rates of geologic processes is usually necessary to show that changes are occurring. However, change is continually occurring. Sediment is carried to the sea (Figure 9.12, p. 238); continents slip along faults (Figure 13.18, p. 352); and volcanoes add new rocks (Figure 12.5, p. 306).

Over millions of years, such slow processes alter the whole appearance of Earth. The presence of large areas of sediments containing fossils of marine organisms shows that shallow seas once covered areas where mountains are found today. Examination of the variety and distribution of fossils

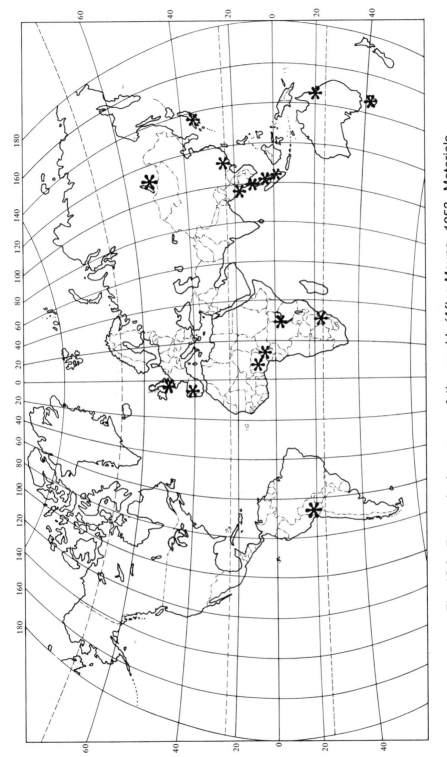

Fig. 16.2 *Tin-producing areas of the world. (After Mayne, 1953, Materials Survey, Tin, Map 4–1. U. S. Dept. of Commerce.)*

shows how animals and plants have evolved throughout Earth's history. Examination of the distribution of ore deposits reveals that the dominant processes of differentiation may also have undergone changes. For example, the fraction of large iron deposits found in the old rocks of the shield areas greatly exceeds the fraction of Earth's surface where shield rocks are exposed (Figure 16.3). Why this is so is not understood; but it must mean that the environment in which they formed was different from what is common on Earth today. Could Earth once have looked more like Venus, or Mars or even Jupiter than it does now?

16.5 ORIGIN OF THE ATMOSPHERE.

The hypothesis of Earth's formation presented above requires that throughout its principal period of accretion there was a steady efflux of the lighter elements, permitting the buildup of a solid body in which hydrogen and helium were minor rather than major constituents. It might be expected, therefore, that the atmosphere initially contained much hydrogen and helium. The present paucity of these two gases would be the result of their loss to space from the upper atmosphere. This loss process would automatically occur if the atmosphere were hot enough. Once the stellar cloud collapsed to form Sun and the present planets, solar radiation would provide one source for such energy.

It is quite likely, however, that Earth went thru a stage when not only hydrogen and helium but all the other gases in its atmosphere were lost. The evidence for this comes from the proportions of the inert gases in Earth's atmosphere as compared to Sun. The elements neon, argon, krypton and xenon are much scarcer on Earth than in Sun, as shown in Table 16.1. This table is based on the assumption that the inert gases are no more prevalent in the whole of Earth than they are in crustal rocks, which seems a reasonable assumption considering that they do not normally enter into chemical combination with other elements, and hence should tend to separate out in any differentiation process. Considering that even argon is less than 1/500,000 as abundant on Earth as in Sun, and the others even scarcer, it seems certain that Earth has not retained more than a small fraction of the inert gases which were once a part of it. Presumably these gases, unable to form solid compounds, separated from the rest of Earth and were lost at an early stage in its history. And if heavy inert gases were lost from the primitive Earth's atmosphere, surely any lighter gaseous elements should also have escaped.

What caused this loss of volatile components remains a mystery. There is no reason to suspect that Sun ever heated the atmosphere to higher temperatures than at present, although high radiation intensity could explain the loss of an original atmosphere. Nor is heat from within Earth a reasonably likely source. The only other large available source is the energy of infalling meteoroids. A falling body gains 6.3×10^7 joules per kilogram dropping thru Earth's present gravitative field. This is enough to vaporize most substances. It is also exactly the energy needed to allow the material

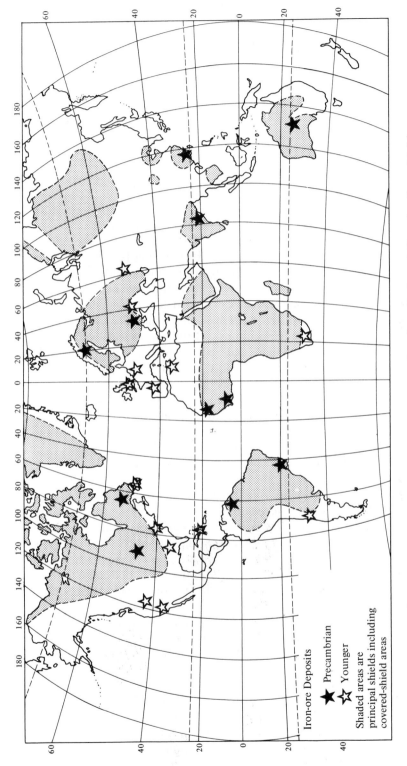

Fig. 16.3 *Iron deposits of the world, showing their relation to the shield areas.* (Shield locations after Umbgrove, The Pulse of the Earth.)

Iron-ore Deposits
★ Precambrian
☆ Younger

Shaded areas are principal shields including covered-shield areas

Table 16.1 *Abundance ratios of inert gases to silica. (After Brown,* Atmospheres of the Earth and Planets *p. 261-3.)*

Ratio of	In Earth	In Sun
Ne/Si	8.0×10^{-11}	2.7
Ar/Si	7.9×10^{-8}	0.14
Kr/Si	1.8×10^{-11}	2.6×10^{-4}
Xe/Si	2.3×10^{-12}	7.0×10^{-4}

to escape again. Thus, an infalling gas molecule must lose part of its kinetic energy in order to be retained by Earth. Any atmospheric gases present are in a prime position to gain a part of the energy lost by any infalling body. During the stage of rapid accretion, this may have provided just the needed extra energy source to drive off all the atmosphere. Once this stage was reached, with no atmosphere left to slow down the infalling meteoroids, any gaseous components of the falling bodies simply evaporated away due to the heat generated by their collision with Earth.

When Earth was small, conditions would have been less extreme. Thus one can picture an early stage of aggradation when light gases might have been retained. Initially, small masses of comet-like assemblages of matter coalesced. At low temperatures, all the components were either solid or liquid. As more and more material streamed into the growing center, this first material was buried. At first, the gravitative pull of the proto-earth was so small that the kinetic energy of the infalling bodies could not melt even the most volatile components. As its size increased, however, a time came when the more volatile materials such as any hydrogen not locked in chemical combination with other elements were evaporated by impact. The composition of material retained by the growing solid Earth would thus have shifted gradually, as heavier and heavier elements became unstable in the impact process. There may have been an intermediate stage when only the heavier volatile elements and compounds were retained and finally a stage when no volatile elements could be kept.

If this speculative view is correct, Earth was built with a composition which varied from undifferentiated cosmic matter at the center to successively less volatile-rich layers as the surface is approached. Throughout the accumulation process, however, differentiation must have been progressing throughout Earth. Volatile materials would have begun working their way to the surface as soon as the interior first melted. During the stage of Earth's history when there was no atmosphere, escaping materials quickly moved on to space. Later, when the rate of infall of meteoric matter decreased, the gases would have begun to reach the surface faster than they were driven off. Soon thereafter, only the largest meteorites would splatter parts of themselves back into space, and the loss of atmospheric gases would be limited to hydrogen and helium as it is at present.

The composition of the material escaping from Earth's interior at any one time in Earth's history is difficult to determine. Even the present efflux is uncertain. The principal gases escaping today from volcanoes (aside from atmospheric contaminants) are water, carbon dioxide and sulfur dioxide (compare Table 12.1, p.305). An unknown amount of gas also escapes in non-volcanic areas. It is known, for example, that radon in the atmosphere increases in periods of low atmospheric pressure and decreases in periods of high atmospheric pressure. Radon is a radioactive gas formed by the decay of uranium and thorium. It disintegrates into polonium and helium and eventually into lead. Its longest-lived isotope has a half life of 3.8 days. Therefore, it must constantly be replenished in the atmosphere or it would quickly disappear. Since radon is found in minute quantities with a wide distribution, it must seep out of the ground. This may be the case for other gases as well.

Even if the present efflux of gases were accurately known, there is no reason to suppose that the composition of material escaping from Earth's interior has always been the same. The gases escaping now consist partly of original material coming from the deep interior to the surface for the first time and partly of reworked material which has been recycled thru crustal processes. If the early history of differentiation of the interior had as a major process the escape of hydrogen and helium, then these elements would have played a more prominent role in the early atmosphere than in the present one. The composition of the earliest permanent atmosphere would have been correspondingly different from the present composition.

Compounds similar to those observed on other planets, particularly methane (CH_4) and ammonia (NH_3) must have been prevalent. Hydrogen rather than oxygen would at one time have been the dominant element. The first stage of the permanent atmosphere, then, is presumed to be one in which oxygen was present only as a component of water (Table 16.2).

As time passed, the steady loss of hydrogen and helium from the upper atmosphere resulted in an increase in the proportion of oxygen and carbon to hydrogen. The composition of the atmosphere changed. Oxides of carbon and sulfur became important components of the air and were dissolved in the oceans. Free hydrogen became scarce. Ammonia was largely dissociated, freeing nitrogen. Methane was oxidized into carbon dioxide and water. The presence of large quantities of oxides, especially carbon dioxide, was the special characteristic of this stage.

Later yet, the continued loss of hydrogen would result in stage 3, an atmosphere in which oxygen was present in excess. But somewhere long before this stage, the presence of large bodies of liquid water at Earth's surface would have begun to play a critical role in events.

16.6 ORIGIN OF THE OCEANS.

The oceans may be presumed to have originated along with the atmosphere. At first, water would have been a minor component compared to hydrogen and helium. As hydrogen was lost, however, the amount of water would eventually have become too great for the atmosphere to hold, and liquid water would have started to condense out. It would have run down-

Table 16.2 *Principal components of atmospheric gases during three stages of its evolution.*

Stage 1 hydrogen rich	Stage 2 CO_2 rich	Stage 3 O_2 rich
H_2	N_2	N_2
CH_4	SO_2	O_2
H_2O	H_2O	H_2O
N_2	CO_2	CO_2
H_2S	CH_4	
NH_3	NH_3	
H_e		

hill and collected into any low spots on Earth's surface. Water-based weathering and erosion of the surface would have replaced wind-dominated processes. The first marine sedimentary rocks were formed.

At this stage there were no plants to hold the soil together or organic acids to help break down exposed minerals. There may have been marked differences from the present in the nature of the sedimentary rocks and of the salts carried in solution to the primordial ocean. With the passage of time and repeated recycling of the water thru evaporation and rainfall and of the rocks thru mountain building, the present pattern of composition was reached.

Water running off the continents into the oceans is not pure, but contains measurable amounts of dissolved material. By measuring stream flow and salinity it is found that about 3×10^{11} kg of dissolved material are carried to the oceans annually. It comes from three sources. Most is obtained by decomposition of the rocks of which the continents are composed. Some is brought to the surface by escaping magmatic water produced as part of the process of volcanism. The rest is precipitated from the atmosphere. A little of the latter is soluble debris from meteoroids striking the atmosphere, but most is recycled material from the oceans. Chlorine and sulfur escape from the oceans as gases and are reprecipitated on land. Oceanic salts are added to the atmosphere when salty spray evaporates. Over half of the sodium in streams is believed to come from air-borne salt crystals which were picked up by the atmosphere.

The two commonest elements in seawater are chlorine and sodium, which make up 55.0 and 30.6 percent respectively of all the dissolved material. Unlike most other elements, very little sodium is precipitated out of the sea. By dividing the present amount of sodium in the oceans by the amount being brought in annually an estimate can be made of how long it would take to salt the oceans. The total volume of seawater is 1.37×10^9 cubic kilometers or 1.37×10^{21} liters. This contains .0106 kg/liter sodium. Thus there are 1.46×10^{19} kg of sodium in the oceans.

About 5.8 percent of the dissolved material in river water is sodium, so 1.7×10^{10} kg of sodium are carried to the sea each year. It would take $1.46 \times 10^{19}/1.7 \times 10^{10}$ or 9×10^8 years for this rate of flow to provide all the sodium in the sea. But over half of the sodium is recycled sodium, so the time required is at least twice this, or roughly two billion years. Furthermore, there is an estimated half as much again sodium entrapped in various sedimentary deposits formed in the sea. This increases the time required for the oceans to form to three billion years. This is over half the estimated age of Earth itself. Since this calculation was made assuming a constant rate of sodium transfer to the sea, as well as other very rough estimates, an error by a factor of two is easily possible.

The most that can be concluded is that the salts found in the sea today could all have originated from the weathering of continental rocks. They need not all have come from this source. Much of the material may have been brought to the oceans directly as part of the dissolved material in submarine volcanic emmanations, or it may have been dissolved from solid igneous material erupted from submarine volcanoes.

The concentrations of other elements are all complicated by their much greater rates of removal from the oceans. Calcium carbonate, the most abundant material dissolved in river water, is removed as limestone deposits. These formations are raised into the continents, where they weather, enriching the water runoff in calcium and returning it to the sea to be redeposited. Similar cycles can be traced for other elements.

The preceding discussion was based on the assumption that the salts of the ocean were obtained from the solid part of Earth. What about the possibility that much of this material was provided directly to the oceans by the influx of meteorites as Earth grew? This possibility cannot be disproven, but seems unlikely. The present rate of influx of meteoritic material is so small that meteoritic matter is rarely recognizable as a component of sediments and never as a major component. Comparison of the present rates of weathering and meteoritic infall as a source of material greatly favors weathering as the source of ocean salts. Since weathering (plus escaping magmatic solutions) adequately explains the composition, there is no reason to call on any other source.

Although the first water to form at Earth's surface is visualized here as having condensed from the atmosphere, most of the total currently present may have reached the oceans more directly. Volcanic processes release large quantities of water. In Section 15.4 it was pointed out that the volume of continental material is 6.7×10^9 cubic kilometers. At 2.7 gm/cc this would be about 1.8×10^{22} kg. The mass of the oceans is about 1.4×10^{21} kg. To provide the water of the oceans from the continental material would require that originally it contained over 7.3 percent water. This is a high water content for an igneous rock, though not an impossible one. Serpentine consists 13 percent of water.

The mass of the mantle is about 4×10^{24} kg. The mass of the ocean water is only 0.0003 of this. The loss of this small fraction of its composition taken equally from all parts of the mantle could leave it with its physical properties hardly altered at all.

A more difficult problem is to estimate the history of the salinity of the oceans. If the salts and the water differentiated from the mantle at the same rate, then the salinity has always been the same as it is at present. On the other hand, if a large part of the water condensed from a primordial atmosphere, then the salinity would have gradually increased throughout the history of Earth.

An intermediate rate seems most reasonable. It was postulated earlier that water is released directly by the differentiation of igneous rocks. The dissolved solids of seawater come from the weathering of rocks. Thus it seems reasonable that the release of water may have preceded the production of the dissolved material. Furthermore, loss of hydrogen from the atmosphere could mean that the volume of surface water today is less than the total amount released from Earth's interior. The salts in seawater, particularly the sodium, suffer no corresponding removal from the marine environment. Thus, these materials would be expected to be increasing in concentration.

The rate of increase in salinity would not be linear. Much of the dissolved material carried to the sea today is derived from marine sedimentary rocks which have been compressed into the continental framework. As a result, the rate of increase in sea salinity may be proportional to the amount of igneous material exposed to erosion for the first time on land. As more and more reworked material becomes available, the fraction of the continents over which new igneous rocks are exposed decreases. A likely result is an asymptotic increase in salinity with the present concentration still well below what it will eventually become.

Another approach is to consider the volume of ocean water as a function of time. Examination of the distribution of sedimentary rocks on the continents shows that the relative elevation of the continents and sea level has varied by no more than a few hundred meters for as far back in time as the record is clear. Shallow-water sediments are everywhere common in the geologic record. Since there appears to have been a steady growth of volcanic material throughout geologic history, this favors a concomitant growth of ocean volume. However, the relative areas of ocean and land are unknown. The processes of continental and oceanic growth seem reasonably well understood, but determination of their relative rates of formation awaits further study.

16.7 ORIGIN OF LIFE.

The changes in the atmosphere and the oceans are both a cause and a result of the evolution of living organisms. The origin of life occurred so far back in geologic time that the events cannot be traced in detail, but it is possible to make a reasonable guess as to what may have happened. When water first began to condense out of the atmosphere as rain it would have carried in solution a wide variety of trace compounds of a more complex nature than the simple molecules listed in Table 16.2. Laboratory experiments have shown that electrical discharges such as lightning have a tendency to stimulate the production of larger molecules from simple compounds. In an inert atmosphere, before life developed, there were no biologic agents to aid in the decomposition of such materials. Solar radiation would tend to break up any complex molecules which stayed in the atmosphere, but any which were precipitated into the primordial oceans or in smaller lakes and pools would be protected, to some degree, from such radiation. Furthermore, this early atmosphere may have been as murky as that of Venus or Jupiter today. Conditions at Earth's surface were warm and, at the proper stage, damp.

One group of compounds which chemists have shown would likely form in trace quantities under these conditions is the amino acids. These constitute important basic components used in many life processes. It requires only a few steps to tie amino acids together with a few additional elements to form virus molecules. Viruses are complex organic molecules which have the ability to act as stable chemical compounds under some conditions, but to crystallize duplicates of themselves in other, more favorable environments.

Two ideal conditions for the reproduction of viruses are the absence of other living creatures, which might look upon them as food, and the presence of the compounds on which the virus itself feeds. Stage 1 of the development of the atmosphere presumably provided these conditions. In pools of water fed by rain, the earliest, most primitive forms of life were created by the chance events of nature. Here, out of inert materials, the first life evolved.

Once compounds capable of reproducing themselves formed, they began to cause changes in their surroundings. They consumed the supply of food as fast as it was formed, and were forced to compete with one another for what was available. Any advantage one variety of virus had over another in utilizing this food supply favored its growth. Thus, the competition of species for survival began. As evolution produced successively more highly developed species, waste products of their metabolism were added to the environment, providing food for other varieties of organisms. This led to an ever increasing multiplicity of interdependent life forms.

The steady change in atmospheric and oceanic composition stimulated evolution by requiring adaptation to change for survival. The shift from an environment of excess hydrogen to one in which carbon dioxide was a more prominent component may have been an important factor in controlling the direction in which organic evolution proceeded. The ability of organisms to use CO_2 in their cellular structure, particularly to lock it into bones and shells as calcium carbonate, was an important advance. It led to widespread deposition of limestones, which was a further change in the nature of the environment.

Finally, plants evolved with their efficient ability to lock atmospheric carbon from CO_2 into more complex organic compounds and to release free oxygen. By this time hydrogen was even further depleted. The locking of carbon into organic compounds then shifted the atmosphere to its present stage of excess oxygen.

The outline of the evolution of Earth presented above leaves many questions unanswered. What role if any did sulfur play in this development?

Sulfur is an important component of the gases given off by igneous activity. Although sulfur enters relatively easily into soluble compounds, high concentrations in natural waters are rare. It appears to be efficiently precipitated in the marine environment.

16.8 THE FUTURE.

One of the most obvious things to be learned from studying the universe is that nothing remains fixed. Change is universal. This applies to Earth as well as every other body.

Man today is disturbing his environment at a rapid rate. He pollutes it with a wide variety of new compounds whose effects are poorly understood. Many of the waste products he flushes into the oceans and underground waters are highly toxic, and some are extremely persistent. Some deadly poisons (e.g. certain insecticides) have been traced thru a whole series of organisms until they were found being consumed by man in fish caught far from where the poisons were used. Artificially created radioactive elements present a special hazard.

The impact of these new substances places a pressure for adaptation on all living organisms from the simplest viruses to the highest mammals. The man who can breath smog and keep on working at high efficiency has an advantage over the individual who is made ill by this experience. If the pressure of changing conditions stimulates the rate of evolution, then presumably new viruses will develop to attack the more resistant men who survive to live in the new atmosphere. Little is yet known of the rate at which these changes occur or the role of trace elements in various life and geochemical cycles.

What is sure is that Sun is warming and its pattern of radiation is slowly changing both secularly over eons of time and cyclically with 22-year and probably longer periods. Will Mars provide a better environment than Earth for our descendents five billion years from now? Or Jupiter in twenty billion years? Should we plant bacteria on one of these planets today to start the evolution of conditions toward a place where man's descendents can live; and, if so, what organisms should we send to prepare the way for us?

Has Venus passed thru a development similar to Earth's and does it represent a stage we have yet to reach, or is it a laggard representing stage two of the development of Earth's atmosphere? We do not yet know the facts of Venus' environment well enough to answer this question.

Once man answers questions such as these, he will be better prepared to decide how to control the evolution of today's environment, and hence himself. Much is said about the danger of upsetting nature's balance and about the need to slow down the rate at which we are producing changes. But change is inevitable. It has always taken place, and is the natural condition of the universe. Man's only choice is how fast and in what directions these changes will proceed. These are the variables man must control if he is to remain the master and not be just another species which will live its span and be replaced.

BIBLIOGRAPHY

Ahrens, L. H., 1965, *Distribution of the Elements in our Planet*. McGraw-Hill Book Co., Inc. 110 pp.

Brancazio, P. H.; A.G.W. Cameron, 1964, *The Origin and Evolution of Atmosphere and Oceans*. John Wiley and Sons, Inc. 314 pp.

Hess, H. H., 1965, Mid-Oceanic Ridges and Tectonics of the Sea-Floor. In *Submarine Geology and Geophysics* pp. 317–334. Butterworths.

Kuiper, G. P. (Ed.), 1952, *The Atmospheres of the Earth and Planets*. University of Chicago Press.

Landsberg, H. E., 1953, *The Origin of the Atmosphere*. Scientific American Offprint 824. W. H. Freeman and Co.

Mason, B., 1966, *Principles of Geochemistry*. John Wiley and Sons, Inc. 399 pp.

Ringwood, A. E., 1966, The Chemical Composition and Origin of the Earth. In *Advances in Earth Sciences* pp. 287–356. M.I.T. Press.

Rubey, W. W., 1955, Development of the Hydrosphere and Atmosphere, with Special Reference to Probable Composition of the Early Atmosphere. Geol. Soc. Am. Special Pap. 62:631–650.

Umbgrove, 1947, *The Pulse of the Earth*, Martinus Nijhoff, 358 pp., 8 plates.

Urey, H. C., 1952, *The Planets*, Yale University, 245 pp.

APPENDIX 1
Metric and English Equivalents

	m.k.s. (meter— kilogram— second)	c.g.s. (centimeter— gram—second)	English
length	1 centimeter	10 millimeters	0.3937 inches
	1 meter	100 centimeters	39.37 inches
	1 kilometer	1000 meters	0.62137 miles
time	1 second	1 second	1 second
mass	1 kilogram	1000 grams	2.20462 pounds
			35.274 ounces
force	1 newton	100,000 dynes	7.233 poundals
Newtonian	6.67×10^{-11}	6.67×10^{-8}	3.44×10^{-8} lb ft^2
constant	$\dfrac{\text{newton m}^2}{\text{kg}^2}$	$\dfrac{\text{dyne-cm}^2}{\text{gm}^2}$	$slug^2$
pressure	10^5 newtons/m^2	10^6 dynes/cm^2	1 bar
energy	1 joule (newton-meter)	10^7 ergs (dyne-cm)	23.73 foot-poundals
	1 kilogram- calorie (1 kg-cal- orie = .00023889 joules)	1000 gram- calories	3.9685 British thermal units
power	1 watt	10^7 ergs/sec	.001341 horse power
temperature	0°C	273.16°K	32°F
temperature difference	1°C	1°K	1.8°F

APPENDIX 2
Some Useful Dimensions

Astronomical unit (distance Earth to Sun) = 1.49599×10^8 km
Day (mean) = 86,400 seconds
Earth's area = 5.101×10^8 sqkm
   density   = 5.51 gm/cc
   heat flux (average) = 1.5×10^{-5} kg-cals/m^2/sec
                 = 470 kg-cals/m^2/year
   gravity (sea level) at equator = 9.780490 newtons/kg
                45° latitude = 9.806294 newtons/kg
                poles   = 9.832213 newtons/kg
   heat loss = 7.9×10^9 kg-cals/sec
           = 2.4×10^{17} kg-cals/year
   mass   = 5.977×10^{24} kg

moment of inertia about polar axis $= 8.08 \times 10^{37}$ kg m^2

equatorial axis $= 8.05 \times 10^{37}$ kg m^2

nutation of the precession $= $ up to $\pm 9.23''$ per year

precession of axis of rotation $= 50.2''$ per average year

radius (average) $= 6371.2$ km

(equatorial) $= 6378.1$ km

(polar) $= 6356.6$ km

temperature gradient (average at surface) $= 3°$ C/100m

volume $= 1.08 \times 10^{21}$ cu meters

Milky Way galaxy's diameter $= 10^5$ light years

mass $= 3.3 \times 10^{41}$ kg

number of stars $= 2 \times 10^{11}$

Moon's density $= 3.34$ gm/cc

distance from Earth (average) $= 384,400$ km

gravity at surface $= 1.62$ newtons/kg

mass $= 7.35 \times 10^{22}$ kg

radius (average) $= 1740$ km

volume $= 2.20 \times 10^{19}$ m^3

Newtonian constant (universal constant of gravitation)

$$= 6.67 \times 10^{-11} \frac{\text{newton-m}^2}{\text{kg}^2}$$

Ocean volume (on Earth) $= 1.37 \times 10^9$ cubic km

Solar constant (radiant energy at Earth) $= 20.0 \pm 0.4$ kg-cals/m^2/min

Sun's density $= 1.41$ gm/cc

gravity $= 273$ newtons/kg

mass $= 1.99 \times 10^{30}$ kg

radius $= 695,600$ km

Velocity of electromagnetic radiation (e.g. light)

$= 299,792.5$ km/sec

$= 9.47 \times 10^{12}$ km/year

Year, anomalistic $= 365.2596$ mean days

siderial $= 365.2564$ mean days

tropical $= 365.2422$ mean days $= 3.156 \times 10^7$ secs

INDEX